Advances in Industrial Control

Other titles published in this Series:

Robust Control of Diesel Ship Propulsion
Nikolaos Xiros

Hydraulic Servo-systems
Mohieddine Jelali and Andreas Kroll

Strategies for Feedback Linearisation
Freddy Garces, Victor M. Becerra, Chandrasekhar Kambhampati and Kevin Warwick

Robust Autonomous Guidance
Alberto Isidori, Lorenzo Marconi and Andrea Serrani

Dynamic Modelling of Gas Turbines
Gennady G. Kulikov and Haydn A. Thompson (Eds.)

Control of Fuel Cell Power Systems
Jay T. Pukrushpan, Anna G. Stefanopoulou and Huei Peng

Fuzzy Logic, Identification and Predictive Control
Jairo Espinosa, Joos Vandewalle and Vincent Wertz

Optimal Real-time Control of Sewer Networks
Magdalene Marinaki and Markos Papageorgiou

Process Modelling for Control
Benoît Codrons

Computational Intelligence in Time Series Forecasting
Ajoy K. Palit and Dobrivoje Popovic

Modelling and Control of mini-Flying Machines
Pedro Castillo, Rogelio Lozano and Alejandro Dzul

Rudder and Fin Ship Roll Stabilization
Tristan Perez

Measurement, Control, and Communication Using IEEE 1588
John Eidson

Hard Disk Drive Servo Systems (2nd Edition)
Ben M. Chen, Tong H. Lee, Kemao Peng and Venkatakrishnan Venkataramanan

Windup in Control
Peter Hippe

Manufacturing Systems Control Design
Stjepan Bogdan, Frank L. Lewis, Zdenko Kovačić and José Mireles Jr.
Publication due May 2006

Nonlinear H_2/H_∞ *Constrained Feedback Control*
Murad Abu-Khalaf, Jie Huang and Frank L. Lewis
Publication due May 2006

Process Control Performance Assessment
Andrzej Ordys, Damien Uduehi and Michael A. Johnson (Eds.)
Publication due August 2006

S.O. Reza Moheimani and Andrew J. Fleming

Piezoelectric Transducers for Vibration Control and Damping

With 203 Figures

Springer

S.O. Reza Moheimani, PhD
Andrew J. Fleming, PhD
School of Electrical Engineering and Computer Science
University of Newcastle
University Drive
Callaghan
NSW 2308
Australia

British Library Cataloguing in Publication Data
Moheimani, S. O. Reza, 1967-
Piezoelectric transducers for vibration control and damping. - (Advances in industrial control)
1.Damping (Mechanics) 2.Piezoelectric materials 3.Vibration transducers
I.Title II.Fleming, Andrew J.
620.3'7
ISBN-13: 9781846283314
ISBN-10: 1846283310

Library of Congress Control Number: 2006921994

Advances in Industrial Control series ISSN 1430-9491
ISBN-10: 1-84628-331-0 e-ISBN 1-84628-332-9 Printed on acid-free paper
ISBN-13: 978-1-84628-331-4

Printed in Germany

9 8 7 6 5 4 3 2 1

Springer Science+Business Media
springer.com

Advances in Industrial Control

Professor Emeritus O.P. Malik
Department of Electrical and Computer Engineering
University of Calgary
2500, University Drive, NW
Calgary
Alberta
T2N 1N4
Canada

Professor K.-F. Man
Electronic Engineering Department
City University of Hong Kong
Tat Chee Avenue
Kowloon
Hong Kong

Professor G. Olsson
Department of Industrial Electrical Engineering and Automation
Lund Institute of Technology
Box 118
S-221 00 Lund
Sweden

Professor A. Ray
Pennsylvania State University
Department of Mechanical Engineering
0329 Reber Building
University Park
PA 16802
USA

Professor D.E. Seborg
Chemical Engineering
3335 Engineering II
University of California Santa Barbara
Santa Barbara
CA 93106
USA

Doctor K.K. Tan
Department of Electrical Engineering
National University of Singapore
4 Engineering Drive 3
Singapore 117576

Doctor I. Yamamoto
Technical Headquarters
Nagasaki Research & Development Center
Mitsubishi Heavy Industries Ltd
5-717-1, Fukahori-Machi
Nagasaki 851-0392
Japan

To Niloofar and Dara
S.O.R.M

To my parents, Graeme and Cathy
A.J.F

Series Editors' Foreword

The series *Advances in Industrial Control* aims to report and encourage technology transfer in control engineering. The rapid development of control technology has an impact on all areas of the control discipline. New theory, new controllers, actuators, sensors, new industrial processes, computer methods, new applications, new philosophies, new challenges. Much of this development work resides in industrial reports, feasibility study papers and the reports of advanced collaborative projects. The series offers an opportunity for researchers to present an extended exposition of such new work in all aspects of industrial control for wider and rapid dissemination.

The piezoelectric effect was discovered in some naturally occurring materials in the 1880s. However it was not until the Second World War that man-made polycrystalline ceramic materials were produced that also showed piezoelectric properties. Quartz and other natural crystals found application in microphones, accelerometers and ultrasonic transducers, whilst the advent of the man-made piezoelectric materials widened the field of applications to include sonar, hydrophones, and piezo-ignition systems. Today, piezoelectric applications include smart materials for vibration control, aerospace and astronautical applications of flexible surfaces and structures and novel applications for vibration reduction in sports equipment (tennis racquets, skis and snowboards), In this entry to the *Advances in Industrial Control* monograph series, Reza Moheimani and Andrew Fleming present a comprehensive tour of the present state of the art in the theory and applications of the piezoelectric property pertaining to vibration control and damping. It is such a well-thought-out presentation that we feel sure the volume will become a classic for the field.

The opening chapters of the volume (Chapters 1 and 2) present interesting overview material and then the fundamentals of piezoelectricity. It is here we learn that piezoelectric transducers can be used as actuators to generate moment or as sensors to measure strain. An important application is the control of a flexible beam and the model for this is found in Chapter 2.

Fundamental theory on the feedback control aspects of piezoelectric systems is found in Chapters 3 to 5; it is useful to note that each new control method in these chapters is supported by experimental verifications and demonstrations. This makes for very enlightening reading of the sequence of increasingly sophisticated control methods used to solve a problem where the process models are of high order and the process modes are lightly damped.

The practical issues of the instrumentation used in piezoelectric systems are dealt with in Chapter 6. Here, emphasis is given to the fact that the quality of the control obtained is highly dependent on the quality of the instrumentation used in piezoelectric systems. This chapter reports the authors' direct experience of constructing the instrumentation to implement their control design work and is a valuable source of "know-how" for use by other researchers in the field.

A key technique used in piezoelectric vibration control systems is that of piezoelectric shunt damping. This is where a piezoelectric transducer, which is bounded to the flexible structure, is connected to an electrical shunt impedance, Z(s). In Chapter 5, Moheimani and Fleming give this design problem a novel twist by reinterpreting the shunt damping structure as a feedback structure, opening the way for the use of control design methods in this problem. Chapters 7 to 10 exploit the feedback interpretation in a number of ways. The spatially distributed nature of flexible structures leads naturally to multivariable problems using multi-port shunts and a generic feedback architecture is ideal for this problem (Chapter 7). To reduce the sensitivity of the shunt circuit design to changes in piezoelectric capacitance and structure resonance frequency, online adaptation is used (Chapter 8). As an alternative to online adaptation, negative capacitor shunt impedance designs are investigated in Chapter 9. Standard feedback control design tools (such as LQG, H_∞ and H_2) are used to compute optimal MIMO shunt damping designs in Chapter 10.

To complete the wide-ranging yet thorough contents of the volume, Chapter 11 looks at hysteresis effects in piezoelectric transducers and Chapter 12 is concerned with the important and fairly recent application of piezoelectric tube scanners in scanning tunnelling microscopes. As with almost all the other chapters in the volume, the discussions in Chapters 11 and 12 are supported by experimental findings and results.

This volume will be an essential reference and source book for all control academics, researchers and engineers working in the field of piezoelectric control systems. It will be equally attractive to those wishing to gain an introduction to the area and to those wishing to find examples to illustrate the use of control in an advanced applications field.

M.J. Grimble and M.A. Johnson
Industrial Control Centre
Glasgow, Scotland, U.K.

Preface

Flexible mechanical systems experience undesirable vibration in response to environmental and operational forces. The very existence of vibrations can limit the accuracy of sensitive instruments or cause significant error in applications where high-precision positioning is essential. In many scientific and engineering applications control of vibrations is a necessity. Piezoelectric transducers have been used in countless applications as sensors, actuators, or both. When traditional passive vibration control techniques fail to meet the requirements, piezoelectric transducers, in conjunction with feedback controllers, can be used to suppress vibrations in an effective way.

The ability of piezoelectric materials to transform mechanical energy into electrical energy and *vice versa* has been known in scientific circles for well over a century. Although a transducer with such capacity was perceived to be ideal for vibration control applications, naturally occurring piezoelectric materials were found to be of little use for such purposes. The discovery of piezoceramic materials, after the second world war, and invention of methods for their high-volume manufacturing enabled researchers to utilize piezoelectric transducers in a variety of applications. In particular, over the past decade there has been significant interest in using piezoelectric transducers as actuators and sensors in applications such as vibration control of flexible structures, micro- and nano-positioning systems, medical instrumentation, and micro electro-mechanical systems.

The purpose of this book is to present, in a unified fashion, some recent developments in the feedback control of vibration using embedded piezoelectric sensors and actuators. The book is intended for researchers, graduate students and practicing engineers with an interest in vibration control systems. The book covers various ways in which active vibration control systems can be designed for piezoelectric laminated structures, paying special attention to how such control systems can be implemented in real time. The text contains numerous examples and experimental results obtained from laboratory-scale apparatus with details of how similar setups can be built.

This book documents research performed in the Laboratory for Dynamics and Control of Smart Structures at the School of Electrical Engineering and Computer Science, University of Newcastle, Australia. The stimulating research environment at the School along with its solid academic culture and long tradition of scholarship was an ideal setting for the development of this monograph. In the production of this book, the authors wish to acknowledge the support they have received from the Australian Research Council (ARC), the University of Newcastle and the ARC Centre for Complex Dynamic Systems and Control (CDSC). Furthermore, the authors wish to acknowledge the contributions made by their colleagues, Sam Behrens, Benjamin Vautier, Bharath Bhikkaji, Dominik Niederberger and Manfred Morari for the research that underlies parts of the material presented in this book. The authors are particularly grateful to Ian Petersen, Hemanshu Pota and Dominik Niederberger who read the book and made valuable suggestions and comments.

Newcastle, Australia
December 2005

S. O. R. Moheimani
A. J. Fleming

Contents

1

Introduction

1.1 Piezoelectric Transducers

Piezoelectric transducers have become increasingly popular in vibration control applications. They are used as sensors and as actuators in structural vibration control systems. They provide excellent actuation and sensing capabilities. The ability of piezoelectric materials to transform mechanical energy into electrical energy and *vice versa* was discovered over a century ago by Pierre and Jacque Curie. These French scientists discovered a class of materials that when pressured, generate electrical charge, and when placed inside an electric field, strain mechanically.

Piezoelectricity, which literally means "electricity generated from pressure" is found naturally in many monocrystalline materials, such as quartz, tourmaline, topaz and Rochelle salt. However, these materials are generally not suitable as actuators for vibration control applications. Instead, man-made polycrystalline ceramic materials, such as lead zirconate titanate (PZT), can be processed to exhibit significant piezoelectric properties. PZT ceramics are relatively easy to produce, and exhibit strong coupling between mechanical and electrical domains. This enables them to produce comparatively large forces or displacements from relatively small applied voltages, or *vice versa.* Consequently, they are the most widely utilized material in manufacturing of piezoelectric transducers.

Piezoelectric transducers are available in many forms and shapes [28]. The most widely used piezoelectric transducers are in the form of thin sheets that can be bonded to or embedded in composite structures. As actuators they are mainly used to generate moment in flexible structures, while as sensors they are used to measure strain. Piezoelectric actuators are also available in the form of "stacks", where many layers of materials and electrodes are assembled together. These stacks generate large forces but small displacements in the direction normal to the top and bottom surfaces.

Piezoelectric transducers are used in many applications such as structural vibration control [11, 14, 73, 161, 176], precision positioning [124, 75, 183], aerospace systems [4, 106, 43], and more recently they have been critical in advancing research in nanotechnology [26, 165, 166].

Properties of piezoelectric transducers are studied in Chapter 2 of this book. This chapter describes piezoelectric constitutive equations, explains piezoelectric coefficients that are of significance, describes the functions of piezoelectric sensing and actuation, and finally derives the dynamics of a flexible structure with piezoelectric sensors and actuators. Specific properties of such systems are studied in this chapter.

1.2 Vibration Control

Piezoelectric transducers have been extensively used in structural vibration control applications. Their wide utilization in this specific application can be attributed to their excellent actuation and sensing abilities which stems from their high electro-mechanical coupling coefficient, as well as their non-intrusive nature.

For vibration control purposes, piezoelectric transducers are bonded to or embedded in a composite structure. The type of structures which lend themselves to piezoelectric transducers are generally flexible in nature. The transfer functions of these systems are of high order, and their poles are very lightly damped. Control problems associated with these systems are by no means trivial.

Control and observation spillover can potentially destabilize a closed-loop system if the controller is only designed for a small number of in-bandwidth modes and then implemented on the full-order system [12, 13]. A number of techniques have been proposed to address this issue, the most promising of which is based on using compatible collocated sensors and actuators [98].

The velocity feedback controller has well-known stability properties when implemented on a flexible structure with collocated actuators and sensors [12, 13, 77]. In particular, although this control is susceptible to spill-over, existence of spill-over does not destabilize the closed-loop system. The foremost difficulty with velocity feedback control is the high controller bandwidth and unnecessary control effort at frequencies far away from the resonance frequencies of the base structure. The controller is often rolled-off at higher frequencies by appending extra low-pass dynamics. Special care must be exercised in adding extra dynamics to the controller since the existence of out-of-bandwidth modes of the structure may result in instabilities [77].

Resonant controllers approximate a velocity feedback controller at and close to the resonance frequencies of a structure [136]. Unlike velocity feedback controllers their control effort at other frequencies is negligible. Furthermore, they are biproper controllers which enjoy similar stability properties to velocity feedback controllers. These controllers are introduced and thoroughly studied in Chapter 3. Relatively similar control techniques have been proposed in the literature (see [161] and references therein). However, they are mostly tailored to applications that involve vibration control of a single structural mode and do not extend to multivariable systems. The presentation in Chapter 3 addresses the problem in its most general form. The same control structures are used in the remainder of the book to construct sensorless controllers and to derive controllers for charge-driven actuators, all in a multivariable setting.

Positive position feedback (PPF) controllers, originally proposed in [58] have been extensively used in structural vibration control applications [162, 176, 85, 183, 32, 69, 113, 52, 57]. These controllers have a structure reminiscent of the system which is being controlled. Compared to resonant controllers they enjoy a quick roll off outside of the controlled bandwidth and they are capable of achieving higher levels of performance.

Stability conditions for PPF controlled systems, as derived in [58], do not allow for the effect of out-of-bandwidth modes of the base structure. Consequently, existence of these modes may destabilize the closed-loop system once the controller is implemented on the structure. Having said this, the natural roll off of this controller at higher frequencies can often be used to alleviate the effect of out-of-bandwidth modes.

Recently it has been demonstrated that the effect of out-of-bandwidth modes of a structure can be captured by adding a feed-through term to the truncated model of the system [34, 134, 133, 138]. The corrected model is a close representation of the true system in the controlled bandwidth. The PPF stability condition, as derived in [58], does not allow for the presence of a feed-through term in the structure model. These stability conditions are re-derived in Chapter 3 and the results are consistent with those reported in [58] when a feed-through term is not present in the structure model.

All control techniques introduced thus far are devised for structures with bonded collocated piezoelectric actuators and sensors. In these systems a piezoelectric transducer performs a single task of actuation, or sensing. These transducers, however, can function simultaneously as actuators and sensors. This property has resulted in emergence of self-sensing circuits which integrate the functions of sensing and actuation in the same piezoelectric transducer [54, 9]. If designed properly, such circuits can reduce the number of requisite piezoelectric transducers by half. However, there are problems associated with this technique which are discussed in Section 3.8.

1.3 Piezoelectric Shunt Damping

Piezoelectric shunt damping is a popular method for vibration suppression in flexible structures. The technique is characterized by the connection of an electrical impedance to a structurally bonded piezoelectric transducer. Such methods do not require an external sensor, and if designed properly, may guarantee stability of the shunted system.

Although first appearing in [68], the concept of piezoelectric shunt damping is mainly attributed to Hagood and von Flotow [82], who demonstrated that a series inductor-resistor network can significantly reduce vibration of a single structural mode. In this technique the network, together with the inherent capacitance of the piezoelectric transducer, is tuned to the resonance frequency of the mode which is to be damped. Similar to a tuned mechanical absorber [95], additional dynamics introduced by the shunt circuit act to increase the effective structural damping [82].

Single-mode damping can be applied to reduce vibration of several structural modes with the use of as many piezoelectric transducers and damping circuits. However, in many cases, this may not be a practical solution since a large number of transducers will be needed if a large number of modes are to be shunt damped. This has encouraged researchers to develop multiple mode shunt damping circuits which use only one piezoelectric transducer.

The first multi-mode shunt circuit is credited to Hollkamp [89], who was able to successfully suppress the second and third modes of a cantilever beam by 19 and 12 dB respectively. The circuit requires as many parallel branches as there are modes to control, and the author proposed a numerical optimization to determine suitable component values. The resulting optimization problem is highly nonlinear, nonconvex and relatively difficult to solve for a large number of modes.

Another multi-mode circuit, proposed by Wu and co-authors, is made of current blocking networks inserted in each parallel branch [198, 200, 199, 203]. Inductive and capacitive components of the proposed circuit can be found in a straightforward manner. A systematic approach for determining effective resistance values is presented in reference [16]. Even in its simplest form [203], the complexity and order of current-blocking circuits restrict their use to a maximum of three modes.

An alternative multi-mode shunt circuit was recently introduced in [17]. The idea is based on inserting current-flowing LC shunts in each parallel branch of the multi-mode shunt, effectively isolating branches from one another at each resonance frequency of the base structure. Besides greatly simplifying the tuning procedure current-flowing shunt circuits require less components and are effective in damping a large number of modes. A dual of this circuit has also been proposed in [67].

Multi-mode shunts discussed thus far are to some extent direct extensions of the original single-mode circuits with additional circuitry that enables the same shunt to damp several modes of a structure. A new paradigm for the design of piezoelectric shunt damping circuits was presented in [140]. By viewing the electrical shunt impedance as parameterizing an equivalent collocated strain feedback controller, a shunt impedance can be found working backwards from an effective feedback controller. In particular, starting from resonant controllers as discussed above, two new passive resonant shunt circuits are proposed in [140] and [137]. The proposed shunts are low in order, easy to tune, and suitable for modally dense systems. The extension of this technique to multivariable, multi-mode shunt design has been reported in [141]. Chapters 5 and 7 cover these topics in a unified way.

All of the shunts introduced so far are passive in nature. This implies that they can be realized using passive components, *i.e.* resistors, inductors and capacitors. However, the requisite inductors are generally so large that they are often implemented using Gyrator circuits [164]. A more practical approach is to simulate the presence of an impedance using a synthetic impedance/admittance circuit as described in [59, 60], and articulated in Chapters 4 and 6.

Once effective tools for the implementation of a given impedance are developed, it is no longer necessary to restrict the shunt impedance to a passive circuit. By allowing the shunt impedance to be active, it is possible to design shunt damping circuits that are capable of delivering higher levels of performance in terms of adding damping to the system. Furthermore, having identified the feedback structure of the shunted system, it is feasible to cast the problem of piezoelectric shunt design as a standard feedback control design problem. Therefore, advanced control design techniques such as LQG, H_∞, H_2, etc. can be employed to design high-performance shunts. Chapter 10 describes a number of ways that the problem of active shunt design can be formulated, and discusses the design of an active H_∞ shunt for a cantilever beam. Further experimental results are reported in [66].

1.4 Hysteresis

The assumption made thus far has been that piezoelectric transducers are linear devices that transduce energy between mechanical and electrical domains, where the coupling mechanism is governed via a linear relationship described by the IEEE Standards on Piezoelectricity [154]. While this is generally true at lower drives when driven by a voltage amplifier, at higher drives piezoelectric actuators display nonlinear behavior known as hysteresis.

The existence of hysteresis in piezoelectric materials is generally attributed to residual misalignment of crystal grains in the poled ceramic [74, 94]. Existence of hysteresis has been shown to have an adverse effect on the stability and closed-loop performance of voltage-controlled piezoelectric actuators [123]. A large number of techniques have been developed that are aimed at reducing the hysteresis associated with voltage-driven piezoelectric actuators, *e.g.* see [71], and references therein. In particular, methods such as inversion-based Preisach modeling [127] and phase control [47] are two examples of the proposed techniques.

It has been argued that hysteresis is an electrical property of piezoelectric materials, which mainly exists between the applied electric field and the resulting electrical charge [78]. Indeed, it has been demonstrated that by controlling electrical charge, or current rather than the applied voltage, the hysteresis effect can be substantially reduced [147]. The above argument clearly suggests that when possible, piezoelectric actuators should be driven by charge, or current sources as this significantly reduces the hysteretic nature of the actuator.

Even though this approach has been known for some time, it has not been widely used due to the perceived difficulty of driving highly capacitive loads such as piezoelectric actuators. The main problem being the existence of offset voltages in the charge or current source circuit, which will eventually charge up the capacitive load. This will then distort the control signal being applied to the piezoelectric load. This issue has been pointed out by a number of authors [122, 124]. Recent research [62] proposes a new structure for charge and current sources capable of regulating the DC profile of the actuator. Due to this development, it is now possible to use electrical charge as the driving control signal for piezoelectric actuators in structural control applications.

Chapter 11 is mainly concerned with control issues related to piezoelectric laminated structures driven by charge amplifiers. Resonant controllers that were originally developed for voltage-driven piezoelectric actuators, and were introduced in Chapter 5 are redesigned to work with charge-driven piezos. It is shown that these modified resonant controllers roll off quickly at higher frequencies and maintain closed-loop stability of the system in presence of out-of-bandwidth vibration modes.

One of the issues with resonant controllers for voltage-driven piezos is the biproper nature of these controllers. Although stability of the closed-loop system under resonant controllers is guaranteed, since these controllers do not roll off at higher frequencies instabilities may occur due to introduction of a phase lag[1], or other imperfections.

Given the specific structure of charge-driven piezoelectric laminated structures, it is possible to construct resonant controllers in a way that they achieve a roll off of 20 dB, or even 40 dB per decade. This makes them comparable to PPF controllers. Furthermore, since the underlying system is free of hysteresis nonlinearity excellent performance can be achieved. Chapter 11 is based on the results reported in [143] and [188].

1.5 Applications

Piezoelectric vibration control has shown promise in a variety of applications ranging from consumer and sporting products to satellite and fighter aircraft vibration control systems.

In the consumer products category, a number of companies such as HEAD and K2 have invested in high-performance and novelty items such as composite piezoelectric tennis racquets, skis, and snowboards [21]. These products typically involve the use of a shunted piezoelectric transducer to decrease vibration. Benefits include increased user comfort, better handling and performance.

---

[1] This may be partially due to the hysteresis

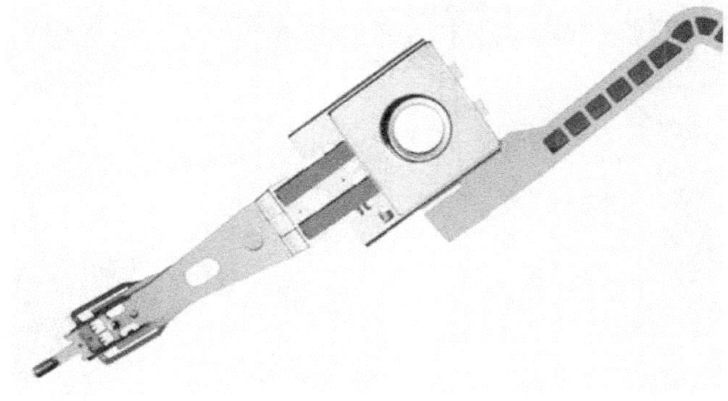

Figure 1.1. A PZT instrumented dual-stage actuator. Picture courtesy of Robert Evans, Hutchinson Technology.

The next generation of hard disk drives may also incorporate piezoelectric vibration control systems in a number of ways. The most promising technology, which is about to be commercialized, appears to be the utilization of dual-stage microactuators in hard disk drives to improve bandwidth and tracking performance of the servo system. In these actuation mechanisms, the read-write head, which is sitting at the end of a small suspension is moved by two piezoelectric actuators that are placed near the root of the suspension, see Figure 1.1. Therefore, the piezoelectric actuators serve as a secondary actuation mechanism, in addition to the voice coil motor.

A considerable research effort has been undertaken on the structural control of military aircraft. In certain modes of flight, buffet loads on wing and stabilizer airfoils can result in high levels of vibration. Such vibration can lead to mechanical fatigue and reduction of the flight envelope and lift performance of an airfoil [172]. Examples include: FA-18 wing, body, and stabilizer control [90], and F-15 panel control [196]. A piezoelectric laminated FA-18 stabilizer is shown in Figure 1.2. Piezoelectric transducers have also been incorporated into helicopter rotor blades for the suppression of lightly damped lag modes in hingeless rotors [106].

Other noise control applications include: suppression of acoustic radiation from underwater submersibles [205], launch vehicle structural and acoustic noise mitigation [153, 51], acoustic transmission reduction panels [105, 171], and active antenna structures [73].

A primary consideration in the design of space structures is the vibration experienced during launch. In future, structures incorporating piezoelectric transducers may form the basis of lightweight, high performance mechanical components for use in space applications [4].

A very successful application for piezoelectric actuators is their use in the scanning stage of scanning probe microscopes, which are used extensively in nanotechnology research. These "piezoelectric tube scanners", first reported

Figure 1.2. A piezoelectric laminate FA-18 vertical stabilizer. Picture courtesy of Dr. Steven Galea, Defence Science and Technology Organization.

in [25], can be used in positioning applications which require high precision (down to sub-nanometer levels), and high bandwidth. Chapter 12 is concerned with problems associated with this specific class of piezoelectric actuators and illustrates how the control design techniques, along with the developed instrumentation discussed in earlier parts of this book, can be employed to significantly improve performance of these nanopositioning systems.

2

Fundamentals of Piezoelectricity

2.1 Introduction

This chapter is concerned with piezoelectric materials and their properties. We begin the chapter with a brief overview of some historical milestones, such as the discovery of the piezoelectric effect, the invention of piezoelectric ceramic materials, and commercial and military utilization of the technology. We will review important properties of piezoelectric ceramic materials and will then proceed to a detailed introduction of the piezoelectric constitutive equations.

The main assumption made in this chapter is that transducers made from piezoelectric materials are linear devices whose properties are governed by a set of tensor equations. This is consistent with the IEEE standards of piezoelectricity [154]. We will explain the physical meaning of parameters which describe the piezoelectric property, and will clarify how these parameters can be obtained from a set of simple experiments.

In this book, piezoelectric transducers are used as sensors and actuators in vibration control systems. For this purpose, transducers are bonded to a flexible structure and utilized as either a sensors to monitor structural vibrations, or as actuators to add damping to the structure. To develop model-based controllers capable of adding sufficient damping to a structure using piezoelectric actuators and sensors it is vital to have models that describe the dynamics of such systems with sufficient precision.

We will explain how the dynamics of a flexible structure with incorporated piezoelectric sensors and actuators can be derived starting from physical principles. In particular, we will emphasize the structure of the models that are obtained from such an exercise. Knowledge of the model structure is crucial to the development of precise models based on measured frequency domain data. This will constitute our main approach to obtaining models of systems studied throughout this book.

2.2 History of Piezoelectricity

The first scientific publication describing the phenomenon, later termed as piezoelectricity, appeared in 1880 [48]. It was co-authored by Pierre and Jacques Curie, who were conducting a variety of experiments on a range of crystals at the time. In those experiments, they cataloged a number of crystals, such as tourmaline, quartz, topaz, cane sugar and Rochelle salt that displayed surface charges when they were mechanically stressed.

In the scientific community of the time, this observation was considered as a significant discovery, and the term "piezoelectricity" was coined to express this effect. The word "piezo" is a Greek word which means "to press". Therefore, piezoelectricity means electricity generated from pressure - a very logical name. This terminology helped distinguish piezoelectricity from the other related phenomena of interest at the time; namely, contact electricity[1] and pyroelectricity[2].

The discovery of the direct piezoelectric effect is, therefore, credited to the Curie brothers. They did not, however, discover the converse piezoelectric effect. Rather, it was mathematically predicted from fundamental laws of thermodynamics by Lippmann [118] in 1881. Having said this, the Curies are recognized for experimental confirmation of the converse effect following Lippmann's work.

The discovery of piezoelectricity generated significant interest within the European scientific community. Subsequently, roughly within 30 years of its discovery, and prior to World War I, the study of piezoelectricity was viewed as a credible scientific activity. Issues such as reversible exchange of electrical and mechanical energy, asymmetric nature of piezoelectric crystals, and the use of thermodynamics in describing various aspects of piezoelectricity were studied in this period.

The first serious application for piezoelectric materials appeared during World War I. This work is credited to Paul Langevin and his co-workers in France, who built an ultrasonic submarine detector. The transducer they built was made of a mosaic of thin quartz crystals that was glued between two steel plates in a way that the composite system had a resonance frequency of 50 KHz. The device was used to transmit a high-frequency chirp signal into the water and to measure the depth by timing the return echo. Their invention, however, was not perfected until the end of the war.

Following their successful use in sonar transducers, and between the two World Wars, piezoelectric crystals were employed in many applications. Quartz crystals were used in the development of frequency stabilizers for vacuum-tube oscillators. Ultrasonic transducers manufactured from piezoelectric crystals were used for measurement of material properties. Many of the classic piezoelectric applications that we are familiar with, applications such

[1] Static electricity generated by friction

[2] Electricity generated from crystals, when heated

as microphones, accelerometers, ultrasonic transducers, etc., were developed and commercialized in this period.

Development of piezoceramic materials during and after World War II helped revolutionize this field. During World War II, significant research was performed in the United States and other countries such as Japan and the former Soviet Union which was aimed at the development of materials with very high dielectric constants for the construction of capacitors. Piezoceramic materials were discovered as a result of these activities, and a number of methods for their high-volume manufacturing were devised. The ability to build new piezoelectric devices by tailoring a material to a specific application resulted in a number of developments, and inventions such as: powerful sonars, piezo ignition systems, sensitive hydrophones and ceramic phono cartridges, to name a few.

2.3 Piezoelectric Ceramics

A piezoelectric ceramic is a mass of perovskite crystals. Each crystal is composed of a small, tetravalent metal ion placed inside a lattice of larger divalent metal ions and O_2, as shown in Figure 2.1.

To prepare a piezoelectric ceramic, fine powders of the component metal oxides are mixed in specific proportions. This mixture is then heated to form a uniform powder. The powder is then mixed with an organic binder and is formed into specific shapes, *e.g.* discs, rods, plates, etc. These elements are then heated for a specific time, and under a predetermined temperature. As a result of this process the powder particles sinter and the material forms a dense crystalline structure. The elements are then cooled and, if needed, trimmed into specific shapes. Finally, electrodes are applied to the appropriate surfaces of the structure.

Above a critical temperature, known as the "Curie temperature", each perovskite crystal in the heated ceramic element exhibits a simple cubic symmetry with no dipole moment, as demonstrated in Figure 2.1. However, at temperatures below the Curie temperature each crystal has tetragonal symmetry and, associated with that, a dipole moment. Adjoining dipoles form regions of local alignment called "domains". This alignment gives a net dipole moment to the domain, and thus a net polarization. As demonstrated in Figure 2.2 (a), the direction of polarization among neighboring domains is random. Subsequently, the ceramic element has no overall polarization.

The domains in a ceramic element are aligned by exposing the element to a strong, DC electric field, usually at a temperature *slightly below* the Curie temperature (Figure 2.2 (b)). This is referred to as the "poling process". After the poling treatment, domains most nearly aligned with the electric field expand at the expense of domains that are not aligned with the field, and the element expands in the direction of the field. When the electric field is removed most of the dipoles are locked into a configuration of near alignment

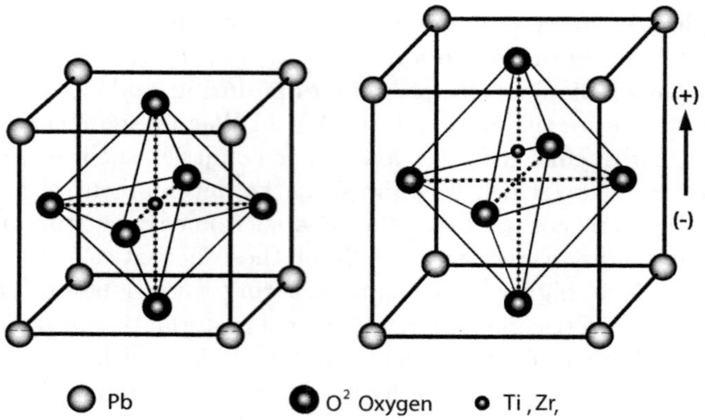

Figure 2.1. Crystalline structure of a piezoelectric ceramic, before and after polarization

(Figure 2.2 (c)). The element now has a permanent polarization, the remnant polarization, and is permanently elongated. The increase in the length of the element, however, is very small, usually within the micrometer range.

Properties of a poled piezoelectric ceramic element can be explained by the series of images in Figure 2.3. Mechanical compression or tension on the element changes the dipole moment associated with that element. This creates a voltage. Compression along the direction of polarization, or tension perpendicular to the direction of polarization, generates voltage of the same polarity as the poling voltage (Figure 2.3 (b)). Tension along the direction of polarization, or compression perpendicular to that direction, generates a voltage with polarity opposite to that of the poling voltage (Figure 2.3 (c)). When operating in this mode, the device is being used as a sensor. That is, the ceramic element converts the mechanical energy of compression or tension into electrical energy. Values for compressive stress and the voltage (or field

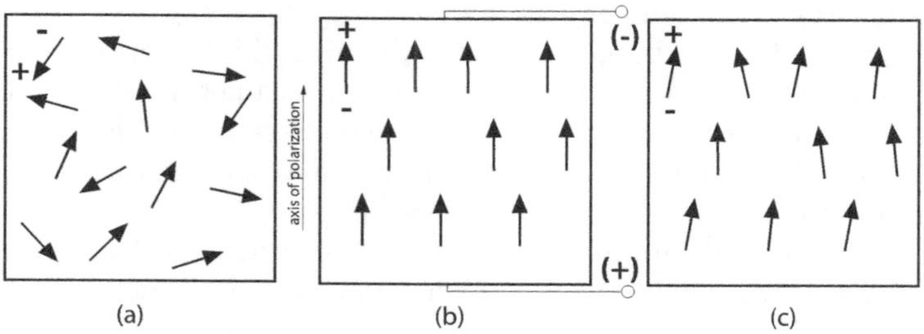

Figure 2.2. Poling process: (a) Prior to polarization polar domains are oriented randomly; (b) A very large DC electric field is used for polarization; (c) After the DC field is removed, the remnant polarization remains.

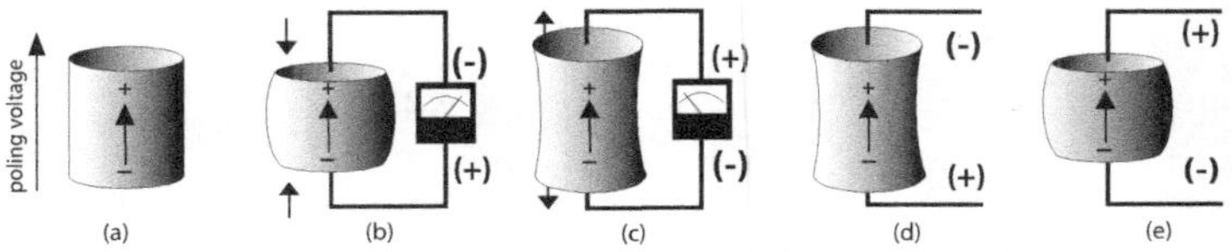

Figure 2.3. Reaction of a poled piezoelectric element to applied stimuli

strength) generated by applying stress to a piezoelectric ceramic element are linearly proportional, up to a specific stress, which depends on the material properties. The same is true for applied voltage and generated strain[3].

If a voltage of the same polarity as the poling voltage is applied to a ceramic element, in the direction of the poling voltage, the element will lengthen and its diameter will become smaller (Figure 2.3 (d)). If a voltage of polarity opposite to that of the poling voltage is applied, the element will become shorter and broader (Figure 2.3 (e)). If an alternating voltage is applied to the device, the element will expand and contract cyclically, at the frequency of the applied voltage. When operated in this mode, the piezoelectric ceramic is used as an actuator. That is, electrical energy is converted into mechanical energy.

2.4 Piezoelectric Constitutive Equations

In this section we introduce the equations which describe electromechanical properties of piezoelectric materials. The presentation is based on the IEEE standard for piezoelectricity [154] which is widely accepted as being a good representation of piezoelectric material properties. The IEEE standard assumes that piezoelectric materials are linear. It turns out that at low electric fields and at low mechanical stress levels piezoelectric materials have a linear profile. However, they may show considerable nonlinearity if operated under a high electric field or high mechanical stress level. In this book we are mainly concerned with the linear behavior of piezoelectric materials. That is, for the most part, we assume that the piezoelectric transducers are being operated at low electric field levels and under low mechanical stress.

When a poled piezoelectric ceramic is mechanically strained it becomes electrically polarized, producing an electric charge on the surface of the material. This property is referred to as the "direct piezoelectric effect" and is the

[3] It should be stressed that this statement is true when the piezoelectric material is being operated under small electric field, or mechanical stress. When subject to higher mechanical, or electrical fields, piezoelectric transducers display hysteresis-type nonlinearity. For the most part, in this monograph, the linear behavior of piezoelectric transducers will be of interest. However, Chapter 11 will briefly review the issues arising when a piezoelectric transducer is operated in the nonlinear regime.

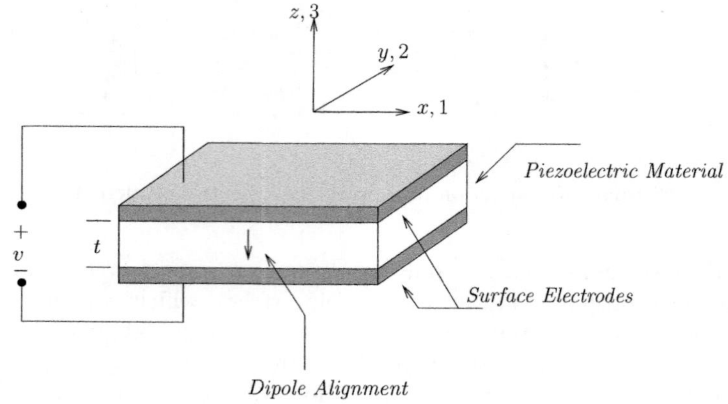

Figure 2.4. Schematic diagram of a piezoelectric transducer

basis upon which the piezoelectric materials are used as sensors. Furthermore, if electrodes are attached to the surfaces of the material, the generated electric charge can be collected and used. This property is particularly utilized in piezoelectric shunt damping applications to be discussed in Chapter 4.

The constitutive equations describing the piezoelectric property are based on the assumption that the total strain in the transducer is the sum of mechanical strain induced by the mechanical stress and the controllable actuation strain caused by the applied electric voltage. The axes are identified by numerals rather than letters. In Figure 2.4, 1 refers to the x axis, 2 corresponds to the y axis, and 3 corresponds to the z axis. Axis 3 is assigned to the direction of the initial polarization of the piezoceramic, and axes 1 and 2 lie in the plane perpendicular to axis 3. This is demonstrated more clearly in Figure 2.5.

The describing electromechanical equations for a linear piezoelectric material can be written as [154, 70]:

$$\varepsilon_i = S^E_{ij}\sigma_j + d_{mi}E_m \tag{2.1}$$

$$D_m = d_{mi}\sigma_i + \xi^\sigma_{ik}E_k, \tag{2.2}$$

where the indexes $i, j = 1, 2, \ldots, 6$ and $m, k = 1, 2, 3$ refer to different directions within the material coordinate system, as shown in Figure 2.5. The above equations can be re-written in the following form, which is often used for applications that involve sensing:

$$\varepsilon_i = S^D_{ij}\sigma_j + g_{mi}D_m \tag{2.3}$$

$$E_i = g_{mi}\sigma_i + \beta^\sigma_{ik}D_k \tag{2.4}$$

where

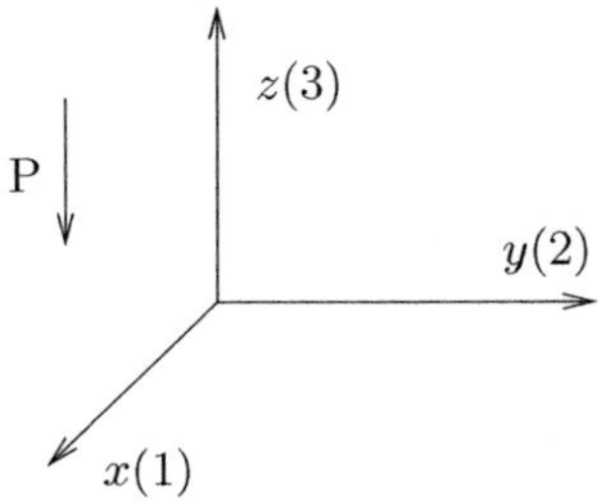

#	Axis
1	x
2	y
3	z
4	Shear around x
5	Shear around y
6	Shear around z

Figure 2.5. Axis nomenclature

σ ... stress vector (N/m^2)
ε ... strain vector (m/m)
E... vector of applied electric field (V/m)
ξ ... permitivity (F/m)
d ... matrix of piezoelectric strain constants (m/V)
S ... matrix of compliance coefficients (m^2/N)
D... vector of electric displacement (C/m^2)
g ... matrix of piezoelectric constants (m^2/C)
β ... impermitivity component (m/F)

Furthermore, the superscripts D, E, and σ represent measurements taken at constant electric displacement, constant electric field and constant stress.

Equations (2.1) and (2.3) express the converse piezoelectric effect, which describe the situation when the device is being used as an actuator. Equations (2.2) and (2.4), on the other hand, express the direct piezoelectric effect, which deals with the case when the transducer is being used as a sensor. The converse effect is often used to determine the piezoelectric coefficients.

In matrix form, Equations (2.1)-(2.4) can be written as:

$$\begin{bmatrix} \varepsilon_1 \\ \varepsilon_2 \\ \varepsilon_3 \\ \varepsilon_4 \\ \varepsilon_5 \\ \varepsilon_6 \end{bmatrix} = \begin{bmatrix} S_{11} & S_{12} & S_{13} & S_{14} & S_{15} & S_{16} \\ S_{21} & S_{22} & S_{23} & S_{24} & S_{25} & S_{26} \\ S_{31} & S_{32} & S_{33} & S_{34} & S_{35} & S_{36} \\ S_{41} & S_{42} & S_{43} & S_{44} & S_{45} & S_{46} \\ S_{51} & S_{52} & S_{53} & S_{54} & S_{55} & S_{56} \\ S_{61} & S_{62} & S_{63} & S_{64} & S_{65} & S_{66} \end{bmatrix} \begin{bmatrix} \sigma_1 \\ \sigma_2 \\ \sigma_3 \\ \tau_{23} \\ \tau_{31} \\ \tau_{12} \end{bmatrix} + \begin{bmatrix} d_{11} & d_{21} & d_{31} \\ d_{12} & d_{22} & d_{32} \\ d_{13} & d_{23} & d_{33} \\ d_{14} & d_{24} & d_{34} \\ d_{15} & d_{25} & d_{35} \\ d_{16} & d_{26} & d_{36} \end{bmatrix} \begin{bmatrix} E_1 \\ E_2 \\ E_3 \end{bmatrix} \tag{2.5}$$

and

$$\begin{bmatrix} D_1 \\ D_2 \\ D_3 \end{bmatrix} = \begin{bmatrix} d_{11} & d_{12} & d_{13} & d_{14} & d_{15} & d_{16} \\ d_{21} & d_{22} & d_{23} & d_{24} & d_{25} & d_{26} \\ d_{31} & d_{32} & d_{33} & d_{34} & d_{35} & d_{36} \end{bmatrix} \begin{bmatrix} \sigma_1 \\ \sigma_2 \\ \sigma_3 \\ \sigma_4 \\ \sigma_5 \\ \sigma_6 \end{bmatrix} + \begin{bmatrix} e^{\sigma}_{11} & e^{\sigma}_{12} & e^{\sigma}_{13} \\ e^{\sigma}_{21} & e^{\sigma}_{22} & e^{\sigma}_{23} \\ e^{\sigma}_{31} & e^{\sigma}_{32} & e^{\sigma}_{33} \end{bmatrix} \begin{bmatrix} E_1 \\ E_2 \\ E_3 \end{bmatrix}. \tag{2.6}$$

Some texts use the following notation for shear strain

$$\gamma_{23} = \varepsilon_4$$
$$\gamma_{31} = \varepsilon_5$$
$$\gamma_{12} = \varepsilon_6$$

and for shear stress

$$\tau_{23} = \sigma_4$$
$$\tau_{31} = \sigma_5$$
$$\tau_{12} = \sigma_6.$$

Assuming that the device is poled along the axis 3, and viewing the piezoelectric material as a transversely isotropic material, which is true for piezoelectric ceramics, many of the parameters in the above matrices will be either zero, or can be expressed in terms of other parameters. In particular, the non-zero compliance coefficients are:

$$S_{11} = S_{22}$$
$$S_{13} = S_{31} = S_{23} = S_{32}$$
$$S_{12} = S_{21}$$
$$S_{44} = S_{55}$$
$$S_{66} = 2(S_{11} - S_{12}).$$

The non-zero piezoelectric strain constants are

$$d_{31} = d_{32}$$

and

$$d_{15} = d_{24}.$$

Finally, the non-zero dielectric coefficients are $e^{\sigma}_{11} = e^{\sigma}_{22}$ and e^{σ}_{33}. Subsequently, the equations (2.5) and (2.6) are simplified to:

$$\begin{bmatrix} \varepsilon_1 \\ \varepsilon_2 \\ \varepsilon_3 \\ \varepsilon_4 \\ \varepsilon_5 \\ \varepsilon_6 \end{bmatrix} = \begin{bmatrix} S_{11} & S_{12} & S_{13} & 0 & 0 & 0 \\ S_{12} & S_{11} & S_{13} & 0 & 0 & 0 \\ S_{13} & S_{13} & S_{33} & 0 & 0 & 0 \\ 0 & 0 & 0 & S_{44} & 0 & 0 \\ 0 & 0 & 0 & 0 & S_{44} & 0 \\ 0 & 0 & 0 & 0 & 0 & 2(S_{11} - S_{12}) \end{bmatrix} \begin{bmatrix} \sigma_1 \\ \sigma_2 \\ \sigma_3 \\ \tau_{23} \\ \tau_{31} \\ \tau_{12} \end{bmatrix} + \begin{bmatrix} 0 & 0 & d_{31} \\ 0 & 0 & d_{31} \\ 0 & 0 & d_{33} \\ 0 & d_{15} & 0 \\ d_{15} & 0 & 0 \\ 0 & 0 & 0 \end{bmatrix} \begin{bmatrix} E_1 \\ E_2 \\ E_3 \end{bmatrix} \tag{2.7}$$

and

$$\begin{bmatrix} D_1 \\ D_2 \\ D_3 \end{bmatrix} = \begin{bmatrix} 0 & 0 & 0 & 0 & d_{15} & 0 \\ 0 & 0 & 0 & d_{15} & 0 & 0 \\ d_{31} & d_{31} & d_{33} & 0 & 0 & 0 \end{bmatrix} \begin{bmatrix} \sigma_1 \\ \sigma_2 \\ \sigma_3 \\ \sigma_4 \\ \sigma_5 \\ \sigma_6 \end{bmatrix} + \begin{bmatrix} e^{\sigma}_{11} & 0 & 0 \\ 0 & e^{\sigma}_{11} & 0 \\ 0 & 0 & e^{\sigma}_{33} \end{bmatrix} \begin{bmatrix} E_1 \\ E_2 \\ E_3 \end{bmatrix} .$$

The "piezoelectric strain constant" d is defined as the ratio of developed free strain to the applied electric field. The subscript d_{ij} implies that the electric field is applied or charge is collected in the i direction for a displacement or force in the j direction. The physical meaning of these, as well as other piezoelectric constants, will be explained in the following section.

The actuation matrix in (2.5) applies to PZT materials. For actuators made of PVDF materials, this matrix should be modified to

$$\begin{bmatrix} 0 & 0 & d_{31} \\ 0 & 0 & d_{32} \\ 0 & 0 & d_{33} \\ 0 & d_{25} & 0 \\ d_{15} & 0 & 0 \\ 0 & 0 & 0 \end{bmatrix} .$$

This reflects the fact that in PVDF films the induced strain is nonisotropic on the surface of the film. Hence, an electric field applied in the direction of the polarization vector will result in different strains in 1 and 2 directions.

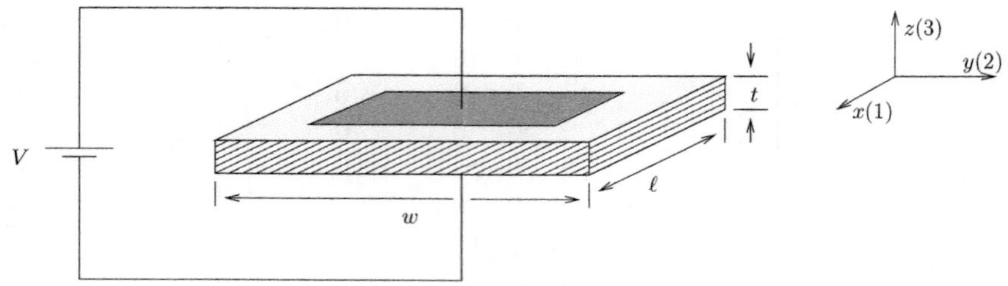

Figure 2.6. A piezoelectric transducer arrangement for d_{31} measurement

2.5 Piezoelectric Coefficients

This section reviews the physical meaning of some of the piezoelectric coefficients introduced in the previous section. Namely d_{ij}, g_{ij}, S_{ij} and e_{ij}.

2.5.1 Piezoelectric Constant d_{ij}

The piezoelectric coefficient d_{ij} is the ratio of the strain in the j-axis to the electric field applied along the i-axis, when all external stresses are held constant. In Figure 2.6, a voltage of V is applied to a piezoelectric transducer which is polarized in direction 3. This voltage generates the electric field

$$E_3 = \frac{V}{t}$$

which strains the transducer. In particular

$$\varepsilon_1 = \frac{\Delta\ell}{\ell}$$

in which

$$\Delta_\ell = \frac{d_{31} V \ell}{t}.$$

The piezoelectric constant d_{31} is usually a negative number. This is due to the fact that application of a positive electric field will generate a positive strain in direction 3.

Another interpretation of d_{ij} is the ratio of short circuit charge per unit area flowing between connected electrodes perpendicular to the j direction to the stress applied in the i direction. As shown in Figure 2.7, once a force F is applied to the transducer, in the 3 direction, it generates the stress

$$\sigma_3 = \frac{F}{\ell w}$$

which results in the electric charge

$$q = d_{33} F$$

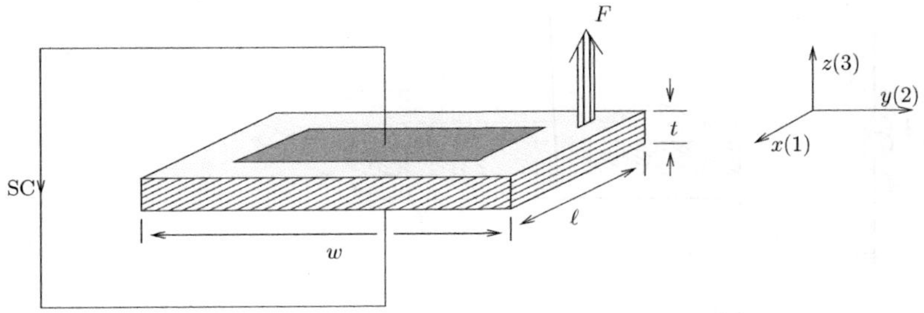

Figure 2.7. Charge deposition on a piezoelectric transducer - An equal, but opposite force, F, is not shown

flowing through the short circuit.

If a stress is applied equally in 1, 2 and 3 directions, and the electrodes are perpendicular to axis 3, the resulting short-circuit charge (per unit area), divided by the applied stressed is denoted by d_p.

2.5.2 Piezoelectric Constant g_{ij}

The piezoelectric constant g_{ij} signifies the electric field developed along the i-axis when the material is stressed along the j-axis. Therefore, in Figure 2.8 the applied force F, results in the voltage

$$V = \frac{g_{31} F}{w}.$$

Another interpretation of g_{ij} is the ratio of strain developed along the j-axis to the charge (per unit area) deposited on electrodes perpendicular to the i-axis. Therefore, in Figure 2.9, if an electric charge of Q is deposited on the surface electrodes, the thickness of the piezoelectric element will change by

$$\Delta \ell = \frac{g_{31} Q}{w}.$$

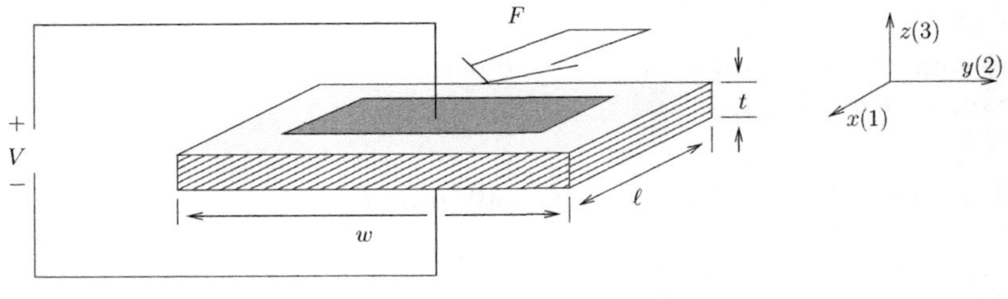

Figure 2.8. An open-circuited piezoelectric transducer under a force in direction 1 - An equal, but opposite force, F, is not shown

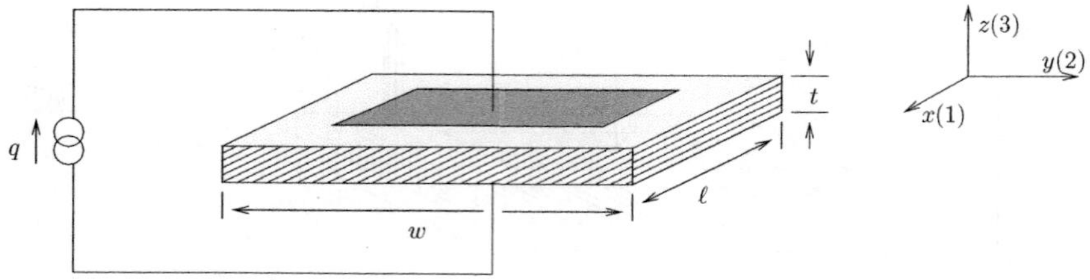

Figure 2.9. A piezoelectric transducer subject to applied charge

2.5.3 Elastic Compliance S_{ij}

The elastic compliance constant S_{ij} is the ratio of the strain the in i-direction to the stress in the j-direction, given that there is no change of stress along the other two directions. Direct strains and stresses are denoted by indices 1 to 3. Shear strains and stresses are denoted by indices 4 to 6. Subsequently, S_{12} signifies the direct strain in the 1-axis when the device is stressed along the 2-axis, and stresses along directions 1 and 3 are unchanged. Similarly, S_{44} refers to the shear strain around the 2-axis due to the shear stress around the same axis.

A superscript "E" is used to state that the elastic compliance S_{ij}^E is measured with the electrodes short-circuited. Similarly, the superscript "D" in S_{ij}^D denotes that the measurements were taken when the electrodes were left open-circuited. A mechanical stress results in an electrical response that can increase the resultant strain. Therefore, it is natural to expect S_{ij}^E to be smaller than S_{ij}^D. That is, a short-circuited piezo has a smaller Young's modulus of elasticity than when it is open-circuited.

2.5.4 Dielectric Coefficient, e_{ij}

The dielectric coefficient e_{ij} determines the charge per unit area in the i-axis due to an electric field applied in the j-axis. In most piezoelectric materials, a field applied along the j-axis causes electric displacement only in that direction. The relative dielectric constant, defined as the ratio of the absolute permitivity of the material by permitivity of free space, is denoted by K. The superscript σ in e_{11}^{σ} refers to the permitivity for a field applied in the 1 direction, when the material is not restrained.

2.5.5 Piezoelectric Coupling Coefficient k_{ij}

The piezoelectric coefficient k_{ij} represents the ability of a piezoceramic material to transform electrical energy to mechanical energy and *vice versa.* This transformation of energy between mechanical and electrical domains is employed in both sensors and actuators made from piezoelectric materials. The

ij index indicates that the stress, or strain is in the direction j, and the electrodes are perpendicular to the i-axis. For example, if a piezoceramic is mechanically strained in direction 1, as a result of electrical energy input in direction 3, while the device is under no external stress, then the ratio of stored mechanical energy to the applied electrical energy is denoted as k_{31}^2.

There are a number of ways that k_{ij} can be measured. One possibility is to apply a force to the piezoelectric element, while leaving its terminals open-circuited. The piezoelectric device will deflect, similar to a spring. This deflection Δ_z, can be measured and the mechanical work done by the applied force F can be determined

$$\mathcal{W}_M = \frac{F\Delta_z}{2}.$$

Due to the piezoelectric effect, electric charges will be accumulated on the transducer's electrodes. This amounts to the electrical energy

$$\mathcal{W}_E = \frac{Q^2}{2C_p}$$

which is stored in the piezoelectric capacitor. Therefore,

$$\begin{aligned} k_{33} &= \sqrt{\frac{\mathcal{W}_E}{\mathcal{W}_M}} \\ &= \frac{Q}{\sqrt{F\Delta_z C_p}}. \end{aligned}$$

The coupling coefficient can be written in terms of other piezoelectric constants. In particular

$$\begin{aligned} k_{ij}^2 &= \frac{d_{ij}^2}{S_{ij}^E e_{ij}^\sigma} \\ &= g_{ij} d_{ij} E_p, \end{aligned} \tag{2.8}$$

where E_p is the Young's modulus of elasticity of the piezoelectric material.

When a force is applied to a piezoelectric transducer, depending on whether the device is open-circuited or short-circuited, one should expect to observe different stiffnesses. In particular, if the electrodes are short-circuited, the device will appear to be "less stiff". This is due to the fact that upon the application of a force, the electric charges of opposite polarities accumulated on the electrodes cancel each other. Subsequently no electrical energy can be stored in the piezoelectric capacitor. Denoting short-circuit stiffness and open-circuit stiffness respectively as K_{sc} and K_{oc}, it can be proved that

$$\frac{K_{oc}}{K_{sc}} = \frac{1}{1 - k^2}.$$

2.6 Piezoelectric Sensor

When a piezoelectric transducer is mechanically stressed, it generates a voltage. This phenomenon is governed by the direct piezoelectric effect (2.2). This property makes piezoelectric transducers suitable for sensing applications. Compared to strain gauges, piezoelectric sensors offer superior signal to noise ratio, and better high-frequency noise rejection. Piezoelectric sensors are, therefore, quite suitable for applications that involve measuring low strain levels. They are compact, easy to embed and require moderate signal conditioning circuitry.

If a PZT sensor is subject to a stress field, assuming the applied electric field is zero, the resulting electrical displacement vector is:

$$\begin{Bmatrix} D_1 \\ D_2 \\ D_3 \end{Bmatrix} = \begin{bmatrix} 0 & 0 & 0 & 0 & d_{15} & 0 \\ 0 & 0 & 0 & d_{15} & 0 & 0 \\ d_{31} & d_{31} & d_{33} & 0 & 0 & 0 \end{bmatrix} \begin{Bmatrix} \sigma_1 \\ \sigma_2 \\ \sigma_3 \\ \tau_{23} \\ \tau_{31} \\ \tau_{12} \end{Bmatrix}.$$

The generated charge can be determined from

$$q = \int\int \begin{bmatrix} D_1 & D_2 & D_3 \end{bmatrix} \begin{bmatrix} dA_1 \\ dA_2 \\ dA_3 \end{bmatrix},$$

where dA_1, dA_2 and dA_3 are, respectively, the differential electrode areas in the 2-3, 1-3 and 1-2 planes. The generated voltage V_p is related to the charge via

$$V_p = \frac{q}{C_p},$$

where C_p is capacitance of the piezoelectric sensor.

Having measured the voltage, V_p, strain can be determined by solving the above integral. If the sensor is a PZT patch with two faces coated with thin electrode layers, *e.g.* the patch in Figure 2.4, and if the stress field only exists along the 1-axis, the capacitance can be determined from

$$C_p = \frac{\ell w e_{33}^{\sigma}}{t}.$$

Assuming the resulting strain is along the 1-axis, the sensor voltage is found to be

$$V_s = \frac{d_{31} E_p w}{C_p} \int_{\ell} \varepsilon_1 dx, \tag{2.9}$$

where E_p is the Young's modulus of the sensor and ε_1 is averaged over the sensor's length. The strain can then be calculated from

$$\varepsilon_1 = \frac{C_p V_s}{d_{31} E_p \ell w}. \tag{2.10}$$

In deriving the above equation, the main assumption was that the sensor was strained only along 1-axis. If this assumption is violated, which is often the case, then (2.10) should be modified to

$$\varepsilon_1 = \frac{C_p V_s}{(1-\nu) d_{31} E_p \ell w},$$

where ν is the Poisson's ratio[4].

2.7 Piezoelectric Actuator

Consider a beam with a pair of collocated piezoelectric transducers bonded to it as shown in Figure 2.10. The purpose of actuators is to generate bending in the beam by applying a moment to it. This is done by applying equal voltages, of 180° phase difference, to the two patches. Therefore, when one patch expands, the other contracts. Due to the phase difference between the voltages applied to the two actuators, only pure bending of the beam will occur, without any excitation of longitudinal waves. The analysis presented in this section follows the research reported in references [42, 11, 76, 53].

When a voltage V is applied to one of the piezoelectric elements, in the direction of the polarization vector, the actuator strains in direction 1 (the x-axis). Furthermore, the amount of free strain is given by

$$\varepsilon_p = \frac{d_{31} V}{t_p}, \tag{2.11}$$

where t_p represents the thickness of the piezoelectric actuator.

Since the piezoelectric patch is bonded to the beam, its movements are constrained by the stiffness of the beam. In the foregoing analysis perfect bonding of the actuator to the beam is assumed. In other words, the shearing effect of the non-ideal bonding layer is ignored [33]. Assuming that the strain distribution is linear across the thickness of the beam[5], we may write

$$\varepsilon(z) = \alpha z. \tag{2.12}$$

The above equation represents the strain distribution throughout the beam, and the piezoelectric patches, if the composite structure were bent, say by an external load, into a downward curvature. Subsequently, the portion of the beam above the neutral axis and the top patch would be placed in

[4] Notice that if $d_{31} \neq d_{32}$, *e.g.* if the sensor is a PVDF film, then this expression for strain must be changed to $\varepsilon_1 = \frac{C_p V_s}{(1-\nu \frac{d_{32}}{d_{31}}) d_{31} E_p \ell w}$.

[5] This is consistent with the Kirchoff hypothesis of laminate plate theory[97].

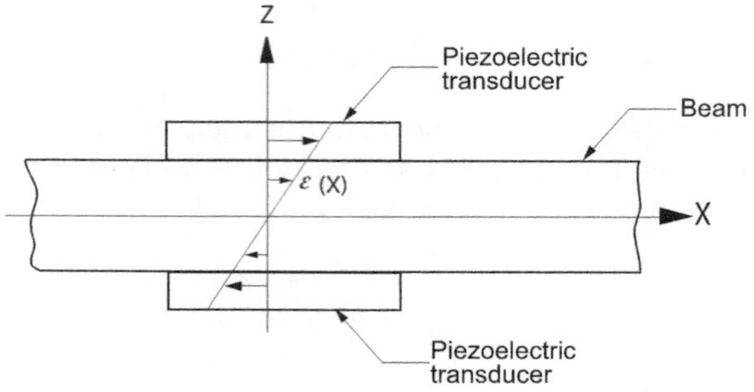

Figure 2.10. A beam with a pair of identical collocated piezoelectric actuators

tension, and the bottom half of the structure and the bottom patch in compression. Although, the strain is continuous on the beam-actuator surface, the stress distribution is discontinuous. In particular, using Hooke's law, the stress distribution within the beam is found to be

$$\sigma_b(z) = E_b \alpha z, \tag{2.13}$$

where E_b is the Young's modulus of elasticity of the beam. Since the two "identical" piezoelectric actuators are constrained by the beam, stress distributions inside the top and the bottom actuators can be written in terms of the total strain in each actuator (the strain that produces stress)

$$\sigma_p^t = E_p(\alpha z - \varepsilon_p) \tag{2.14}$$

$$\sigma_p^b = E_p(\alpha z + \varepsilon_p), \tag{2.15}$$

where E_p is the Young's modulus of elasticity of the piezoelectric material and the superscripts t and b refer to the top and bottom piezoelectric patches respectively. Applying moment equilibrium about the center of the beam[6] results in

$$\int_{-\frac{t_b}{2}-t_p}^{-\frac{t_b}{2}} \sigma_p^b(z) z dz + \int_{-\frac{t_b}{2}}^{\frac{t_b}{2}} \sigma_p(z) z dz + \int_{\frac{t_b}{2}}^{\frac{t_b}{2}+t_p} \sigma_p^t(z) z dz = 0. \tag{2.16}$$

After integration α is determined to be

$$\alpha = \frac{3E_p\left(\left(\frac{t_b}{2}+t_p\right)^2 - \left(\frac{t_b}{2}\right)^2\right)}{2\left(E_p\left\{\left(\frac{t_b}{2}+t_p\right)^3 - \left(\frac{t_b}{2}\right)^3\right\} + E_b\left(\frac{t_b}{2}\right)^3\right)}\varepsilon_p. \tag{2.17}$$

[6] Due to the symmetrical nature of the stress field, the integration need only be carried out starting from the centre of the beam, *i.e.* $\int_0^{\frac{t_b}{2}} \sigma_p(z) z dz + \int_{\frac{t_b}{2}}^{\frac{t_b}{2}+t_p} \sigma_p^t(z) z dz = 0$.

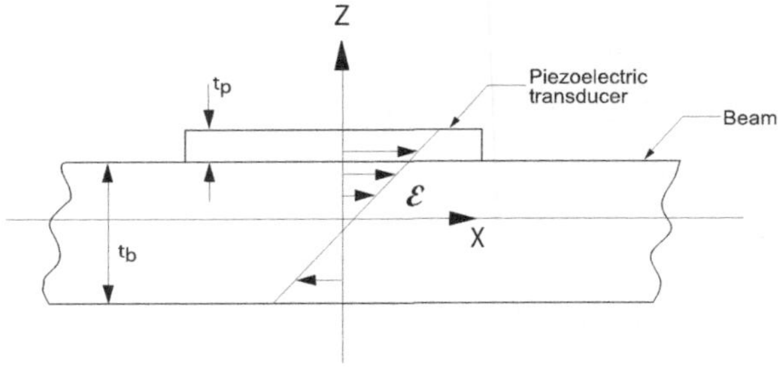

Figure 2.11. A beam with a single piezoelectric actuator

The induced moment intensity[7], M in the beam is then determined by integrating the triangular stress distribution across the beam:

$$M = E_b I \alpha, \tag{2.18}$$

where I is the beam's moment of inertia. Knowledge of M is crucial in determining the dynamics of the piezoelectric laminate beam.

If only one piezoelectric actuator is bonded to the beam, such as shown in Figure 2.11, then the strain distribution (2.12) needs to be modified to

$$\varepsilon(z) = (\alpha z + \varepsilon_0). \tag{2.19}$$

This expression for strain distribution across the beam thickness can be decomposed into two parts: the flexural component, αz and the longitudinal component, ε_0. Therefore, the beam extends and bends at the same time. This is demonstrated in Figure 2.12. The stress distribution inside the piezoelectric actuator is found to be

$$\sigma_p(z) = E_p(\alpha z + \varepsilon_0 - \varepsilon_p). \tag{2.20}$$

The two parameters, ε_0 and α can be determined by applying the moment equilibrium about the centre of the beam

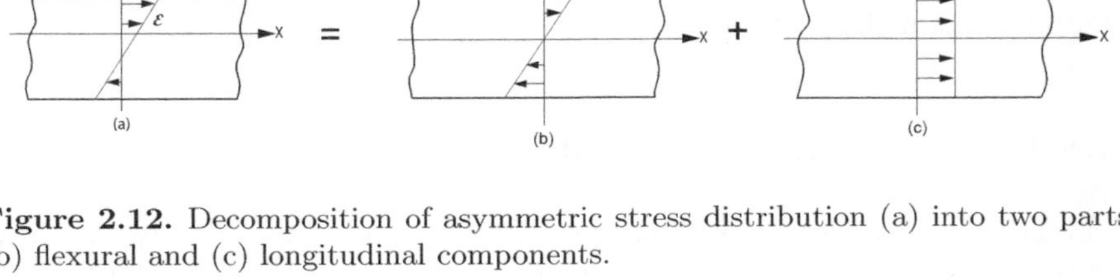

Figure 2.12. Decomposition of asymmetric stress distribution (a) into two parts: (b) flexural and (c) longitudinal components.

[7] moment per unit length

$$\int_{-\frac{t_b}{2}}^{\frac{t_b}{2}} \sigma_b(z) z dz + \int_{\frac{t_b}{2}}^{\frac{t_b}{2}+t_p} \sigma_p(z) z dz = 0 \tag{2.21}$$

and the force equilibrium along the x-axis

$$\int_{-\frac{t_b}{2}}^{\frac{t_b}{2}} \sigma_b(z) dz + \int_{\frac{t_b}{2}}^{\frac{t_b}{2}+t_p} \sigma_p(z) dz = 0. \tag{2.22}$$

Unlike the symmetric case, the force equilibrium condition (2.22) needs to be applied. This is due to the asymmetric distribution of strain throughout the beam. Solving (2.21) and (2.22) for ε_0 and α we obtain

$$\alpha = \frac{6E_b E_p t_b t_p (t_b + t_p)}{E_b^2 t_b^4 + E_p E_b (4t_b^3 t_p + 6t_b^2 t_p^2 + 4t_b t_p^3) + E_p^2 t_p} \varepsilon_p \tag{2.23}$$

and

$$\varepsilon_0 = \frac{\{E_b t_p^3 + E_p t_p^3\} E_p (t_b/2)}{E_b^2 t_b^4 + E_p E_b (4t_b^3 t_p + 6t_b^2 t_p^2 + 4t_b t_p^3) + E_p^2 t_p} \varepsilon_p. \tag{2.24}$$

The response of the beam to this form of actuation consists of a moment distribution

$$M_x = E_b I \alpha \tag{2.25}$$

and a longitudinal strain distribution

$$\varepsilon_x = \varepsilon_0. \tag{2.26}$$

It can be observed that the moment exerted on the beam by one actuator is not exactly half of that applied by two collocated piezoelectric actuators driven by 180° out-of-phase voltages. This arises from the fact that the Expressions (2.25) and (2.23) do not include the effect of the second piezoelectric actuator. However, if this effect is included by allowing for the stiffness of the second actuator in the derivations, while ensuring that the voltage applied to this patch is set to zero, then it can be shown that the resulting moment will be exactly half of that predicted by (2.18) and (2.17). The collocated situation is often used in vibration control applications, in which one piezoelectric transducer is used as an actuator while the other one is used as a sensor. This configuration is appealing for feedback control applications for reasons that will be explained in Chapter 3.

2.8 Piezoelectric 2D Actuation

This section is concerned with the use of piezoelectric actuators for excitation of two-dimensional structures, such as plates in pure bending. The analysis is similar to that presented in the previous section. A typical application is

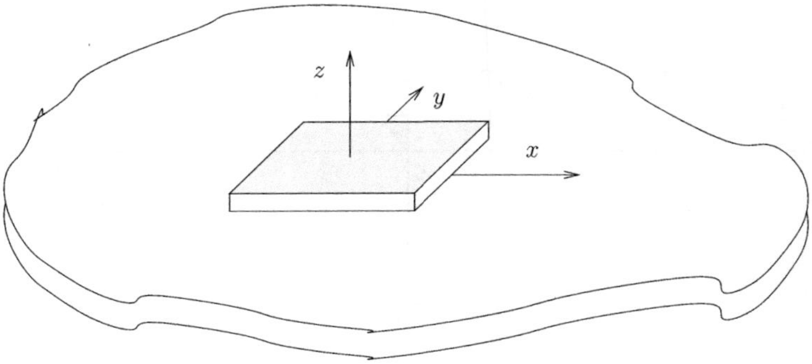

Figure 2.13. A piezoelectric actuator bonded to a plate

shown in Figure 2.13, which demonstrates a piezoelectric transducer bonded to the surface of a plate. It is also assumed that another identical transducer is bonded to the opposite side of the structure in a collocated fashion. If the two patches are driven by signals that are 180° out of phase, the resulting strain distribution, across the plate, will be linear as shown in Figure 2.14 a and b. That is,

$$\varepsilon_x = \alpha_x z \tag{2.27}$$

$$\varepsilon_y = \alpha_y z, \tag{2.28}$$

where α_x and α_y represent the strain distribution slopes in the $x-z$ and $y-z$ planes respectively.

Assuming that the piezoelectric material has similar properties in the 1 and 2 directions, *i.e.* $d_{31} = d_{32}$, the unconstrained strain associated with the actuator in both the x and y directions, under the voltage V, is given by

$$\varepsilon_p = \frac{d_{31}V}{t_p}.$$

Now the resulting stresses in the plate, in the x and y directions are

$$\sigma_x = \frac{E}{1-\nu^2}(\varepsilon_x + \nu\varepsilon_y)$$

and

$$\sigma_y = \frac{E}{1-\nu^2}(\varepsilon_y + \nu\varepsilon_x),$$

where ν is the Poisson's ratio of the plate material. Representing the stresses in the top piezoelectric patch as σ_x^p and σ_y^p, and the stresses in the bottom patch as $\tilde{\sigma}_x^p$ and $\tilde{\sigma}_y^p$, we may write

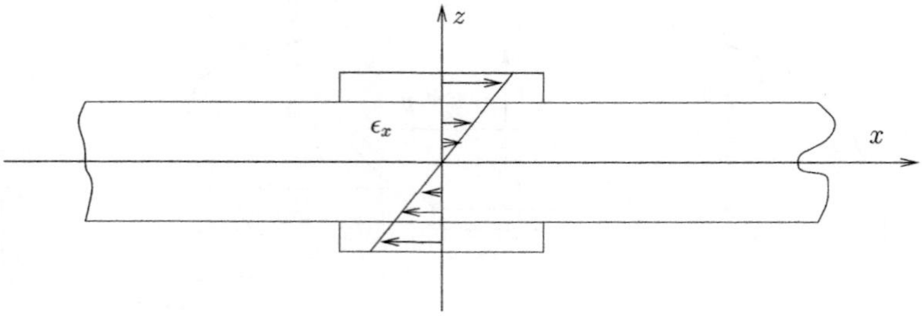

(a)

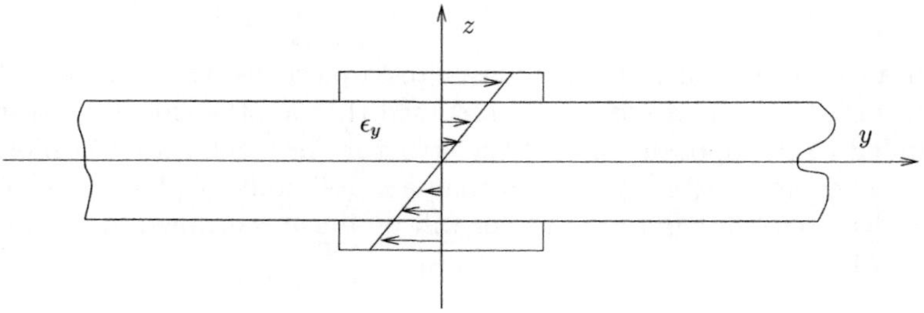

(b)

Figure 2.14. Two dimensional strain distribution in a plane with two collocated anti-symmetric piezoelectric actuators

$$\sigma_x^p = \frac{E_p}{1-\nu_p^2}\left\{\varepsilon_x + \nu_p\varepsilon_y - (1+\nu_p)\varepsilon_p\right\} \tag{2.29}$$

$$\tilde{\sigma}_x^p = \frac{E_p}{1-\nu_p^2}\left\{\varepsilon_x + \nu_p\varepsilon_y + (1+\nu_p)\varepsilon_p\right\} \tag{2.30}$$

$$\sigma_y^p = \frac{E_p}{1-\nu_p^2}\left\{\varepsilon_y + \nu_p\varepsilon_x - (1+\nu_p)\varepsilon_p\right\} \tag{2.31}$$

$$\tilde{\sigma}_y^p = \frac{E_p}{1-\nu_p^2}\left\{\varepsilon_y + \nu_p\varepsilon_x + (1+\nu_p)\varepsilon_p\right\}, \tag{2.32}$$

where ν_p is the Poisson's ratio of the piezoelectric material.

Given that ε_p is the same in both directions and that the plate is homogeneous, we may write

$$\varepsilon_x = \varepsilon_y = \varepsilon.$$

Subsequently, the strain distribution across the plate thickness can be written as

$$\varepsilon = \alpha_x z = \alpha_y z = \alpha z.$$

The condition of moment of equilibrium about the x and y axes can now be applied. That is,

$$\int_0^{\frac{t}{2}} \sigma_x z dz + \int_{\frac{t}{2}}^{\frac{t}{2}+t_p} \sigma_x^p z dz = 0$$

and

$$\int_0^{\frac{t}{2}} \sigma_y z dz + \int_{\frac{t}{2}}^{\frac{t}{2}+t_p} \sigma_y^p z dz = 0,$$

where t represents the plate's thickness. Integrating and solving for α gives

$$\alpha = \frac{3E_p\{(\frac{t_b}{2}+t_p)^2 - (\frac{t_b}{2})^2\}(1-\nu)}{2E_p\{(\frac{t_b}{2}+t_p)^3 - (\frac{t_b}{2})^3\}(1-\nu) + 2E(\frac{t_b}{2})^3(1-\nu_p)}\varepsilon_p.$$

The resulting moments in x and y directions are

$$M_x = M_y = EI\alpha. \tag{2.33}$$

For the symmetric case, *i.e.* when only one piezoelectric actuator is bonded to the plate, similar derivations to the previous section can be made.

2.9 Dynamics of a Piezoelectric Laminate Beam

In this section we explain how the dynamics of a beam with a number of collocated piezoelectric actuator/sensor pairs can be derived. At this stage we do not make any specific assumptions about the boundary conditions since we wish to keep the discussion as general as possible. However, we will explain how the effect of boundary conditions can be incorporated into the model.

Let us consider a setup as shown in Figure 2.15, where m identical collocated piezoelectric actuator/sensor pairs are bonded to a beam. The assumption that all piezoelectric transducers are identical is only adopted to simplify the derivations, and can be removed if necessary. The ith actuator is exposed to a voltage of $v_{a_i}(t)$ and the voltage induced in the ith sensor is $v_{p_i}(t)$.

We assume that the beam has a length of L, width of W, and thickness of t_b. Corresponding dimensions of each piezoelectric transducer are L_p, W_p, and t_p. Furthermore, we denote the transverse deflection of the beam at point x and time t by $z(x,t)$. The dynamics of such a structure are governed by the Bernoulli-Euler partial differential equation

$$E_b I \frac{\partial^4 z(x,t)}{\partial x^4} + \rho A_b \frac{\partial^2 z(x,t)}{\partial t^2} = \frac{\partial^2 M_x(x,t)}{\partial x^2}, \tag{2.34}$$

where ρ, A_b, E_b and I represent density, cross-sectional area, Young's modulus of elasticity and moment of inertia about the neutral axis of the beam respectively. The total moment acting on the beam is represented by $M_x(x,t)$, which is the sum of moments exerted on the beam by each actuator, *i.e.*

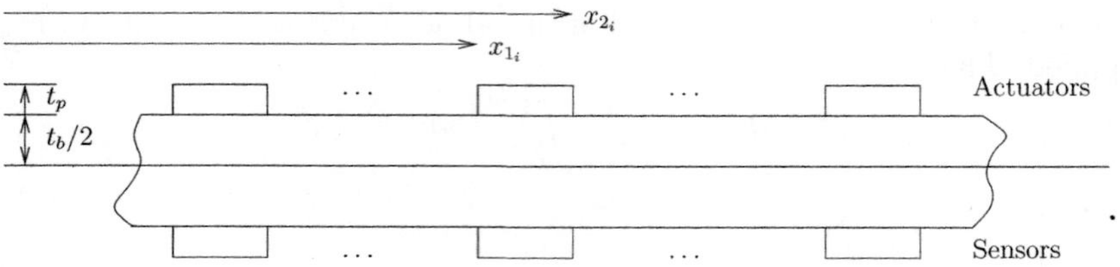

Figure 2.15. A beam with a number of collocated piezoelectric actuator/sensor pairs

$$M_x(x,t) = \sum_{i=1}^{m} M_{x_i}(x,t). \tag{2.35}$$

The moment exerted on the beam by the ith actuator, $M_{x_i}(x,t)$ can be written as

$$M_{x_i}(x,t) = \bar{\kappa} v_{a_i}(t)\{u(x - x_{1_i}) - u(x - x_{2_i})\}, \tag{2.36}$$

where $u(x)$ represents the unit step function, *i.e.* $u(x) = 0$ for $x < 0$ and $u(x) = 1$ for $x \geq 0$. The term $\{u(x - x_{1_i}) - u(x - x_{2_i})\}$ is incorporated into (2.36) to account for the spatial placement of the ith actuator. The constant $\bar{\kappa}$ can be determined from (2.11), (2.17) and (2.18). The forcing term in (2.34) can now be determined from Expressions (2.36) and (2.35), and using the following property of Dirac delta function

$$\int_{-\infty}^{\infty} \delta^{(n)}(t - \theta)\phi(t)dt = (-1)^n \phi^{(n)}(\theta), \tag{2.37}$$

where $\delta^{(n)}$ is the nth derivative of δ, and ϕ is a continuous function of θ [111]. Having determined the expression for the forcing function in (2.34) we can now proceed to solving the partial differential equation. One approach to solving this PDE is based on using the modal analysis approach [130]. In this technique the solution of the PDE is assumed to be of the form

$$z(x,t) = \sum_{k=1}^{\infty} w_k(x) q_k(t). \tag{2.38}$$

Here $w_k(x)$, known as the modeshape, is the eigenfunction which is determined from the eigenvalue problem obtained by substituting (2.38) into (2.34) and using the following orthogonality properties [130]

$$\int_0^L w_k(x) w_p(x) dx = \delta_{kp} \tag{2.39}$$

$$\int_0^L \frac{E_b I}{\rho A_b} \frac{d^4 w_k(x)}{dx^4} w_p(x) dx = \omega_k^2 \delta_{kp}, \tag{2.40}$$

where ω_k describes the kth natural frequency of the beam and δ_{kp} is the Kroneker delta function, *i.e.*

$$\delta_{kp} = \begin{cases} 1, & k = p \\ 0, & \text{otherwise} \end{cases}$$

A solution to the eigenvalue problem requires precise knowledge of boundary conditions. For specific boundary conditions, *e.g.* simply-supported and cantilevered, mode shapes and resonance frequencies can be determined analytically from the eigenvalue problem. For further details, the reader is referred to [130] and [50].

A set of uncoupled ordinary differential equations can be obtained from (2.34) using the orthogonality properties (2.39) and (2.40) and the property (2.37), as well as (2.38). It can be shown that the ordinary differential equations are of the form:

$$\ddot{q}_k(t) + \omega_k^2 q_k(t) = \frac{\bar{\kappa}}{\rho A_b} \sum_{i=1}^{m} \psi_{k_i} v_{a_i}(t), \tag{2.41}$$

where $k = 1, 2, \ldots$ and $q_k(t)$ is the generalized coordinate of the kth mode. Furthermore, the parameter ψ_{k_i} is found to be:

$$\psi_{k_i} = \int_0^L w_k(x) \left\{ \delta'(x - x_{1_i}) - \delta'(x - x_{2_i}) \right\} dx \tag{2.42}$$

$$= w_k'(x_{2_i}) - w_k'(x_{1_i}), \tag{2.43}$$

where $f'(x)$ represents the first derivative of the function f with respect to x, and we have used (2.37) to obtain (2.43) from (2.42).

To this end we point out that the differential equation (2.41) does not contain a term to account for the natural damping associated with the beam. The presence of damping can be incorporated into (2.41) by adding the term $2\zeta_k \dot{q}_k(t)$ to (2.41). This results in the differential equation

$$\ddot{q}_k(t) + 2\zeta_k \omega_k \dot{q}_k(t) + \omega_k^2 q_k(t) = \frac{\bar{\kappa}}{\rho A_b} \sum_{i=1}^{m} \psi_{k_i} v_{a_i}(t). \tag{2.44}$$

Applying the Laplace transform to (2.44), assuming zero initial conditions, we obtain the following transfer function from the vector of applied actuator voltages $V_a(s) = [v_{a_1}(s), \ldots, v_{a_m}(s)]$ to the beam deflection $z(x, s)$ at location x

$$G(x, s) = \gamma \sum_{i=1}^{\infty} \frac{w_k(x) \bar{\psi}_k'}{s^2 + 2\zeta_k \omega_k s + \omega_k^2}, \tag{2.45}$$

where $\gamma = \frac{\bar{\kappa}}{\rho A_b}$ and

$$\bar{\psi}_k = \begin{bmatrix} \psi_{k_1} \\ \vdots \\ \psi_{k_m} \end{bmatrix}.$$

The piezoelectric voltage induced in the ith sensor can be obtained using Expression (2.9). That is,

$$v_{p_i}(t) = \frac{d_{31} E_p W_p}{C_p} \int_0^L \varepsilon_{x_i} dx.$$

The expression for the mechanical strain in the ith sensor patch can be obtained from

$$\varepsilon_{x_i} = -\left(\frac{t_b}{2} + t_p\right) \frac{\partial^2 z}{\partial x^2}.$$

Now v_{p_i} is found to be

$$v_{p_i}(t) = \frac{-d_{31} E_p W_p \left(\frac{t_b}{2} + t_p\right)}{C_p} \sum_{k=1}^{\infty} \psi_{k_i} q_k(t). \tag{2.46}$$

Therefore, the transfer function matrix relating the voltages applied to the piezoelectric actuators $V_a(s) = [v_{a_1}(s), \ldots, v_{a_m}(s)]$ to the voltages measured at the piezoelectric sensors $V_p(s) = [v_{p_1}(s), \ldots, v_{p_m}(s)]$ is found to be:

$$G_{vv}(s) = \bar{\gamma} \sum_{k=1}^{\infty} \frac{\bar{\psi}_k \bar{\psi}_k'}{s^2 + 2\zeta_k \omega_k s + \omega_k^2}, \tag{2.47}$$

where

$$\bar{\gamma} = \frac{-d_{31} E_p W_p \left(\frac{t_b}{2} + t_p\right) \bar{\kappa}}{C_p \rho A_b}.$$

To this end we need to explain how the systems represented by Infinite Series (2.47) and (2.45) can be approximated with finite dimensional models. In any controller design scenario we may only be interested in designing a controller for a finite bandwidth. If N modes of the structure lie within that bandwidth of interest, the series (2.47) and (2.45) are often truncated to obtain a finite dimensional model of the structure, which is of minimum dimensions. While the truncation may not be of concern in non-collocated models, *e.g.* (2.45), it can be a serious problem for a collocated system, *e.g.* (2.47). Truncation of a collocated model can result in perturbations in open-loop zeros of the system, which in the worst case can cause closed-loop instabilities, and in the best case will contribute to the loss of closed-loop performance [34].

A number of techniques have been proposed to compensate for the effect of truncated out-of-bandwidth modes on collocated structural models. The reader is referred to [34, 207, 138, 135, 84] and references therein for an overview of such techniques. We do, however, point out that the truncation

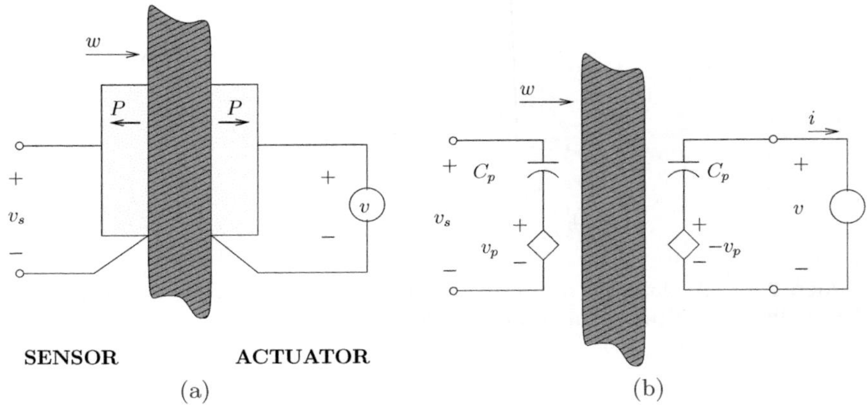

Figure 2.16. (a) A flexible structure with a pair of collocated piezoelectric transducers, and (b) its electrical equivalent model

error can be minimized by appending the truncated model by a feed-through term, *i.e.* to approximate (2.47) with

$$G_{vv}(s) = \bar{\gamma} \sum_{k=1}^{N} \frac{\bar{\psi}_k \bar{\psi}'_k}{s^2 + 2\zeta_k \omega_k s + \omega_k^2} + D.$$

If chosen properly, the addition of this feed-through term to the truncated model can result in an acceptable approximation. In this book, in most cases, we will use system identification to directly identify this parameter, as well as the rest of the dynamics of the system.

We conclude this chapter with an important observation that will be utilized in the forthcoming chapters. Figure 2.16 (a) illustrates a flexible structure with a collocated piezoelectric actuator/sensor pair. The piezoelectric transducer on the right functions as an actuator, and the one on the left as a sensor. What is of importance here is the electrical model of the system depicted in Figure 2.16 (b). Each piezoelectric transducer is modeled as a capacitor in series with a strain-dependent voltage source. The transfer function from the voltage applied to the actuator, v to the voltage induced in the sensor, $v_s = v_p$ is given by (2.47). This observation is particularly important in designing piezoelectric shunts, and to identify the connection between piezoelectric shunt damping and feedback control of a collocated system. This will be discussed in more detail in Chapter 5.

2.10 Active and Macro Fiber Composite Transducers

Active Fiber Composites (AFCs) are an alternative to traditional monolithic piezoelectric transducers. First proposed in 1992 [20], longitudinally polarized

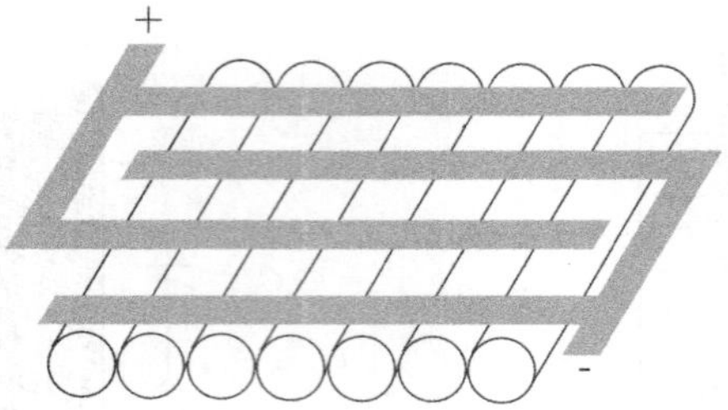

Figure 2.17. A piezoelectric Active Fiber Composite (AFC) comprised of piezoelectric fibers with interdigitated electrodes top and bottom

piezoelectric fibers, as shown in Figure 2.17, are encased in an epoxy resin with interdigitated electrodes laminated onto the top and bottom surfaces of the transducer. With an applied voltage, the interdigitated electrodes induce longitudinal electric fields along the length of each fiber. The original motivation was to increase the electromechanical coupling by utilizing the high d_{33} piezoelectric strain constant rather than the lesser d_{31} constant.

Active fiber composites have a number of practical advantages over traditional monolithic transducers [19]:

- The fibers are encapsulated by the printed polymer electrodes and epoxy resin thus increasing the reliability and service-life in harsh environments.
- The short length and diameter or the fibers together with their alignment along the length of the transducer increases the conformability of AFC transducers. They can be laminated onto structures with complex geometries and curvatures.
- AFC transducers are more robust to mechanical failure than monolithic transducers. In addition to their conformability, they can also tolerate local and incremental damage. If some of the fibers are fractured, the transducer will not be substantially damaged, in contrast, monolithic actuators will fracture and fail if they are stressed beyond their yield limit.
- AFCs have been reported to develop greater strains than monolithic actuators [19]. The strain actuation is also unidirectional.

Macro Fiber Composites (MFCs) [174] are similar in nature to AFCs as they utilize the direct d_{33} piezoelectric effect through the use of interdigitated electrodes. Rather than individual fibers, a monolithic transducer is simply cut into a number of long strips. The resulting transducer is conformable in one dimension and more robust to mechanical failure than monolithic patches.

The greatest disadvantages of AFC and MFC transducers is their high present cost, and the large voltages required to achieve the same actuation strain as monolithic transducers. The equivalent piezoelectric capacitance is also much lesser making them unsuitable as low-frequency strain sensors (see Section 6.2).

The low capacitance of AFC and MFC transducers also causes difficulties in the implementation of piezoelectric shunt damping systems, to be discussed in Chapter 4. Device capacitances of less than 50 nF have been deemed impractical for shunt damping [18]. A performance comparison of monolithic, AFC, and MFC transducers in a passive shunt damping application can be found in references [18] and [148].

Although the piezoelectric transducers used throughout this book are exclusively monolithic, all of the techniques discussed in the proceeding chapters are equally as applicable to AFC and MFC variants. Indeed, from the control engineers viewpoint, transducer physics is usually lumped into a simplified electrical model, or identified as part of the structural system.

3

Feedback Control of Structural Vibration

3.1 Introduction

This chapter is concerned with the problem of feedback control system design for vibration reduction in flexible structures. In the previous chapter we studied some important properties of flexible structures. We paid special attention to those structures in which piezoelectric transducers were used in the form of actuators and sensors.

Transfer functions of flexible structures have interesting properties. For example, different transfer functions associated with the same structure have identical poles. These systems have a large number of poles, and these poles are very close to the $j\omega$ axis. The phase of a collocated system is between 0 and -180 degrees. This property implies that poles and zeros of the system interlace.

Some characteristics of flexible structures make the task of controlling them rather difficult. For example, the fact that these transfer functions are of very high orders means that if a controller is designed with a view to control only a small number of in-bandwidth modes of the system, the existence of out-of-bandwidth modes could potentially destabilize the closed-loop system. This problem has been investigated in the literature in the context of spill-over effect [12, 13].

Despite the fact that control of flexible structures is a difficult task, it is possible to design and implement high-performance controllers capable of reducing structural vibrations and ensuring closed-loop stability of the controlled system. In particular, when actuators and sensors are collocated and compatible, *e.g.* collocated piezoelectric transducers, very effective control design techniques can be proposed.

All of the control design techniques introduced in this chapter assume a collocated structure. Collocation is a property of flexible structures that enables the control designer to devise feedback structures that are guaranteed to be closed-loop stable in the presence of out-of-bandwidth dynamics. Although in some applications it may be difficult to collocate the sensor and

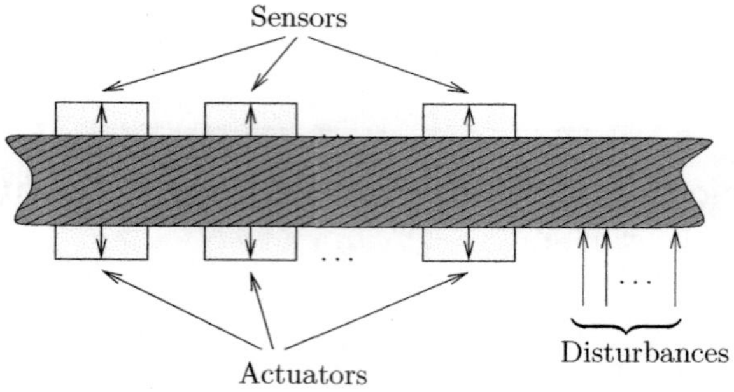

Figure 3.1. A flexible structure with several collocated piezoelectric actuator/sensor paris

actuator, if piezoelectric transducers are to be used for both actuation and sensing collocation should not be viewed as a difficulty. Collocated piezoelectric actuator/sensor packs are commercially available in the market.

In this chapter we will also introduce the problem of self-sensing piezoelectric actuators. The idea is to use a piezoelectric transducer simultaneously as a sensor and an actuator, thereby reducing the number of requisite piezoelectric transducers by half. This is done by simulating the presence of a collocated piezoelectric sensor using properties of the piezoelectric actuated system and an electronic circuit. Some of the techniques and procedures are described and complications associated with the proposed methods are discussed. A more practical approach will be introduced in the next chapter.

3.2 Structural Properties of Resonant Systems

As explained in the previous chapter resonant systems, such as flexible structures are continuous distributed systems described by hyperbolic partial differential equations. However, for practical considerations, it is common to approximate these systems by lumped models. Such models can be obtained through a number of ways, *e.g.* finite element modeling, modal analysis or system identification.

In this book we are mainly concerned with flexible structures with embedded, or bonded piezoelectric transducers. Such a system is shown in Figure 3.1. In the previous chapter we demonstrated that the multivariable collocated transfer function of the system in Figure 3.1 can be represented as

$$G_{vv}(s) = \sum_{i=1}^{M} \frac{\psi_i \psi_i'}{s^2 + 2\zeta_i \omega_i s + \omega_i^2}, \tag{3.1}$$

where ψ_i is an $m \times 1$ vector (assuming m piezoelectric actuator/sensor pairs), and $M \to \infty$. In practice, however, the integer M is finite, but possibly a very

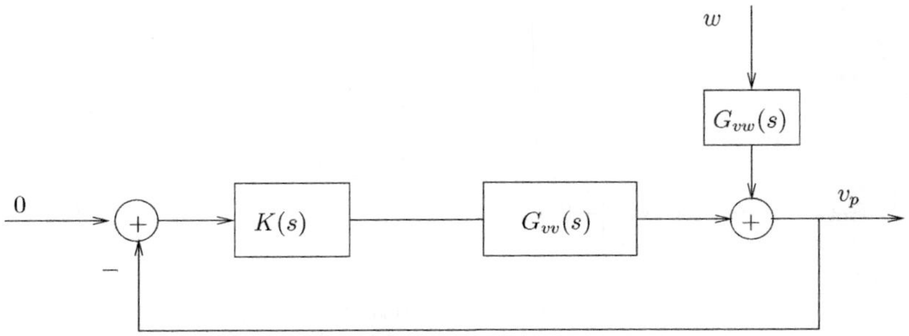

Figure 3.2. Feedback control of vibration with a collocated piezoelectric actuator/sensor pair

large number which represents the number of modes that sufficiently describe the elastic properties of the structure under excitation [93, 100].

Also shown in Figure 3.1 is a number of disturbances acting on the structure. These may represent point forces, distributed forces (such as a wind gust) or torques acting on the structure. Nevertheless, the transfer function matrix relating the disturbance vector W to the vector of voltages measured at the sensors V_p can be written as

$$G_{vw}(s) = \sum_{i=1}^{M} \frac{\psi_i \gamma_i'}{s^2 + 2\zeta_i \omega_i s + \omega_i^2}, \tag{3.2}$$

where γ_i is a $\ell \times 1$ matrix, assuming there are ℓ disturbances acting on the structure.

For the SISO case, *i.e.* when $G_{vv}(s)$ represents the transfer function associated with only one collocated actuator/sensor pair, this system is known to possess interesting properties [126]. In particular, it is known that the system is minimum phase, and furthermore, poles and zeros of the system interlace. This ensures that phase of the collocated transfer function will be within the 0 to $-180°$ range.

Given the highly resonant nature of flexible structures it is natural to investigate ways of adding damping to the system. One approach to adding damping to the structure is to use a feedback controller, as illustrated in Figure 3.2.

The transfer function matrix (3.1) consists of a very large number of parallel second order terms. This is demonstrated in block diagram form in Figure 3.3, which illustrates the situation when such a system is being controlled. In most scenarios, only control of a limited bandwidth is of importance. Typically N modes of the structure would fit within this bandwidth while modes $N+1$ and above are left uncontrolled. The uncontrolled modes, however, do exist and have the potential to destabilize the closed-loop system. Therefore, the existence of these modes should be taken into account, and a controller

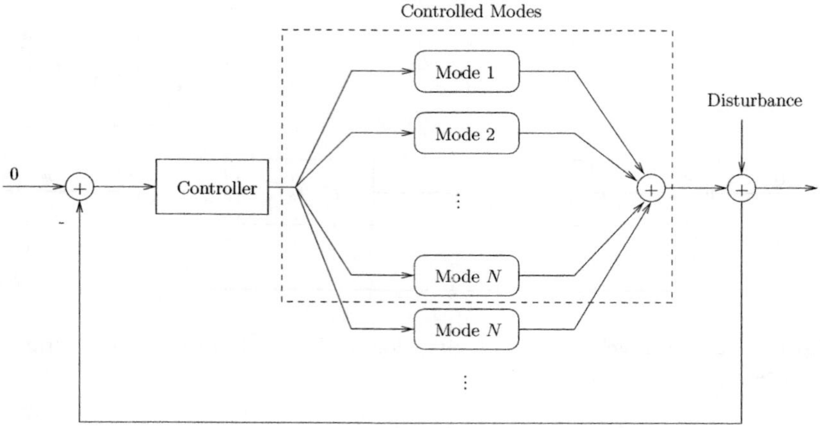

Figure 3.3. Feedback control of flexible structure consisting of a large number of elastic modes

should be designed to ensure adequate damping performance, as well as stability in the presence of these out-of-bandwidth modes.

A number of control design techniques which satisfy the above requirements will be introduced in the forthcoming sections.

3.3 Modeling and System Identification

During the analysis and design of structural vibration control systems it is of great benefit to possess a dynamic system model. That is, a state-space or transfer-function model describing outputs such as sensor voltages and displacements in response to external forces and control inputs. A model allows the designer to utilize model-based design tools and to simulate the characteristics of a closed-loop system before experimental implementation. A brief discussion follows on the three most common techniques for procuring models of flexible structures. It is concluded that the last technique, system identification, is the most suitable for the applications described in this book.

3.3.1 Analytic Modeling

Analytic modeling, typically involving the assumed modes approach [130], requires distinct models for both structural dynamics and piezoelectric transducers [70]. The modal analysis procedure has been used extensively throughout the literature for obtaining models of structural [130] and acoustic systems [87]. Its major disadvantage is the requirement for detailed physical information regarding the sensors, actuators, and the underlying mechanical system. Practical application typically involves the use of experimental data and a non-linear optimization to identify unknown parameters such as modal amplitudes, resonance frequencies and damping ratios. Even in this case, the descriptive partial differential equations must still be solved (as functions of the unknown parameters) to obtain the mode shapes. This may be difficult or impossible for realistic structural or acoustic systems with complicated boundary conditions.

As discussed in Section 2.9, the Lagrangian or modal expansion represents a small deflection of a distributed parameter system as an infinite summation of modes [130]. The modes are a product of two functions, one of the spatial co-ordinate x, and another of the temporal t, *i.e.*

$$z(x,t) = \sum_{k=1}^{\infty} w_k(x) q_k(t), \tag{3.3}$$

where the $q_k(t)$s are the modal displacements, $z(x,t)$ is the displacement at a point, and the $w_k(x)$s are the system eigenfunctions or mode-shapes in the context of mechanical systems analysis. The mode shapes $w_k(x)$ must form a complete coordinate basis for the system, satisfy the geometric boundary conditions, and for analytic analysis be differentiable over the spatial domain to at least the degree required by the describing partial differential equations. Many practical systems also obey certain orthogonality conditions. The reader is referred to Section 2.9 for an in-depth discussion of the Partial Differential Equations (PDEs) governing flexible structures and their solution by modal analysis.

3.3.2 Finite Element Analysis

Another popular technique for obtaining structural models is that of finite element (FE) analysis [40]. This is an approximate method that results in high-order spatially discrete systems. If sensors and actuator dynamics are known, the integrated model can be cast in a state-space form to facilitate control design and analysis [115].

In situations involving non-trivial boundary conditions or irregular geometry, finite element modeling is generally regarded as a better alternative to analytic modeling. Due to the approximate nature of finite element modeling, there is no requirement to solve a partial differential equation, thus, the process is significantly more straightforward. Unfortunately, detailed information regarding the structures' material properties and boundary conditions is still required. Similar to the modal analysis procedure, errors in these parameters are accounted for by correcting the FE models with experimental data [56].

3.3.3 System Identification

System identification can be employed to procure a composite structural and piezoelectric model directly from experimental data. Although the field of system identification is extremely diverse [120], the range of possibilities is significantly reduced if we restrict ourselves to techniques capable of identifying systems with multiple inputs and outputs, that do not require an explicit parameterization or non-linear optimization, two rather undesirable traits. The majority of the residue comprises the so-called subspace class of system identification algorithms. Such methods identify state-space models by exploiting geometric properties of the input and output sequences. Methods exist for both time and frequency domain data, see [189] for a summary of time domain methods.

Before selecting a system identification algorithm it is worth considering the various options for data collection and processing. The traditional approach has been to record input and output sequences then apply a time-domain system identification technique directly. When considering systems with highly resonant modes a number of problems arise:

1. Unless the system dynamics are known *a-priori*, it is impossible to excite all of the modes of interest equally, or to achieve a reasonably constant signal to noise ratio over the entire bandwidth. Typically the first mode will dominate the energy of the measurement.
2. Data records can be prohibitively long. As an example consider a typical mechanical system - a cantilever beam as described in Section 8.3.2 with a first-mode resonance frequency of $11Hz$ and a fourth-mode resonance frequency of $400Hz$. Allowing for the roll-off of anti-aliasing and reconstruction filters, a suitable sampling frequency would be around 4-$10kHz$. To record only 10 cycles of the lowest frequency mode, approximately 10,000 samples would be required. Records of this size typically result in numerical instabilities, out-of-memory problems, and large amounts of computing time. This problem is exacerbated if the system under consideration has multiple inputs or outputs.

Frequency domain data is a more suitable format for highly resonant systems. Frequency domain data can be acquired directly by swept sine analysis or by computing the ratio of input and output Fourier transforms. A full discussion on obtaining empirical transfer function estimates (frequency responses) from time domain data can be found in [159] and [120]. The most straightforward approach is to excite the system with a periodic disturbance, *e.g.* a periodic chirp, and to record exactly an integer number of periods. This avoids Fourier transform leakage and negates the need for window functions introducing systematic error. Even though a frequency response obtained by Fourier transform still requires relatively long time histories, these can be heavily truncated before performing the identification. Furthermore, the data points of interest can be chosen arbitrarily, although this is impossible in the time domain.

The best technique for measuring frequency response data from highly resonant mechanical systems is by swept sine analysis. A spectrum analyzer such as the Hewlett-Packard 35670A will provide extremely high-quality data with logarithmically spaced frequencies and auto-leveled inputs to achieve an excellent signal-to-noise ratio over the entire bandwidth. Each frequency point is estimated from a large amount of time domain data, hence the superior noise performance.

Of the various frequency domain system identification algorithms treated in [159], subspace methods stand out, they are simple and have proven to be effective in identifying resonant systems of high-order [129, 128]. The multivariable discrete-time algorithm of McKelvey [128] is used extensively throughout this book. Continuous time systems can be identified by applying a straightforward frequency pre-warping to the data and a bilinear transform to the resulting model [128]. Commercial implementations can be found in the System Identification Toolbox for MATLAB® [119] and the Frequency Domain System Identification Toolbox for MATLAB® [158]. Another algorithm used in this book is that of Van Overschee and De Moor [186]. This is a continuous time method that avoids the usual numerical conditioning problems associated with a basis constructed from the powers of s.

A further benefit of procuring models in the frequency domain is that the model quality of wide-band systems can be easily assessed by overlaying the data and model response. If a suitable model cannot be obtained it is also possible to perform some design and analysis procedures with only frequency response data. Non-parametric controllers such as resonant shunts can be designed with only frequency response data. It is also straightforward to estimate the closed-loop response with only frequency response data and a controller transfer function.

3.4 Velocity Feedback

A controller that guarantees unconditional closed-loop stability of a collocated system is the velocity feedback controller. This feedback controller is defined as [12, 13, 10]

$$K(s) = \gamma' \gamma s, \tag{3.4}$$

where γ is a $1 \times m$ matrix with m the number of collocated actuator sensor pairs.

Closed-loop stability of (3.1) under (3.4) can be verified in a number of ways. However, in the SISO case, this can be explained graphically. Being a collocated transfer function, the phase of $G_{vv}(s)$ is between 0 and -180 degrees. This, along with the fixed 90 degree phase of the differentiator and the negative feedback gain of -1, ensures that Nyquist plot of $sG_{vv}(s)$ will not circle -1+j0 point in the complex plane.

To prove closed-loop stability in the multivariable case the following theorem is needed.

Theorem 3.1. *Consider the following second order multivariable dynamical system*

$$\ddot{x}(t) + \mathcal{D}\dot{x}(t) + \mathcal{K}x(t) = 0, \tag{3.5}$$

where $\mathcal{D}, \mathcal{K} \in \mathbf{R}^{N\times N}$ *and* $x \in \mathbf{R}^{N\times 1}$. *Furthermore, assume that* $\mathcal{D} = \mathcal{D}' > 0$. *Then (3.5) is exponentially stable if and only if* $\mathcal{K} = \mathcal{K}' > 0$.

Proof: The second order system (3.5) can be re-written as a first order system by defining $X(t) = [x(t) \quad \dot{x}(t)]$:

$$\dot{X}(t) = AX(t), \tag{3.6}$$

where

$$A = \begin{bmatrix} 0 & I \\ -\mathcal{K} & -\mathcal{D} \end{bmatrix}. \tag{3.7}$$

Now, let

$$P = \begin{bmatrix} \mathcal{K} + \tau\mathcal{D} & \tau I \\ \tau I & I \end{bmatrix},$$

where

$$0 < \tau < \lambda_{\min}(\mathcal{D}). \tag{3.8}$$

Then, it is straightforward to verify that

$$A'P + PA = \begin{bmatrix} -2\tau\mathcal{K} & 0 \\ 0 & 2(\tau I - D) \end{bmatrix} < 0.$$

This is a necessary and sufficient condition for exponential stability of (3.6). ■

The next theorem establishes that (3.4) is a stabilizing controller for (3.1).

Theorem 3.2. *Negative feedback connection of (3.1) and (3.4) is exponentially stable.*

The following theorem proves closed-loop stability of the multivariable system.

Proof. The transfer function materix $G_{vv}(s)$ can be represented by the following second-order state-space equation

$$\begin{aligned} \ddot{x}(t) + 2Z\Omega\dot{x}(t) + \Omega^2 x(t) &= \Psi' u(t) \\ y(t) &= \Psi x(t), \end{aligned} \tag{3.9}$$

where

$$Z = \begin{bmatrix} \zeta_1 & & & \\ & \zeta_2 & & \\ & & \ddots & \\ & & & \zeta_M \end{bmatrix}$$

$$\Omega = \begin{bmatrix} \omega_1 & & & \\ & \omega_2 & & \\ & & \ddots & \\ & & & \omega_M \end{bmatrix}$$

$$\Psi = \begin{bmatrix} \psi_1 \; \psi_2 \; \dots \; \psi_M \end{bmatrix}.$$

Also, the feedback controller (3.4) incorporating the negative sign can be represented as follows

$$u(t) = -\gamma'\gamma\dot{y}(t). \tag{3.10}$$

Hence, the closed-loop system dynamics can be described by

$$\ddot{x}(t) + (2Z\Omega + \Psi'\gamma'\gamma\Psi)\,\dot{x}(t) + \Omega^2 x(t) = 0. \tag{3.11}$$

Clearly, $2Z\Omega + \Psi'\gamma'\gamma\Psi$ is a symmetric, positive-definite matrix. Therefore, according to Theorem 3.5 we can conclude that the closed-loop system represented by (3.11) is exponentially stable.

Although velocity feedback allows for spill-over, existence of spill-over does not destabilize the closed-loop system. This is a very interesting property. However, despite its appealing stability properties, velocity feedback as described by (3.4) is of little use in the context of piezoelectric laminated structures. The main problem is due to the requirement for a differentiator, which is not possible. Some extra dynamics would have to be added to the compensator to ensure that it will roll off at high frequencies. Due to the high-bandwidth nature of the underlying system these additional dynamics have the potential to destabilize the closed-loop system.

Another problem with velocity feedback, which is perhaps of a greater concern, is high control effort at all frequencies. For vibration damping purposes, the control effort should ideally be restricted to regions around the resonance frequencies of the base structure. This calls for different classes of controller that will be introduced in the remainder of this chapter.

3.5 Resonant Controllers

As indicated earlier, the structure of a collocated system, as expressed by (3.1), allows for the design of feedback controllers with specific structures that guarantee unconditional stability of the closed-loop system. Such controllers are of interest due to their ability to avoid closed-loop instabilities arising from the spill-over effect [13]. Resonant controllers have the distinctive property that closed-loop stability of the system is guaranteed in presence of out-of-bandwidth modes.

Two possible resonant controllers are:

$$K_v^\alpha = \sum_{i=1}^{\tilde{N}} \frac{\alpha_i \alpha_i' s^2}{s^2 + 2d_i \tilde{\omega}_i s + \tilde{\omega}_i^2}, \tag{3.12}$$

and

$$K_v^\beta = \sum_{i=1}^{\tilde{N}} \frac{\beta_i \beta_i' s(s + 2d_i \tilde{\omega}_i)}{s^2 + 2d_i \tilde{\omega}_i s + \tilde{\omega}_i^2}, \tag{3.13}$$

where typically $\tilde{N} \ll M$. Note that both α_i and β_i are $m \times 1$ vectors. Furthermore, $\tilde{\omega}_1 < \tilde{\omega}_2 < \ldots < \tilde{\omega}_{\tilde{N}}$.

The typical feedback control problem associated with the system (3.1) and a resonant controller K_v is illustrated in Figure 3.2. Here, v_p is the vector of voltages measured at the piezoelectric sensors, while w is the vector of disturbances acting on it. The purpose of the controller is to add damping to the structure, hence reducing the effect of disturbances. This is done by shifting closed-loop poles of the system deeper into the left half of the complex plane.

Closed-loop stability of the multivariable collocated system (3.1) under (3.12) and (3.13) can be proved in a number of ways. We will first prove stability of the feedback connection of (3.1) and (3.13).

Theorem 3.3. *Feedback connection of (3.1) and (3.13) is stable.*

Proof: Stability of (3.1) and (3.13) is equivalent to the stability of

$$\tilde{G}_{vv}(s) = \sum_{i=1}^{M} \frac{\psi_i \psi_i' s}{s^2 + 2\zeta_i \omega_i s + \omega_i^2}$$

and

$$\tilde{K}_v^\beta = \sum_{i=1}^{\tilde{N}} \frac{\beta_i \beta_i' (s + 2d_i \tilde{\omega}_i)}{s^2 + 2d_i \tilde{\omega}_i s + \tilde{\omega}_i^2}.$$

Now, it can be proved that $\tilde{G}_{vv}(s)$ is a positive-real (PR) transfer function matrix, while $\tilde{K}_v^\beta$ is a strictly positive-real (SPR) transfer function matrix. Negative feedback connection of a PR and a SPR system is known to be stable [104]. ■

Closed-loop stability of (3.1) under (3.12) can be established by constructing a Lyapunov function.

Theorem 3.4. *Feedback connection of (3.1) and (3.12) is exponentially stable.*

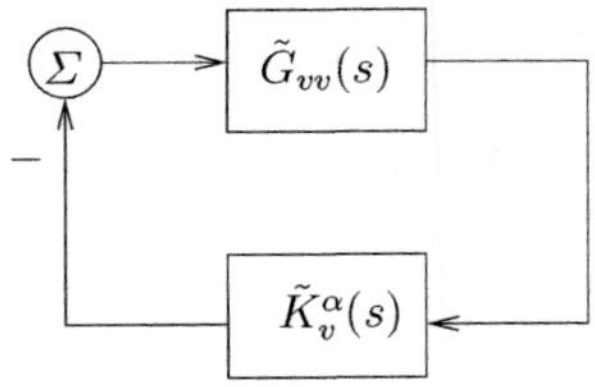

Figure 3.4. Equivalent system for study of closed-loop stability of K_v^α and G_{vv}

Proof: Feedback connection of (3.1) and (3.12) is exponentially stable if and only if the closed-loop system depicted in Figure 3.4 is stable, where

$$\tilde{G}_{vv}(s) = \sum_{i=1}^{M} \frac{\psi_i \psi_i' s}{s^2 + 2\zeta_i \omega_i s + \omega_i^2}$$

and

$$\tilde{K}_v^\alpha = \sum_{i=1}^{\tilde{N}} \frac{\alpha_i \alpha_i' s}{s^2 + 2d_i \tilde{\omega}_i s + \tilde{\omega}_i^2}.$$

Now, $\tilde{G}_{vv}(s)$ can be represented by

$$\ddot{x}(t) + 2Z\Omega\dot{x}(t) + \Omega^2 x(t) = \Psi' u(t) \tag{3.14}$$
$$y(t) = \Psi\dot{x}(t), \tag{3.15}$$

where

$$Z = \begin{bmatrix} \zeta_1 & & & \\ & \zeta_2 & & \\ & & \ddots & \\ & & & \zeta_N \end{bmatrix}$$
$$\Omega = \begin{bmatrix} \omega_1 & & & \\ & \omega_2 & & \\ & & \ddots & \\ & & & \omega_N \end{bmatrix}$$
$$\Psi = \begin{bmatrix} \psi_1 \; \psi_2 \; \dots \; \psi_N \end{bmatrix}.$$

Furthermore, $\tilde{K}_v^\alpha(s)$ can be written as[1]

$$\ddot{\tilde{x}}(t) + 2\Delta\tilde{\Omega}\dot{\tilde{x}}(t) + \tilde{\Omega}^2 \tilde{x}(t) = \Gamma' y(t) \tag{3.16}$$
$$u(t) = -\Gamma\dot{\tilde{x}}(t), \tag{3.17}$$

where

[1] Note that the negative sign is incorporated in the feedback controller.

$$\Delta = \begin{bmatrix} d_1 & & & \\ & d_2 & & \\ & & \ddots & \\ & & & d_N \end{bmatrix}$$

$$\tilde{\Omega} = \begin{bmatrix} \tilde{\omega}_1 & & & \\ & \tilde{\omega}_2 & & \\ & & \ddots & \\ & & & \tilde{\omega}_N \end{bmatrix}$$

$$\Gamma = \begin{bmatrix} \alpha_1 \ \alpha_2 \ \ldots \ \alpha_N \end{bmatrix}.$$

Combining (3.14)-(3.15), the closed-loop system dynamics can be obtained

$$\begin{bmatrix} \ddot{x}(t) \\ \ddot{\tilde{x}}(t) \end{bmatrix} + \begin{bmatrix} 2Z\Omega & \Psi'\Gamma \\ -\Gamma'\Psi & \Delta\tilde{\Omega} \end{bmatrix} \begin{bmatrix} \dot{x}(t) \\ \dot{\tilde{x}}(t) \end{bmatrix} + \begin{bmatrix} \Omega^2 & 0 \\ 0 & \tilde{\Omega}^2 \end{bmatrix} \begin{bmatrix} x(t) \\ \tilde{x}(t) \end{bmatrix} = \begin{bmatrix} 0 \\ 0 \end{bmatrix}. \tag{3.18}$$

Now, a Lyapunov function, $V(x(t), \tilde{x}(t))$ can be defined as

$$V(x(t), \tilde{x}(t)) = \begin{bmatrix} \dot{x}(t) \\ \dot{\tilde{x}}(t) \end{bmatrix}' \begin{bmatrix} \dot{x}(t) \\ \dot{\tilde{x}}(t) \end{bmatrix} + \begin{bmatrix} x(t) \\ \tilde{x}(t) \end{bmatrix}' \begin{bmatrix} \Omega^2 & 0 \\ 0 & \tilde{\Omega}^2 \end{bmatrix} \begin{bmatrix} x(t) \\ \tilde{x}(t) \end{bmatrix}. \tag{3.19}$$

Clearly, $V(x(t), \tilde{x}(t)) > 0$ for all nontrivial x, $\dot{x}$, $\tilde{x}$ and $\dot{\tilde{x}}$. Differentiating V with respect to time yields

$$\dot{V}(x(t), \tilde{x}(t)) = \begin{bmatrix} \dot{x}(t) \\ \dot{\tilde{x}}(t) \end{bmatrix}' \begin{bmatrix} -2Z\Omega & -\Psi'\Gamma \\ \Gamma'\Psi & -2\Delta\tilde{\Omega} \end{bmatrix} \begin{bmatrix} \dot{x}(t) \\ \dot{\tilde{x}}(t) \end{bmatrix} \tag{3.20}$$

which implies

$$\dot{V}(x(t), \tilde{x}(t)) < 0$$

for nontrivial values of $\dot{x}$ and $\dot{\tilde{x}}$. Therefore the closed-loop system is asymptotically stable. ∎

3.6 Positive Position Feedback

One of the difficulties associated with the implementation of resonant controllers is that the frequency responses of (3.12) and (3.13) do not roll off at higher frequencies. Although under ideal assumptions the closed-loop stability of the collocated system (3.1) under (3.12) and (3.13) is guaranteed, in real world applications due to issues such as phase contributions of the antialiasing filters, etc. the existence of out-of-bandwidth dynamics may destabilize the closed-loop system. It is, therefore, of interest to have a controller which possesses a similar structure to (3.12) and (3.13), but whose frequency response rolls off at higher frequencies. Positive position feedback, proposed by Caughey and co-authors [77, 58] has such properties.

For a system of the form (3.1), a positive position feedback controller is defined as

$$K_{pp}(s) = \sum_{i=1}^{\tilde{N}} \frac{-\gamma_i \gamma_i'}{s^2 + 2\delta_i \tilde{\omega}_i s + \tilde{\omega}_i^2}, \tag{3.21}$$

where $\gamma_i \in \mathbf{R}^{m \times 1}$ and $\delta_i > 0$ for $i = 1, 2, \ldots, \tilde{N}$.

Due to the existence of the negative sign in all terms of (3.21) the overall system resembles a positive feedback loop. Also, the transfer function matrix (3.1) is similar to that of the force to displacement transfer function matrix associated with a flexible structure, hence the terminology positive position feedback.

To derive stability conditions for this control loop, the series in (3.1) is first truncated by keeping the first N modes ($N < M$) that lie within the bandwidth of interest, and then incorporating the effect of truncated modes by adding a feed-through term to the truncated model [34, 135]. That is to approximate (3.1) by

$$G_{vv}^N(s) = \sum_{i=1}^{N} \frac{\psi_i \psi_i'}{s^2 + 2\zeta_i \omega_i s + \omega_i^2} + D. \tag{3.22}$$

The following theorem gives the necessary and sufficient conditions for closed-loop stability under positive position feedback.

Theorem 3.5. *The negative feedback connection of (3.22) and (3.21), with* $Z \geq 0$ *is stable if and only if*

$$\tilde{\Omega}^2 - \Gamma' D \Gamma > 0 \tag{3.23}$$

and

$$\Omega^2 - \Psi' \Gamma (\tilde{\Omega}^2 - \Gamma' D \Gamma)^{-1} \Gamma' \Psi > 0, \tag{3.24}$$

where

$$Z = \begin{bmatrix} \zeta_1 & & & \\ & \zeta_2 & & \\ & & \ddots & \\ & & & \zeta_N \end{bmatrix}$$

$$\Omega = \begin{bmatrix} \omega_1 & & & \\ & \omega_2 & & \\ & & \ddots & \\ & & & \omega_N \end{bmatrix}$$

$$\Psi = \begin{bmatrix} \psi_1 \ \psi_2 \ \ldots \ \psi_N \end{bmatrix}$$

and

$$\Delta = \begin{bmatrix} \delta_1 & & & \\ & \delta_2 & & \\ & & \ddots & \\ & & & \delta_{\tilde{N}} \end{bmatrix}$$

$$\tilde{\Omega} = \begin{bmatrix} \tilde{\omega}_1 & & & \\ & \tilde{\omega}_2 & & \\ & & \ddots & \\ & & & \tilde{\omega}_{\tilde{N}} \end{bmatrix}$$

$$\Gamma = \begin{bmatrix} \gamma_1 \; \gamma_2 \; \ldots \; \gamma_{\tilde{N}} \end{bmatrix}.$$

Proof: Using the above notation, (3.22) and (3.21) can be represented in second-order state-space form as

$$\ddot{x}(t) + 2Z\Omega\dot{x}(t) + \Omega^2 x(t) = \Psi' u(t) \tag{3.25}$$

$$y(t) = \Psi x(t) + Du(t) \tag{3.26}$$

and

$$\ddot{\tilde{x}}(t) + 2\Delta\tilde{\Omega}\dot{\tilde{x}}(t) + \tilde{\Omega}^2\tilde{x}(t) = \Gamma' y(t) \tag{3.27}$$

$$u(t) = \Gamma\tilde{x}(t). \tag{3.28}$$

The closed-loop system can then be written as

$$\begin{bmatrix} \ddot{x}(t) \\ \ddot{\tilde{x}}(t) \end{bmatrix} + \begin{bmatrix} 2Z\Omega & 0 \\ 0 & 2\Delta\tilde{\Omega} \end{bmatrix} \begin{bmatrix} \dot{x}(t) \\ \dot{\tilde{x}}(t) \end{bmatrix} + \begin{bmatrix} \Omega^2 & -\Psi'\Gamma \\ -\Gamma'\Psi & \tilde{\Omega}^2 - \Gamma' D\Gamma \end{bmatrix} \begin{bmatrix} x(t) \\ \tilde{x}(t) \end{bmatrix} = \begin{bmatrix} 0 \\ 0 \end{bmatrix}. \tag{3.29}$$

Clearly

$$\begin{bmatrix} 2Z\Omega & 0 \\ 0 & 2\Delta\tilde{\Omega} \end{bmatrix} > 0,$$

therefore, according to Theorem 3.1 the closed-loop system is stable, if and only if

$$\begin{bmatrix} \Omega^2 & -\Psi'\Gamma \\ -\Gamma'\Psi & \tilde{\Omega}^2 - \Gamma' D\Gamma \end{bmatrix} > 0. \tag{3.30}$$

The Schur complement [27] implies that this inequality holds if and only if

$$\tilde{\Omega}^2 - \Gamma' D\Gamma > 0$$

and

$$\Omega^2 - \Psi'\Gamma(\tilde{\Omega}^2 - \Gamma' D\Gamma)^{-1}\Gamma'\Psi > 0$$

which proves the theorem. ■

It should be pointed out that the stability condition (3.23), (3.24) can be represented by a linear matrix inequality in variables Δ, $\tilde{\Omega}^2$ and Ψ, by applying the Schur complement as indicated in the following lemma:

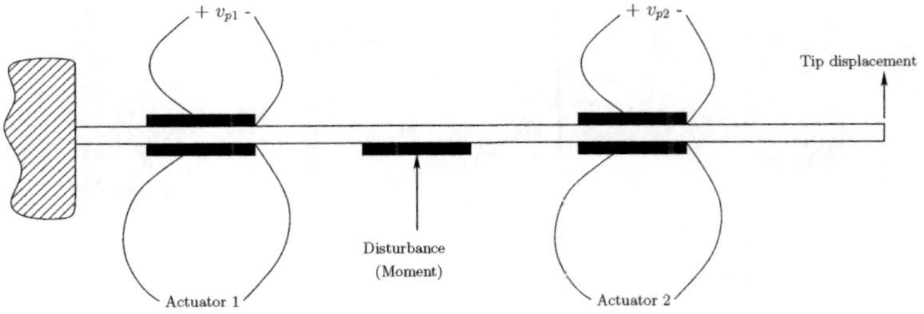

Figure 3.5. Beam layout as used in the experiments

Lemma 3.6. *The set of positive position feedback controllers is a convex set characterized by the following linear matrix inequalities:*

$$\Delta > 0 \tag{3.31}$$

and

$$\begin{bmatrix} \Omega^2 & -\Psi'\Gamma & 0 \\ -\Gamma'\Psi & \tilde{\Omega}^2 & \Gamma' \\ 0 & \Gamma & D^{-1} \end{bmatrix} > 0 \tag{3.32}$$

Proof: This can be proved using the Schur complement as follows:

$$\begin{bmatrix} \Omega^2 & -\Psi'\Gamma \\ -\Gamma'\Psi & \tilde{\Omega}^2 - \Gamma' D \Gamma \end{bmatrix} > 0$$

$$\Updownarrow$$

$$\begin{bmatrix} \Omega^2 & -\Psi'\Gamma \\ -\Gamma'\Psi & \tilde{\Omega}^2 \end{bmatrix} - \begin{bmatrix} 0 \\ \Gamma' \end{bmatrix} D \begin{bmatrix} 0 \; \Gamma \end{bmatrix} > 0$$

$$\Updownarrow$$

$$\begin{bmatrix} \omega^2 & -\Psi'\Gamma & 0 \\ -\Gamma'\Psi & \tilde{\Omega}^2 & \Gamma' \\ 0 & \Gamma & D^{-1} \end{bmatrix} > 0.$$

■

An implication of Lemma 3.6 is that all admissible positive position feedback controllers must be stable, as asserted by Condition (3.31).

Positive position feedback controllers are generally capable of ensuring higher levels of performance compared with resonant controllers (3.12) and (3.13). However, in making a comparison one has to be careful about the assumptions made in proving stability of the closed-loop systems associated with two different classes of controllers. To prove stability under Feedback Controllers (3.12) and (3.13) no simplifying assumption on the system had to made. Indeed it was proved that stability of the full order system was guaranteed under these two controllers. However, to obtain stability conditions for

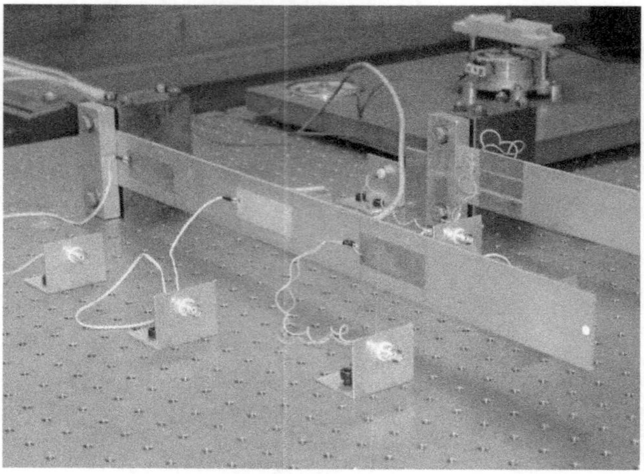

Figure 3.6. Picture of the cantilever beam

the positive position feedback controller, we had to assume that a truncated model of the system with only N modes adequately represented the dynamics of the system within the bandwidth of interest. Although, from a mathematical point of view this may be problematic, in most realistic applications where sensors and actuators have limited bandwidths, this assumption is unlikely to be of concern. Indeed many applications for positive position feedback control have been reported in the literature. To this end the reader is referred to [8, 3, 170, 109, 177] and references therein.

To conclude this section we point out that the stability condition (3.23), (3.24) does allow for a non-zero feed through term in the model of the system (3.22). As discussed in Section 2.9, inclusion of a feed through term in the model of a flexible structure may be essential. Particularly, when sensors and actuators are collocated. It should be noted that this analysis is missing from the original work of Caughy and Fanson [58]. Nevertheless, the stability condition reported in [58] can be derived from (3.23), (3.24) by setting D to zero.

3.7 Experimental Implementation of PPF Control on an Active Structure

3.7.1 Experimental Setup

All discussions on resonant systems so far were based on Analytic Models (3.1), (3.2) and (3.22). The model structures of the resonant controllers and PPF controllers, see (3.12), (3.13) and (3.21), strongly resemble the system model structures (3.1), (3.2) and (3.22). Here, a cantilever beam representing a physical resonant system is considered. This beam, which is clamped at one end and free at the other end, is susceptible to high amplitude vibrations when disturbed. In what follows, a PPF controller will be designed to damp these highly resonant vibration modes of the beam. It is worth noting that cantilever beams of the type used here are known to have models of the form (3.1), (3.2) and (3.22), as articulated in Section 4, and resonant controllers designed for them are known to have damped the vibrations to a reasonable extent, see [187] and [144].

In this section a description of the experimental setup meant for both modeling and control of the beam is presented. This setup will remain the same for all the experiments to be performed on the beam.

As mentioned above, the cantilever beam is clamped at one end and free at the other. Two pairs of piezoelectric patches are attached to this beam, one pair located close to the clamped end and the other pair is located close to the free end of the beam. For each pair, one piezoelectric patch will be used as an actuator (where input signals are applied) and the other patch will act as a sensor (where output signals are recorded). Another solitary piezoelectric patch is attached to the center of the beam and will be driven by a voltage source w. This voltage w represents the disturbance acting on the beam. See Figures 3.5 and 3.6 for a schematic beam setup and an actual picture of the beam respectively.

Not surprisingly, the mechanical properties (including the damping and the resonance frequencies) will depend on the dimensions and material properties of the beam and the piezoelectric patches. In Figure 3.7, Table 3.1 and Table 3.2 the dimensions of the beam setup and the material properties of the beam and the piezoelectric transducers are given.

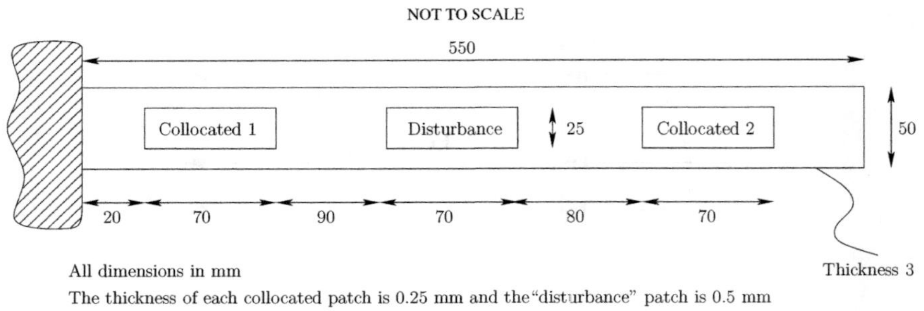

Figure 3.7. Beam setup and dimensions

Table 3.1. Beam Properties

Length, L	550 mm
Thickness, h	3 mm
Width, W	50 mm
Density, ρ	2.77×10^3 kg/m^3
Young's Mod., E	7.00×10^{10} N/m^2

Table 3.2. PIC 151 Ceramic Properties

Length, L_{pz}	50 mm
Thickness, h_{pz}	0.25 mm
Width, W_{pz}	25 mm
Charge Constant, d_{31}	-210×10^{-12} m/V
Voltage Constant, g_{31}	-11.5×10^{-3} Vm/N
Coupling Coefficient, k_{31}	0.34
Capacitance, C_p^S	115 nF

3.7.2 System Identification

In order to obtain a model for the beam, the experimental setup is treated as a three-input-three-output multivariable system, see Figure 3.8. The inputs (v_1 and v_2) in Figure 3.8 are the voltages applied to the actuators of the collocated piezo patches and the outputs v_{p1} and v_{p2} are the voltages induced at the corresponding sensors. The third input w is the disturbance on the beam and the output y_{tip} is the displacement of the the tip of the beam. To be precise, y_{tip} is the displacement of the top vertex point of the beam at the free end.

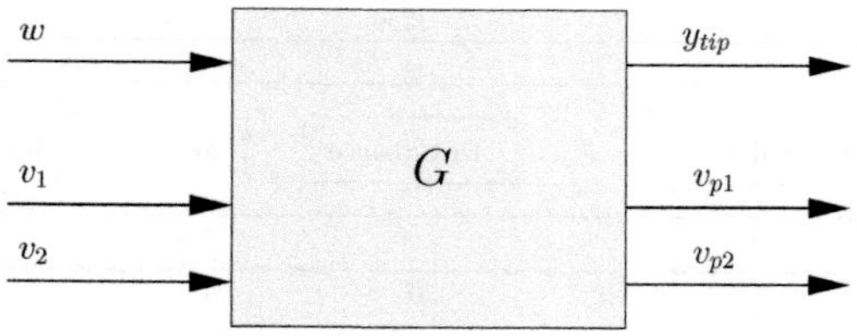

Figure 3.8. Augmented MIMO plant

Since the system is modeled as a three-input-three-output system, the frequency response function (FRF) $G(\mathrm{i}\omega)$ is a 3×3 matrix with each element $G_{ij}(\mathrm{i}\omega), i, j = 1, 2$ and 3, corresponding to a particular combination of the input and the output; *i.e.*,

$$G_{ij}(\mathrm{i}\omega) = G_{y_i u_j} = Y_i(\mathrm{i}\omega)/U_j(\mathrm{i}\omega), \tag{3.33}$$

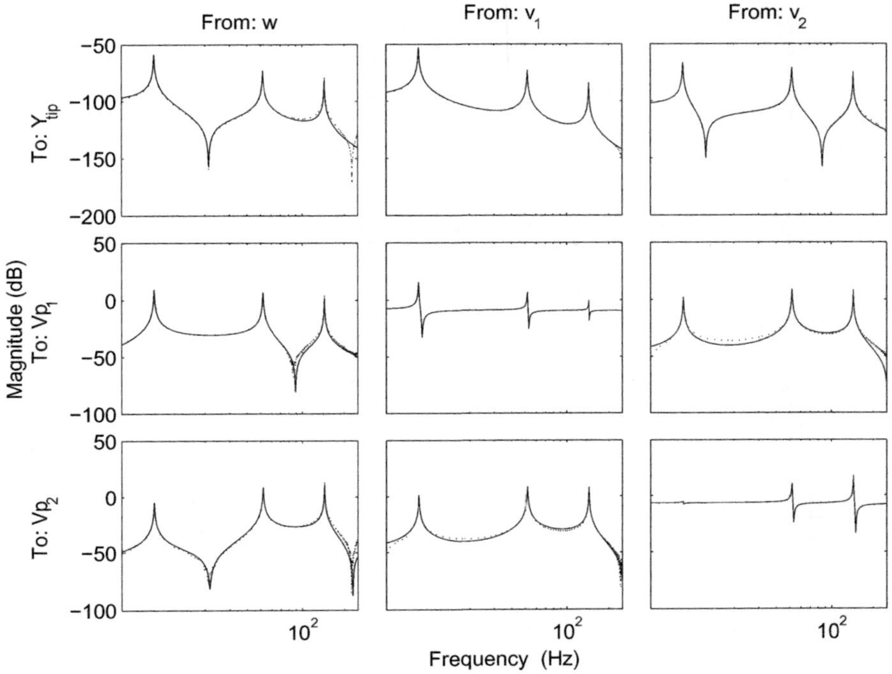

Figure 3.9. Identified model (solid) with measured data (dotted)

where $y_1 = y_{tip}$, $y_2 = v_{p1}$ and $y_3 = v_{p2}$, and $u_1 = w$, $u_2 = v_1$ and $u_3 = v_2$. $Y_i(\mathrm{i}\omega)$ and $U_j(\mathrm{i}\omega)$ are the Fourier transforms of y_i and u_j respectively. These FRFs are determined by applying a sinusoidal chirp of varying frequency (from $5 - 250$Hz) to the piezoelectric actuators (including the central patch corresponding to the disturbance term w) and measuring the corresponding output signals y_{tip}, v_{p1} and v_{p2}. The inputs are generated using a standard HP signal generator and the output signals y_{tip}, v_{p1} and v_{p2} were measured using a Polytec laser scanning vibrometer (PSV-300). The vibrometer PSV-300 also includes software which computes the ratio of input and output FFTs to provide $G(\mathrm{i}\omega)$.

In Figure 3.9, the FRFs $G_{ij}(\mathrm{i}\omega), i, j = 1, 2$ and 3 are plotted. It is apparent from the plots that all the FRFs have three resonance frequencies in the plotted frequency region, and the resonance frequencies are more or less the same for all the FRFs.

The goal here is to have good fits of the form (3.22) for the collocated FRFs $G_{v_{p1}v_1}$ and $G_{v_{p2}v_2}$, and good fits of the form (3.2) for all the other FRFs. It is known, see [142] and [187], that the beam under consideration has a multivariable state space model of the form

$$\dot{x}(t) = Ax(t) + B_w w(t) + B_v V(t) \tag{3.34}$$

$$y_{tip} = C_y x(t) + D_{yw} w(t) + D_{yv} V(t) \tag{3.35}$$

$$V_p(t) = C_v x(t) + D_{vw} w(t) + D_{vv} V(t), \tag{3.36}$$

where

$$A = \begin{bmatrix} 0 & 1 & 0 & 0 \\ -\omega_1^2 & -2\zeta_1\omega_1 & 0 & 0 \\ & & \ddots & \\ 0 & 0 & 0 & 1 \\ 0 & 0 & -\omega_N^2 & -2\zeta_N\omega_N \end{bmatrix} \tag{3.37}$$

$$B = \begin{bmatrix} B_w & B_{v_1} & B_{v_2} \end{bmatrix} = \begin{bmatrix} 0 & 0 & 0 \\ \beta_1 & \Psi_1^{v_1} & \Psi_1^{v_2} \\ \vdots & \vdots & \vdots \\ 0 & 0 & 0 \\ \beta_N & \Psi_N^{v_1} & \Psi_N^{v_2} \end{bmatrix} \tag{3.38}$$

$$C = \begin{bmatrix} C_y \\ C_{v_1} \\ Cv_2 \end{bmatrix} = \begin{bmatrix} \gamma_1 & 0 \dots & \gamma_N & 0 \\ \Psi_1^{v_1} & 0 \dots & \Psi_N^{v_1} & 0 \\ \Psi_1^{v_2} & 0 \dots & \Psi_N^{v_2} & 0 \end{bmatrix} \tag{3.39}$$

$$V(t) = \begin{bmatrix} v_1 \; v_2 \end{bmatrix}' \tag{3.40}$$

$$V_p(t) = \begin{bmatrix} v_{p1} \; v_{p2} \end{bmatrix}'. \tag{3.41}$$

The terms D_{yw}, D_{vw}, D_{yv} and D_{vv} are matrices of appropriate dimensions.

Using the Laplace transform it can be verified that (3.34)-(3.36) enforces the structure (3.22) on the collocated FRFs $G_{v_{p1}v_1}$ and $G_{v_{p2}v_2}$, and the structure (3.2) on the other non-collocated FRFs. The model structure (3.34)-(3.36) also forces all the FRFs to have the same set of poles.

Note that the given FRF data has only three resonance frequencies, see Figure 3.9. In other words the given FRF data takes into account only the first three modes of vibrations. The other higher order modes have been discarded (or not taken into account) as they are beyond the frequency region of interest. Since the given data includes only three modes, it suffices to set $N = 3$, or consider a third order model of the form (3.34)-(3.36). A standard method to determine the model parameters $\{\Psi_k^{v_1}, \Psi_k^{v_2}, \beta_k, \gamma_k, \omega_k, \zeta_k\}_{k=1}^N$ and the feed-through terms D_{yw}, D_{yv}, D_{vw} and D_{vv} is to choose them as the minimizers of the cost function,

$$\mathcal{M} = \sum_{i,j=1}^{3} \sum_{k=1}^{M} \left| \frac{G_{ij}^N(\mathrm{i}\omega_k) - G_{ij}(\mathrm{i}\omega_k)}{G_{ij}(\mathrm{i}\omega_k)} \right|^2, \tag{3.42}$$

where $G_{ij}^N(\mathrm{i}\omega)$ are the FRFs corresponding to the multivariable state space model (3.34)-(3.36) and ω_k denotes the frequency points where the non-parametric FRFs $G_{ij}(\mathrm{i}\omega)$ are measured.

Minimizing the cost function $\mathcal{M}$, (3.42), involves using a computationally complex non-linear search. Due to the enormity in the number of parameters to be estimated, the cost function $\mathcal{M}$ may have numerous local minimas. Therefore a poor initial guess for the parameters $\{\Psi_k^{v_1}, \Psi_k^{v_2}, \beta_k, \gamma_k, \omega_k, \zeta_k\}_{k=1}^N$, D_{yw}, D_{yv}, D_{vw} and D_{vv} (*i.e.*, parameter values which give poor fits for the FRF data), the non-linear search may land in a local minima which also gives a poor fit for the FRF data. An initialization which in itself gives a decent fit to the FRF data would lead the non-linear search to the parameter values that give a good fit for the FRF data. One approach to get such initial values is to fit the non-parametric data first using subspace methods, [128]. Using subspace methods one can obtain a multivariable state-space model, of the same dimensions as (3.34)-(3.36), for the FRFs $G_{ij}(\mathrm{i}\omega)$. Even though subspace methods are black box methods, not conforming to any particular model structure, as the cantilever beam is known to possess a model of the form (3.34)-(3.36), the subspace fit would also resemble the models (3.34)-(3.36) to a large extent. From the subspace fit one can then extract good initial estimates for parameters $\{\Psi_k^{v_1}, \Psi_k^{v_2}, \beta_k, \gamma_k, \omega_k, \zeta_k\}_{k=1}^N$, D_{yw}, D_{yv}, D_{vw} and D_{vv}. For details of subspace methods the reader is referred to [187] or [128].

In Figure 3.9 the parametric model of the form (3.34)-(3.36) estimated from the FRF data is plotted along with the measured data. It is apparent from the plots that the estimated parametric model gives a good fit for the non-parametric data.

3.7.3 PPF Controller Design

The tip displacement y_{tip} of the beam gives a measure of the amplitude of vibrations in the beam. As w represents disturbance, the FRF $G_{y_{tip}w}(\mathrm{i}\omega) = Y_{tip}(\mathrm{i}\omega)/W(\mathrm{i}\omega)$ is a good indicator of the effect of noise on the beam. A well damped FRF $G_{y_{tip}w}(\mathrm{i}\omega)$ would imply a well damped system. Hence, a controller is designed such that the closed-loop FRF $G_{Cl,yw}(\mathrm{i}\omega)$ corresponding to the input w and output y_{tip} is well damped. Alternatively stated, a controller is designed such that the poles of the closed-loop FRF $G_{Cl,yw}(\mathrm{i}\omega)$ are well inside the left half plane.

The PPF controller to be designed is a two-input two-output multivariable controller, which forms a feedback loop connecting the outputs $V_p(t) = [v_{p1} \;\; v_{p2}]'$ to the inputs $V(t) = [v_1 \;\; v_2]'$ of the system. Adhering to the notations in Section 3.6 the PPF controller to be designed is of the form

$$\dot{\tilde{x}}(t) = \tilde{A}\tilde{x}(t) + \tilde{\Gamma} V_p(t) \tag{3.43}$$

$$V(t) = \tilde{\Gamma}'\tilde{x}(t), \tag{3.44}$$

where

$$\tilde{A} = \begin{bmatrix} 0 & 1 & 0 & 0 \\ -\tilde{\omega}_1^2 & -2\delta_1\tilde{\omega}_1 & 0 & 0 \\ & & \ddots & \\ 0 & 0 & 0 & 1 \\ 0 & 0 & -\tilde{\omega}_3^2 & -2\delta_3\tilde{\omega}_3 \end{bmatrix} \tag{3.45}$$

and

$$\tilde{\Gamma} = \begin{bmatrix} 0 & 0 \\ \Gamma_1^{v_{p1}} & \Gamma_1^{v_{p2}} \\ \vdots & \vdots \\ 0 & 0 \\ \Gamma_3^{v_{p1}} & \Gamma_3^{v_{p2}} \end{bmatrix}. \tag{3.46}$$

Note that (3.43)-(3.44) is (3.27)-(3.28) written in first order form with $\tilde{N} = 3$.

Using (3.43)-(3.44) in (3.34)-(3.36), the closed-loop system can be written in the form

$$\dot{X}(t) = \bar{A}(\tilde{\Omega}, \Delta, \Gamma)X(t) + \bar{B}_w w(t) \tag{3.47}$$

$$y_{tip}(t) = \bar{C}(\Gamma)X(t) + \bar{D}_{yw}w(t), \tag{3.48}$$

where $X = [x' \;\; \tilde{x}']'$ and

$$\Gamma = \begin{bmatrix} \Gamma_1^{v_{p1}} & \Gamma_2^{v_{p1}} & \Gamma_3^{v_{p1}} \\ \Gamma_1^{v_{p2}} & \Gamma_2^{v_{p2}} & \Gamma_3^{v_{p2}} \end{bmatrix}. \tag{3.49}$$

Derivations of the expressions for $\bar{A}(\tilde{\Omega}, \Delta, \Gamma), \bar{B}, \bar{C}$ and $\bar{D}_{yw}$ from (3.43)-(3.44) and (3.34)-(3.36) involve standard (and straightforward) calculations and are hence omitted.

The FRF $G_{Cl,yw}(\mathrm{i}\omega)$ corresponding to the closed-loop system (3.47)-(3.48) is the one that is to be damped by appropriately choosing $\tilde{\Omega}$, Δ and Γ. A sufficiency condition for the stability of the closed-loop (3.47)-(3.48) is provided by Lemma 3.6 in Section 3.6. Lemma 3.6 states that for the closed-loop system $G_{Cl,yw}(\mathrm{i}\omega)$ to be stable the LMI condition (3.32) has to be satisfied. Hence the PPF controller design problem can be posed as a constrained optimization problem:

$$\min_{\tilde{\Omega}, \Delta, \Gamma} \| G_{Cl,yw}(\mathrm{i}\omega) \| \tag{3.50}$$

subject to the constraint

$$\begin{bmatrix} \Omega^2 & -\Psi'\Gamma & 0 \\ -\Gamma'\Psi & \tilde{\Omega}^2 & \Gamma' \\ 0 & \Gamma & D^{-1} \end{bmatrix} > 0, \tag{3.51}$$

where

$$\Psi = \begin{bmatrix} \Psi_1^{v_1} & \Psi_2^{v_1} & \Psi_3^{v_1} \\ \Psi_1^{v_2} & \Psi_2^{v_2} & \Psi_3^{v_2} \end{bmatrix}, \tag{3.52}$$

Ω is the 3×3 diagonal matrix with resonance frequencies of the beam as its diagonal elements, and D denotes the feed-through term D_{vv} of the model, see (3.36). The design variables Γ, Δ and $\tilde{\Omega}$ are 2×3, 3×3 and 3×3 matrices respectively, with Γ as defined in (3.49) and the matrices Δ and $\tilde{\Omega}$ being diagonal. In (3.50) $\| \cdot \|$ denotes a norm. Here the H_∞-norm is considered as a performance measure for the controller design. An H_2 optimal PPF controller has also been designed and tested.

Another approach to design the PPF controller would be to place the poles $\{P_k^c\}^{4N}$ of $G_{Cl,yw}(s)$ in a desired region of the left half plane. One way to design such a controller is to choose (Ω, Δ, Γ) such that

$$V = \alpha \sum_{k=1}^{4N} | P_k^d - Re(P_k^c) |^2 + \beta \sum_{i=2,\ i\ even}^{4N} | P_k^c - P_{k-1}^c |^2 \tag{3.53}$$

is minimal under the constraint (3.51). In (3.53), $\{P_k^d\}^{4N}$ denotes a set of prespecified negative real numbers, $Re(\cdot)$ denotes the real part of a complex number, and α and β are positive scalars. The first term in the right-hand side of the cost function V, (3.53), tries to bring the real part of the closed-loop poles $\{P_k^c\}^{4N}$ close to the set of prespecfied real values $\{P_k^d\}^{4N}$ in the left half plane. The second term keeps the pair of closed-loop poles that correspond to the same open-loop pole close to each other (see Figure 3.10).

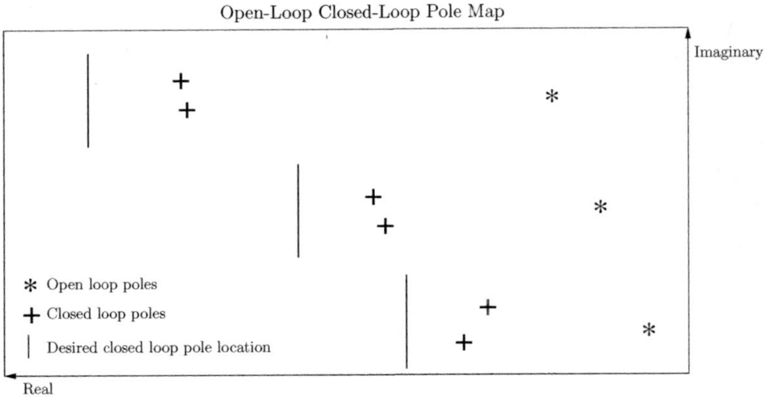

Figure 3.10. Pole optimisation procedure

As in the case of identification, both optimization problems (3.50) and (3.53) involve using a non-linear search, which needs to be initialized amicably. One method to initialize the non-linear search would be to solve the LMI (3.51) using well known LMI solvers like SeDuMi, or the MATLAB® LMI tool box, to obtain a feasible solution. Although such an initial condition would be valid, its use may result in a controller gain $\| \Gamma\Gamma' \|$ which is too large. A prohibitively large controller gain may result in implementation difficulties. A simple way to obtain initial values is to solve for Γ by solving the LMI (3.51) for certain fixed values of $\tilde{\Omega}$ and Δ. Past research (see [177] and [58]) and experience with PPF controllers suggest that good PPF designs have their resonance frequencies $\tilde{\Omega}$ close to the system resonance frequencies Ω. Taking this cue, we let

$$\tilde{\Omega} = \kappa\Omega \tag{3.54}$$

$$\Delta = \frac{\eta}{\kappa}, \tag{3.55}$$

where κ is a scalar reasonably close to one (for example $\kappa = \sqrt{2}$), as our initial guesses for $\tilde{\Omega}$ and Δ. Using the initial values (3.54) and (3.55) a meaningful initial guess for Γ can be obtained in the following fashion: Let $\Psi_\perp^{v_1}$ and $\Psi_\perp^{v_2}$ denote vectors orthonormal to $\Psi^1 = [\Psi_1^{v_1} \;\; \Psi_1^{v_2}]'$ and $\Psi^2 = [\Psi_2^{v_1} \;\; \Psi_2^{v_2}]'$, see (3.52). Note that due the linear independence of Ψ^1 and Ψ^2 (which is assumed) there exists C_1 and C_2 such that $\Psi^3 = [\Psi_3^{v_1} \;\; \Psi_3^{v_2}]'$ is equal to $C_1\Psi^1 + C_2\Psi^2$. Let $\tilde{\Gamma}_C = [\alpha_1\Psi_\perp^1 \;\; \alpha_2\Psi_\perp^2 \;\; 0]$ denote a candidate Γ for the initial value, where α_1 and α_2 are real valued parameters. It is easy to note that

$$\Psi'\Gamma_C = \begin{bmatrix} \alpha_1 & 0 & 0 \\ 0 & \alpha_2 & 0 \\ C_1\alpha_1 & C_2\alpha_2 & 0 \end{bmatrix}. \tag{3.56}$$

Since the feed-thorough term D_{vv} is in general very small, we drop the quadratic term $\Gamma' D\Gamma$ in (3.24) and approximate it by

$$\Omega^2 - \Psi'\Gamma_C\tilde{\Omega}^{-2}\Gamma_C'\Psi > 0. \tag{3.57}$$

Due to the chosen structure of Γ_C, (3.57) can be further simplified to

$$\Omega^2\tilde{\Omega}^2 > M'M, \tag{3.58}$$

where

$$M = \begin{bmatrix} \alpha_1 & 0 & 0 \\ 0 & \alpha_2 & 0 \\ C_1\alpha_1\frac{\tilde{\Omega}_3}{\tilde{\Omega}_1} & C_2\alpha_2\frac{\tilde{\Omega}_3}{\tilde{\Omega}_2} & 0 \end{bmatrix}. \tag{3.59}$$

Note that the right-hand side of (3.58) is a constant diagonal matrix, therefore choosing α_1 and α_2 such that the bound (3.58) is satisfied would give an initial guess for Γ_C. However, it must be stressed that the approximation introduced in (3.57) could fail for large values of α_1 and α_2. Hence, it is worth checking if the LMI (3.24) is satisfied for the chosen initial guesses.

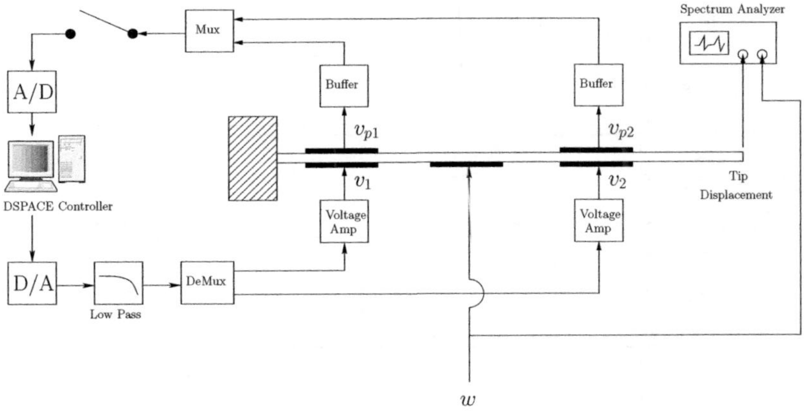

Figure 3.11. Schematics of the experimental setup

Results of Pole Placement Controller Design

In this subsection, the results pertaining to the effectiveness of the pole optimized PPF controller are presented. In Figure 3.12, the two-input two-output multivariable pole optimized PPF controller is plotted. Note from the plots that the FRFs of the PPF controller have a good roll-off in the high frequency regions, and they also do not have large gains. These twin properties of the PPF make them easily implementable. By construction, see (3.21), the diagonal entries of the PPF controller correspond to collocated FRFs and the cross diagonal terms refer to the non-collocated FRFs. It can be noted from Figure 3.12 the non-collocated FRFs are identical, which is expected due to (3.21).

Schematics of the experimental setup used to implement the pole optimized PPF controller, and all subsequent controllers, is illustrated in Figure 3.11. The controller was designed, and simulations were performed, in MATLAB® Simulink®. The controller was then downloaded onto a dSPACE DS-1103 rapid prototyping system. Sampling frequency of the DSP system was set to 20 KHz, and low-pass antialiasing and reconstruction filters with cut-off frequencies of 10 KHz were added to the system as shown in Figure 3.11. Voltages induced in the two piezoelectric sensors were measured through high-input-impedance buffer circuits, to reduce the low-frequency distortions caused by the finite impedance of the measurement device.

To determine a controller, the desired real parts of closed-loop poles in (3.53) were placed at -8, -66 and -120 respectively. The real parts of the actual closed-loop poles were found to be -9.7494, -60.8264 and -105.9135 respectively.

To evaluate the damping introduced in the system due to the PPF controller, both the open-loop and closed-loop FRFs $G_{y_{tip}w}(i\omega)$ and $G_{Cl,y_{tip}w}(i\omega)$

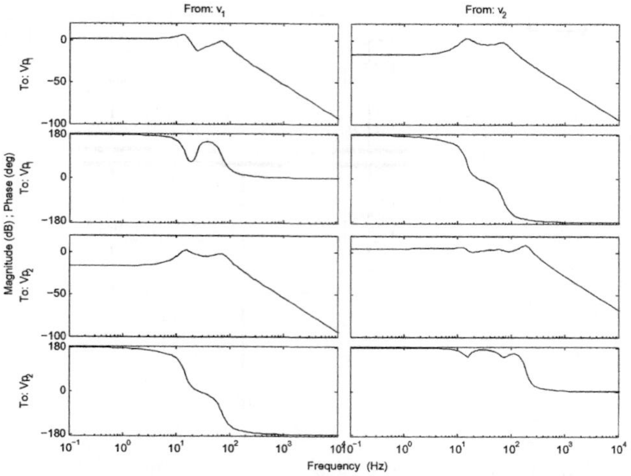

Figure 3.12. PPF Controller using the pole optimization

are plotted in Figure 3.13 and 3.14. In Figure 3.13 simulated values of $G_{y_{tip}w}(\mathrm{i}\omega)$ and $G_{Cl,y_{tip}w}(\mathrm{i}\omega)$ are plotted, *i.e.*, $G_{y_{tip}w}(\mathrm{i}\omega)$ obtained from the model (3.34)-(3.36) and $G_{Cl,y_{tip}w}(\mathrm{i}\omega)$ obtained from (3.47) and (3.48) with the parameters $\tilde{\Omega}, \Delta$ and Γ set to their optimal values. In Figure 3.14 experimentally determined $G_{y_{tip}w}(\mathrm{i}\omega)$ and $G_{Cl,y_{tip}w}(\mathrm{i}\omega)$ are plotted. Experimentally determined FRFs refer to the case where the piezoelectric patch corresponding disturbance input w in the cantilever beam setup is driven by a chirp signal and the corresponding output y_{tip} is recorded for both the closed- and the open-loop systems. The FRFs $G_{y_{tip}w}(\mathrm{i}\omega)$ and $G_{Cl,y_{tip}w}(\mathrm{i}\omega)$ are then calculated from the recorded input-output data, as done in the non-parametric identification. Simulations and experimental results are fairly similar, thus validating the use of the model (3.34)-(3.36) for the beam. The plots also suggest a satisfactory damping in the magnitude of the closed-loop FRF $G_{Cl,y_{tip}w}(\mathrm{i}\omega)$ at the resonance frequencies.

In Figure 3.15 the open-loop poles and the closed-loop poles are plotted. It is evident that $G_{Cl,y_{tip}w}(\mathrm{i}\omega)$ is much better damped compared to $G_{y_{tip}w}(\mathrm{i}\omega)$.

In practice, due to wear and tear and also due to other factors such as surrounding temperature, etc. the material properties of the structure and the piezoelectric patches tend to change. This leads to shifts in the resonance frequencies of the beam. A good controller design must be robust enough to provide good damping even under changed circumstances. In order to check the robustness of the PPF design artificial shifts in the resonance frequencies of the beam were brought about by adding an extra mass at the free end of the beam. In Figure 3.16 the measured open-loop and the closed-loop FRFs $G_{y_{tip}w}(\mathrm{i}\omega)$ and $G_{Cl,y_{tip}w}(\mathrm{i}\omega)$ for the loaded system are plotted. It must be

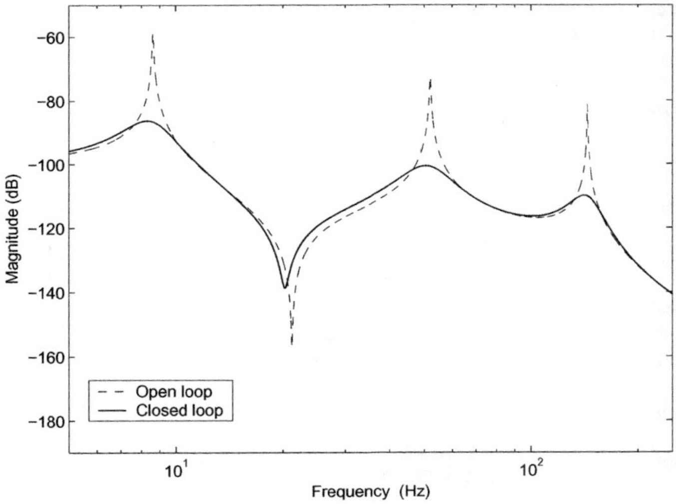

Figure 3.13. Simulated open- and closed-loop frequency response using the pole optimization

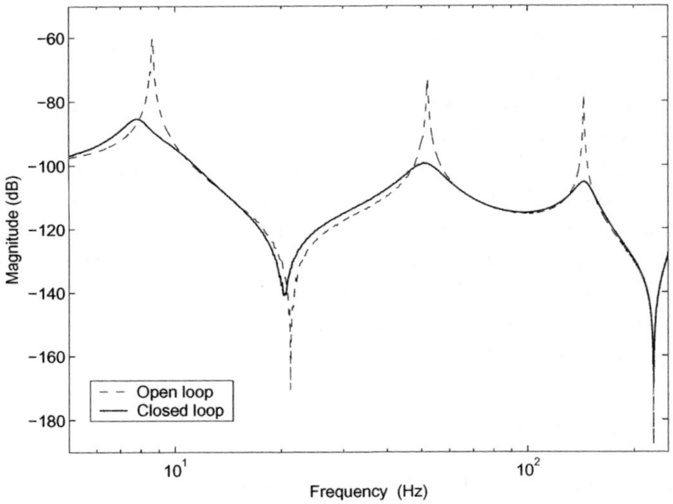

Figure 3.14. Measured open- and closed-loop frequency response using the pole optimization

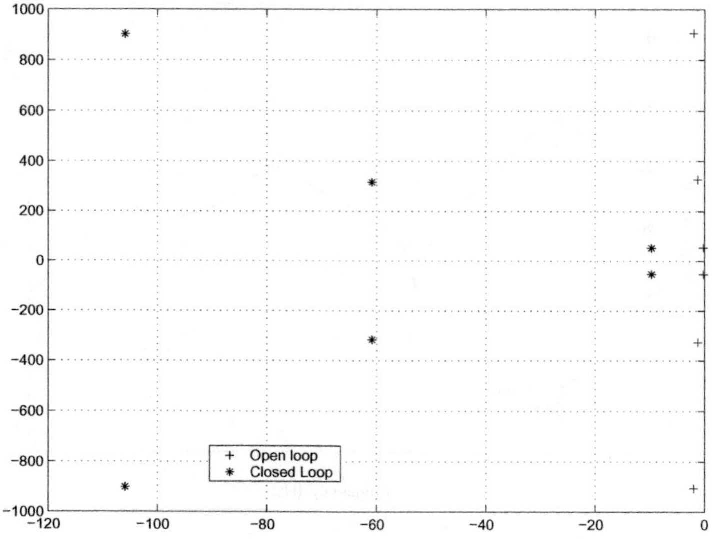

Figure 3.15. Open- and closed-loop pole map for the pole placement optimization

stressed that, the PPF controller used in the closed-loop system is the same PPF controller that was used for the unloaded Cantilever beam. The plots suggest that the PPF controller designed using pole optimization is robust with respect to shifts in the resonance frequencies.

Finally to illustrate controller performance in the time-domain, a pulse-shaped disturbance voltage was applied to the disturbance piezoelectric patch. The resulting tip displacements in open- and closed-loop were recorded for the beam with and without the tip mass. These time-domain results are plotted in Figure 3.17 and clearly demonstrate the effectiveness of the controller in terms of disturbance rejection and its relative immunity to uncertainty in the structural dynamics of the system.

H_∞ PPF Controller

Finally, this section is concluded with the results of using an H_∞ optimized PPF controller. FRF of the H_∞ optimized PPF controller is plotted in Figure 3.18. As in the previous case the PPF controller designed by minimizing the H_∞ norm also has a good roll-off in the high frequency regions. In order to limit the level of control signals applied to the two piezoelectric actuators, the exogenous output y_{tip} is augmented with weighted versions of v_1 and v_2. The weights are chosen carefully to trade off between the achievable damping and the necessary control signals. This is a standard approach [206], hence no more details are included here.

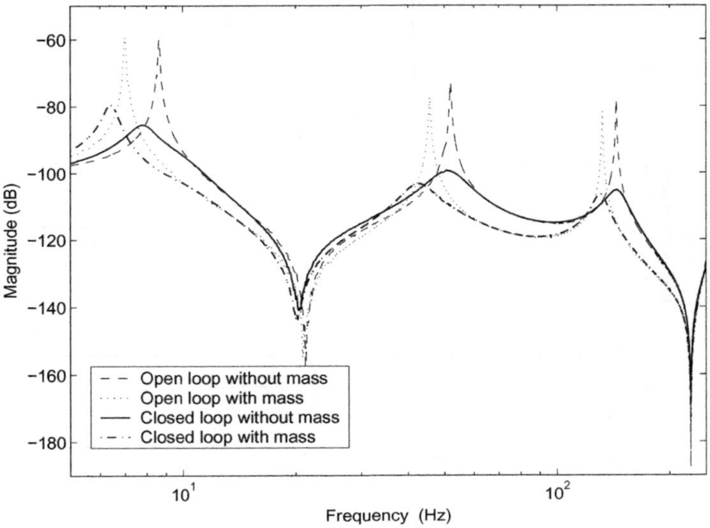

Figure 3.16. Measured open- and closed-loop frequency responses using the pole optimization, with and without the mass

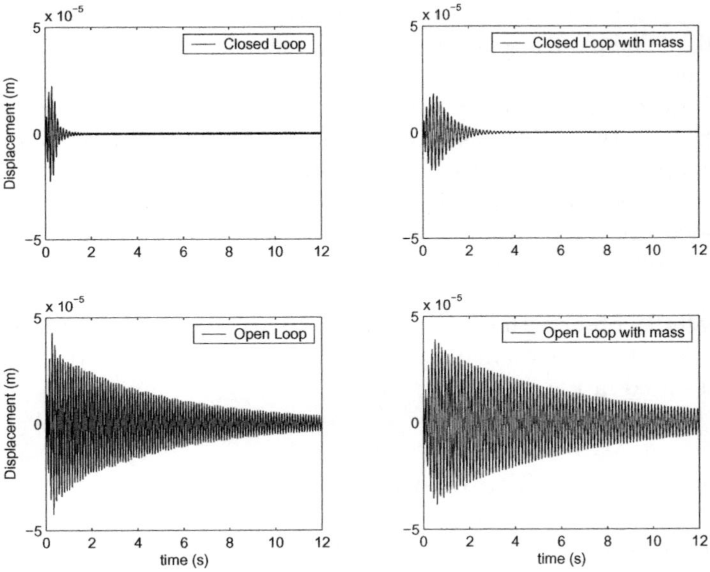

Figure 3.17. Measured disturbance response using pole optimization, with and without the mass

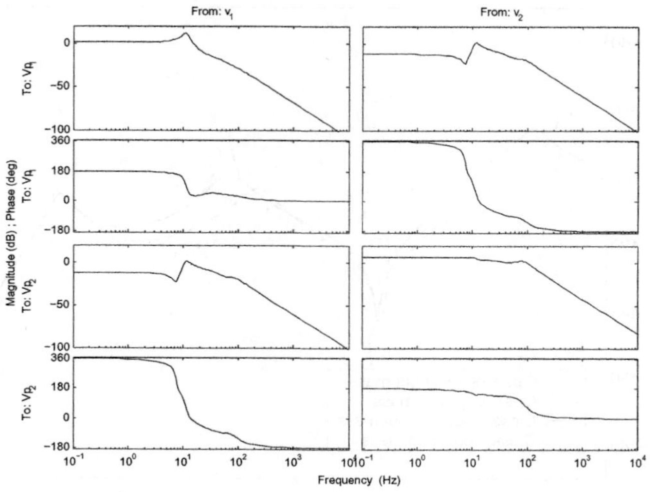

Figure 3.18. PPF Controller using the H_∞ optimization

In Figures 3.19 and 3.20 the closed-loop and the open-loop FRFs are plotted. The plots suggest a good damping of the closed-loop due to the PPF controller. The plots also suggest good agreement between the simulated and experimental FRFs.

In Figure 3.21 the poles of the open- and the closed-loop systems are plotted. As in the earlier case the plot suggests a good deal of damping in the closed-loop system due to the PPF controller.

Robustness of the H_∞ optimized PPF controller is tested by using it on the loaded cantilever beam. The plots 3.22 suggest that the H_∞ minimized PPF controller is also robust to changes in the resonance frequencies of the beam.

Finally, to observe the performance of the H_∞ optimized PPF in the time-domain, similar time-domain tests are performed on the system. The results are illustrated in Figure 3.23.

3.8 Self-sensing Techniques

In a typical active vibration control application, piezoelectric elements are often used as actuators, or sensors. In this case, the piezoelectric device performs a single function; either sensing, or actuation. The piezoelectric self-sensing actuator, or sensori-actuator, on the other hand, is a piezoelectric transducer used simultaneously as a sensor and an actuator. This technique was developed concurrently by Dosch, Inman, and Garcia [54]; and Anderson, Hagood and Goodlife [9], who made the observation that with the capacitance of the

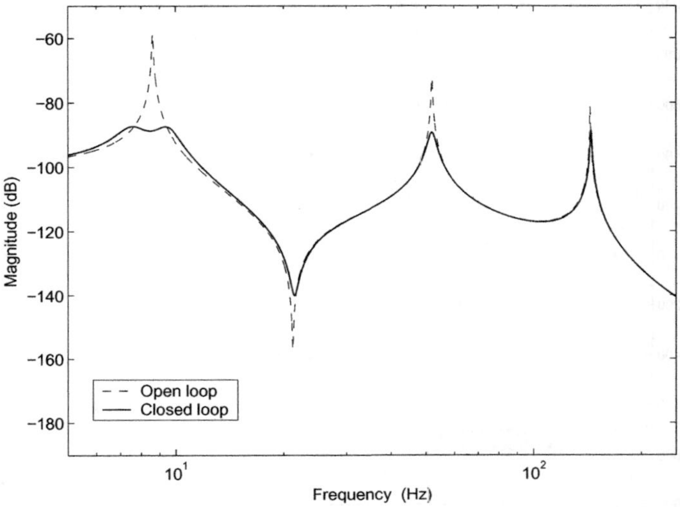

Figure 3.19. Simulated open- and closed-loop frequency responses using H_∞ optimization

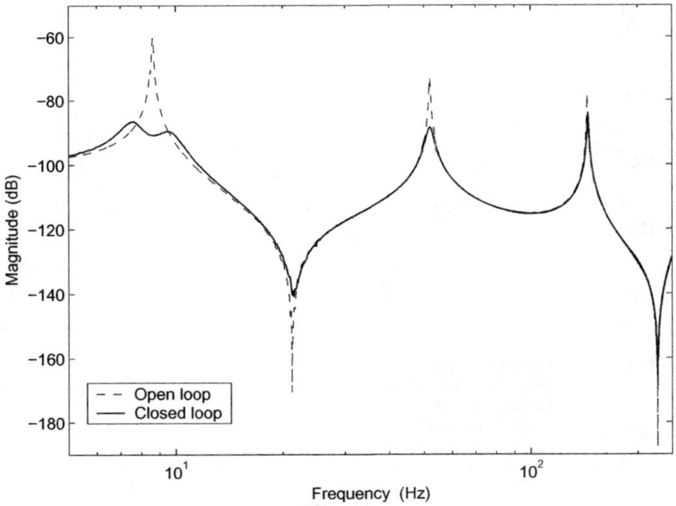

Figure 3.20. Measured open- and closed-loop frequency responses using H_∞ optimization

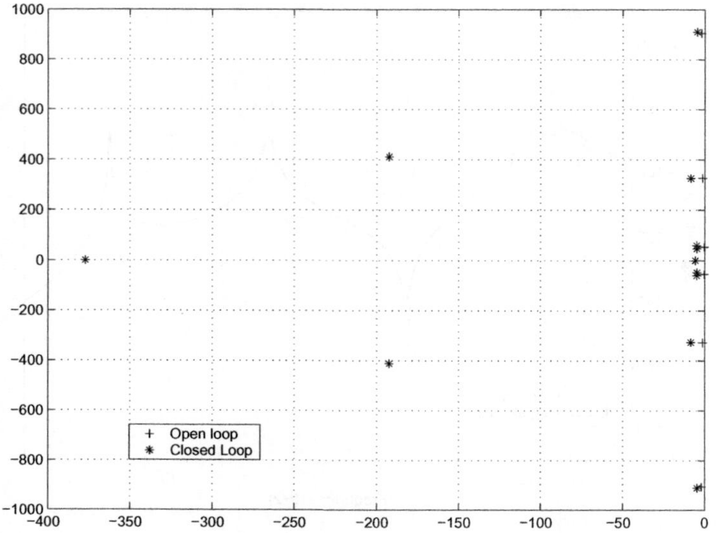

Figure 3.21. Open- and closed-loop pole map for the H_∞ optimization

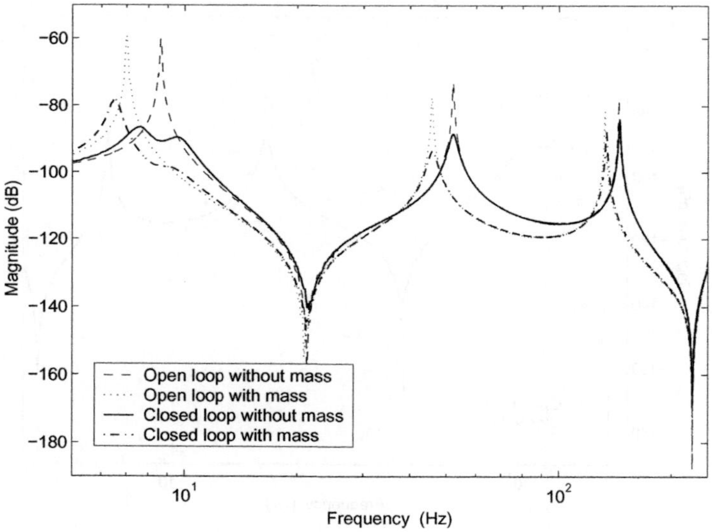

Figure 3.22. Measured open- and closed-loop frequency responses using the H_∞ optimization, with and without the mass

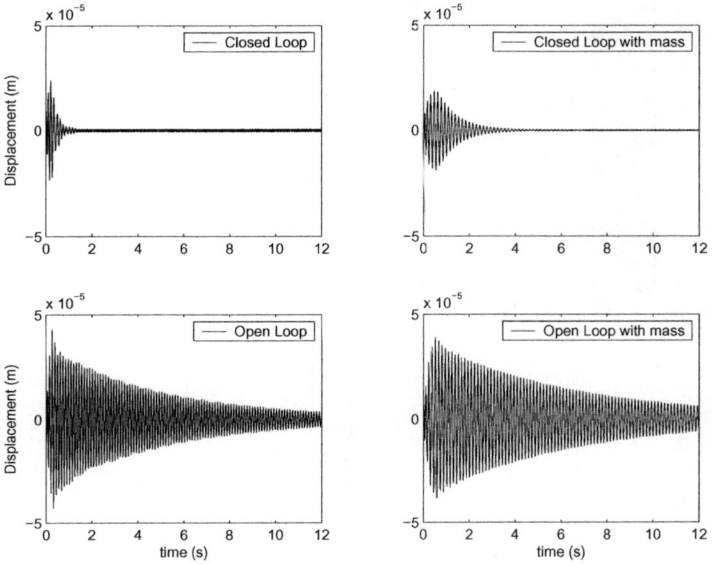

Figure 3.23. Measured disturbance responses using the H_∞ optimization, with and without the mass

piezoelectric device known, one can simply apply the same voltage across an "identical" capacitor and subtract the electrical response from that of the sensori-actuator to resolve the mechanical response of the structure.

The key idea, here, is to replace the function of a sensor in the feedback loop by estimating the voltage induced inside the piezoelectric transducer, v_p. Since this voltage is proportional to the mechanical strain in the base structure, the estimated signal would provide a meaningful measurement for a feedback compensator. Furthermore, by estimating v_p, one would effectively replace the role of the collocated piezoelectric transducer in Figure 3.1 by the additional electronic circuitry. In this way one would expect to design feedback controllers that possess appealing properties associated with compensated collocated systems, such as those discussed in previous sections.

Two realizations for the piezoelectric sensori-actuator, as proposed in [9], are sketched in Figures 3.24 and 3.25. The two circuits have rather similar functions; they use a signal proportional to the electrical charge or current and subtract that from the signal proportional to the total charge or current to produce a signal proportional to the mechanical strain, or its derivative. This signal is then used for feedback.

In the strain measurement circuit of Figure 3.24, assuming that the leakage resistors R_1 and R_2 are very large and that the gain of each op-amp voltage follower is unity, we may write:

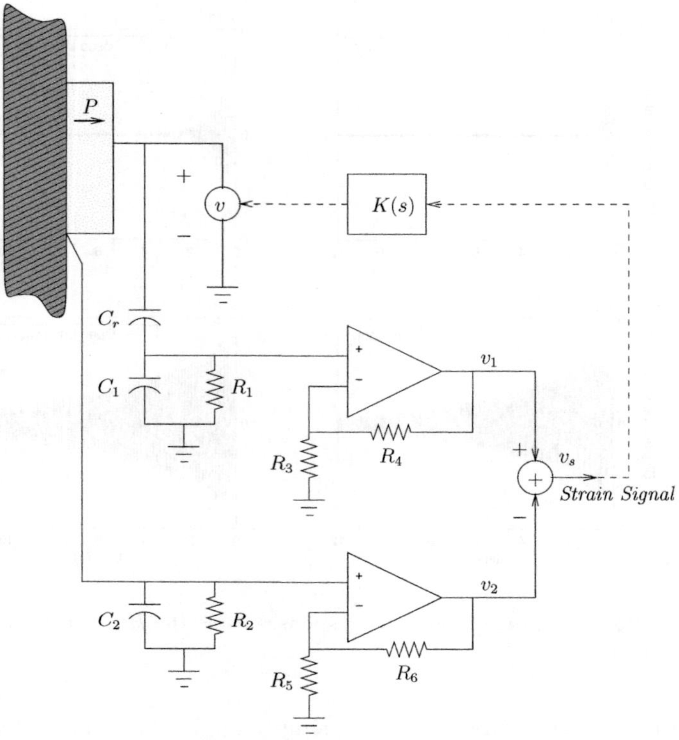

Figure 3.24. The piezoelectric-based sensori-actuator, generating an estimate of the mechanical strain

$$v_1 = \frac{C_r}{C_1 + C_r} v$$
$$v_2 = \frac{C_p}{C_2 + C_p} (v - v_p),$$

where v_p and C_p are respectively, the voltage induced in and the capacitance of the piezoelectric transducer (refer to Figure 2.16 (b)). The voltage v_p is proportional to the mechanical strain. Subtracting v_2 from v_1, we obtain:

$$\begin{aligned} v_s &= v_1 - v_2 \\ &= \left(\frac{C_r}{C_1 + C_r} - \frac{C_p}{C_2 + C_p} \right) v + \frac{C_p}{C_2 + C_p} v_p. \end{aligned} \tag{3.60}$$

Hence, if $C_1 = C_2$ and $C_r = C_p$, equation (3.60) reduces to

$$v_s = \frac{C_p}{C_2 + C_p} v_p. \tag{3.61}$$

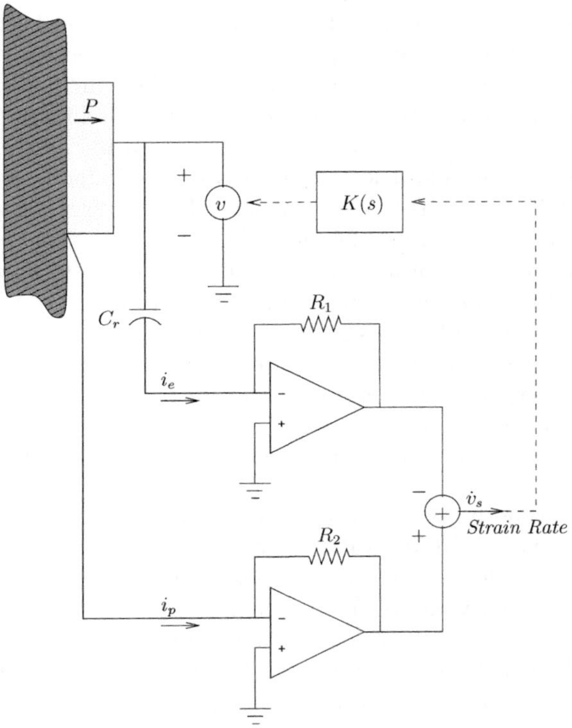

Figure 3.25. The piezoelectric-based sensori-actuator, generating an estimate of the strain rate

Therefore, under the above "ideal assumptions", the estimated voltage is proportional to v_p. Now, consider the sensori-actuator in Figure 3.25. The voltage v is applied to both the piezoelectric transducer and the reference capacitor, C_r. A current, i_e, flows through the upper path, while i_p flows through the lower path. Each signal is converted to a voltage using an op-amp. The two signals are differenced, resulting in a voltage proportional to the derivative of v_p; *i.e.*, the strain rate. To be more precise, if the two resistors R_1 and R_2 are both equivalent to R, v_s is found to be:

$$v_s = R(C_r - C_p)sv + RC_p s v_p.$$

Again, it can be observed that if $C_r = C_p$, the first term will disappear, and v_s will be proportional to the strain rate. For practical reasons, however, very often the capacitive and resistive elements are chosen differently; see [9] and [54] for more details.

The two sensori-actuator schemes in Figures 3.24 and 3.25 should perform well under ideal assumptions. Having estimated v_p, or perhaps $\dot{v}_p$, the signal produced by the sensori-actuator can now be used for feedback. Several applications for this method have appeared throughout literature (see for example [81, 200, 106, 21, 185, 4]. In practice, however, there are a number factors that limit the performance of the sensori-actuator, the foremost being the choice of the reference capacitor C_r that is directly related to the size of piezoelectric capacitance C_p. The piezoelectric properties are influenced by variations in environmental conditions and operation. This requires a continual effort to tune the circuits in Figures 3.24 and 3.25.

A primary obstacle for implementation of the piezoelectric sensori-actuator is the difficulty in obtaining an accurate estimation of the capacitance of the piezoelectric device, C_p. This may not severely affect the open-loop performance of the sensori-actuator, however, if v_s is used as measurement for feedback, such variations may destabilize the closed-loop system[2]. An attempt to address this problem was made in [38, 191, 36, 1], where the authors suggest an adaptive sensori-actuator implementation based on the LMS algorithm [195, 55].

The sensori-actuator is a linear estimator that generates an estimate of the strain signal, or its derivative. The structure of the estimator is rather crude and largely dependent on the added electronic circuitry. Often a nominal model for the underlying system is at hand. Therefore, it should be possible to construct better estimates of the required signals using an optimal estimation method such as a Kalman Filter [114, 7]. The issue of uncertainty associated with the varying piezoelectric capacitance can then be addressed using the recent advances in robust state estimation and Kalman Filtering (see [156] and references therein).

[2] In fact, it can be shown that the transfer function estimated by the self-sensing circuit is $G_{vv}(s) + \delta$, where δ is proportional to $C_p - C_r$. This additional feedthrough term does not alter poles of the open-loop system. However, it does perturb the open-loop zeros, and this could be detrimental to the closed-loop performance and stability of the system.

4

Piezoelectric Shunt Damping

4.1 Introduction

Piezoelectric shunt damping is a popular technique for vibration suppression in smart structures. As illustrated in Figure 4.1, techniques encompassed in this broad description are characterized by the connection of an electrical impedance to a structurally bonded piezoelectric transducer. Such methods do not require an external sensor, may guarantee stability of the shunted system and do not require parametric models for design purposes.

The first goal of this chapter is to provide a brief review of the various techniques categorized as piezoelectric shunt damping. One particular sub-

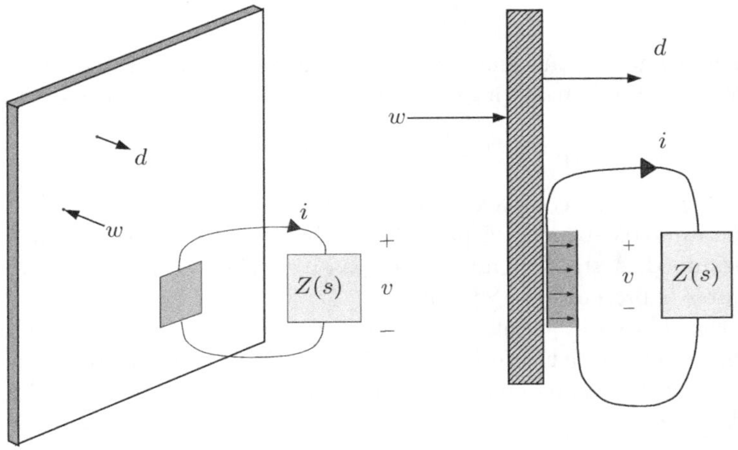

Figure 4.1. A piezoelectric laminate structure disturbed by an external force w. The resulting vibration d is suppressed by the presence of a shunt impedance $Z(s)$. The poling direction of the transducer is indicated by the shaded arrows.

category, the so-called resonant shunts, has been the focus of significant development over the past decade. Resonant shunt damping circuits, comprised of inductors, capacitors, and resistors, are simple to design and can significantly augment the damping of lightly-damped flexible structures. The foremost difficulty associated with resonant shunt circuits is the requirement for impractically large inductance values. In Section 4.4 the synthetic impedance is introduced as a simplified implementation technique.

In the next chapter, the structural influence of a shunted piezoelectric transducer is studied and expressions revealing the dynamics of a piezoelectric laminate structure are derived. The technique of piezoelectric shunt damping is shown to be equivalent to collocated strain feedback control. A simple relationship is shown to exist between a connected shunt impedance and an equivalent strain feedback controller, and *vice versa*.

4.2 Passive Shunt Damping

In the following subsections, the various passive shunt damping techniques depicted in Figure 4.2 are discussed in detail.

4.2.1 Passivity Defined

An electrical shunt impedance is said to be passive if and only if it does not supply power to the system. In mathematical terms, this means that [6]

$$\int_0^\infty v(t) \cdot i(t) \geq 0, \tag{4.1}$$

where $v(t)$ and $i(t)$ are the voltage and current as illustrated in Figure 4.1. For linear systems, passivity of the impedance $Z(j\omega)$ is guaranteed if [6]

$$\Re\left(v(j\omega) \cdot i^*(j\omega)\ \right) \geq 0 \ \text{ or } \ \Re\left(Z(j\omega)\right) \geq 0 \ \ \forall \omega, \tag{4.2}$$

where $i^*(j\omega)$ is the complex conjugate of $i(j\omega)$.

The foremost benefit of passivity is that stability of the shunted system is guaranteed. A stability analysis of passively shunted piezoelectric laminate structures is presented in Section 5.5.

In the context of shunt damping and vibration control, the reader should be aware that the definition of electrical passivity can vary between authors. In some discussions, passivity not only implies that the above conditions hold, but also that the impedance is constructed solely from physical components with no external power supply. This meaning is occasionally used by some authors while discussing the characteristics of particular shunt damping circuits; for example, a shunt circuit that operates autonomously without a power supply is commonly referred to as "passive". In this book we adhere to the standard definition as discussed in [6]. For example, in Section 4.4 op-amp circuits

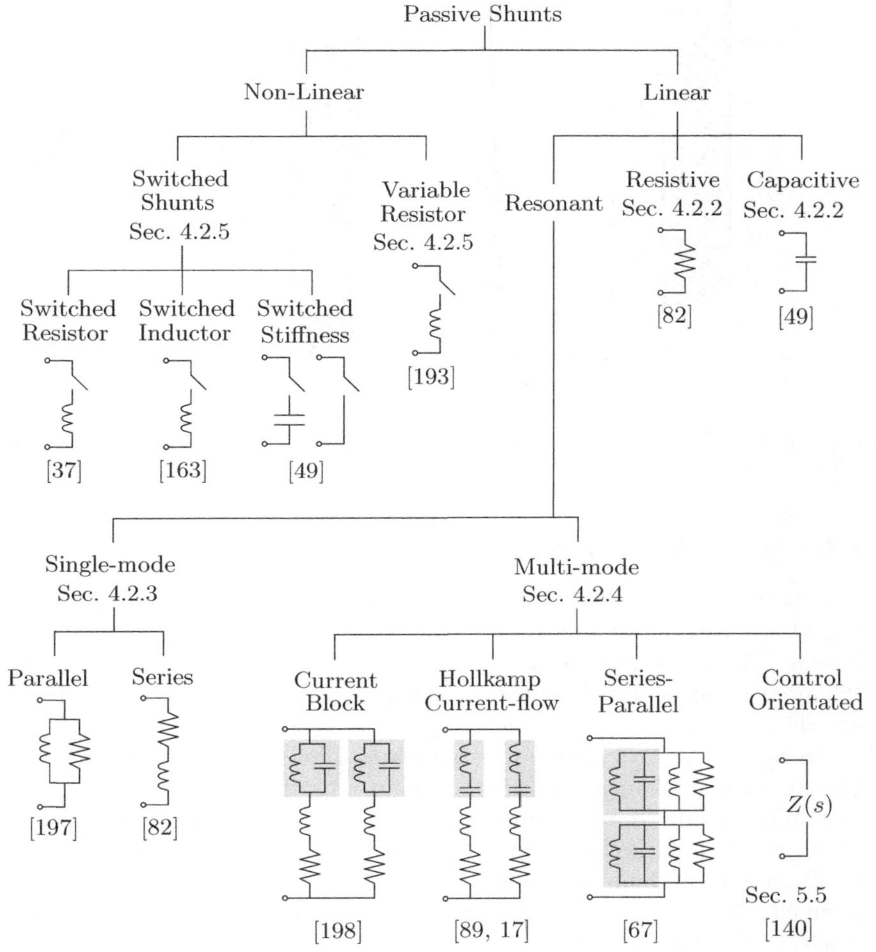

Figure 4.2. Passive piezoelectric shunt damping techniques discussed in Section 4.2

are introduced as a means for implementing impractically large inductance values. Although the impedance is synthesized using active components, the terminal voltage and current satisfy the conditions in Equations (4.1) and (4.2), thus it is referred to as "passive".

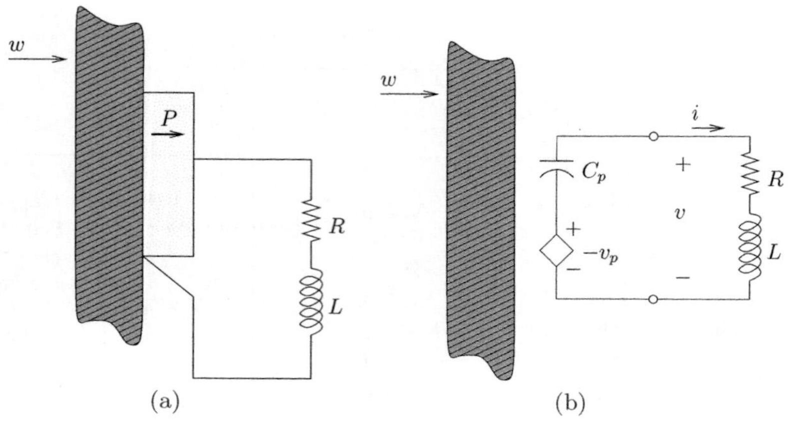

Figure 4.3. A piezoelectric transducer with an RL shunt and the equivalent electrical circuit

4.2.2 Linear Techniques

For the purposes of this discussion, a linear shunt circuit is defined as any impedance with a linear current-to-voltage relationship over the bandwidth of interest. By this definition, slowly varying impedances such as adaptive shunts (Chapter 8) and shunts implemented by switched mode amplifier (Section 6.7) are classed as linear.

In reference [82] a simple resistance was applied to a structurally bonded piezoelectric transducer in order to damp vibration. Its effect on structural dynamics was shown to be equivalent to that of a light visco-elastic damping treatment [82]. Although simple, resistors are rarely used as they offer only a small amount of damping; typically only a few dB when applied to a lightly damped Aluminum structure. Future transducers utilizing high d_{33} electromechanical coupling factors may provide better performance.

Capacitive shunting [49] is another simple but poorly performing technique. Adding capacitance to the terminals of a piezoelectric transducer varies its effective stiffness. Although capacitive shunts have not proven useful in vibration control, they may find application where a slight adjustment in resonance frequency is required. Capacitive shunts and their equivalent strain-feedback interpretation are discussed further in Section 5.3.

4.2.3 Resonant Single-mode Shunt Circuits

The first shunt circuit of any kind should be credited to Forward [68], who in 1979 proposed the idea of inductive (LC) shunting for narrow band reduction of resonant mechanical response. In particular, he demonstrated that the effect of inductive shunting was to cancel the inherent capacitive reactance

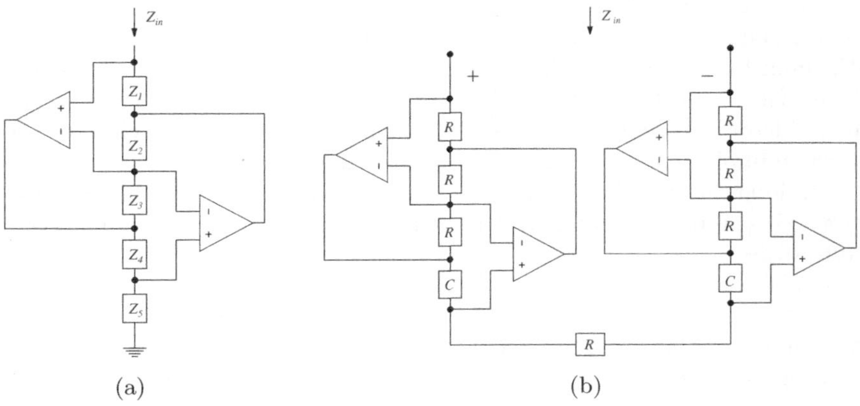

Figure 4.4. Circuit implementation of (a) a grounded and (b) a floating inductor

of a piezoelectric transducer. Later, Hagood and von Flotow [82] interpreted the operation of a resonant shunted piezoelectric transducer as analogous to a tuned mass damper, in which a relatively small second order system is appended to the dynamics of a larger system. Moreover, they addressed the situation in which a resistive element is added to the shunt network, resulting in an RLC tuned circuit. This system, and its electrical equivalent are depicted in Figure 4.3. The resulting RLC circuit is tuned to a specific resonance frequency of the composite system. That is, if the vibration associated with the ℓth mode is to be reduced, then L is chosen as:

$$L = \frac{1}{\omega_\ell^2 C_p}.$$

By adopting a proper value for R, the resonant response at and in the vicinity of ω_ℓ can be reduced. However, we should keep in mind that due to the passive nature of the shunt there will be hard constraints on the level of performance that this approach can achieve.

A number of techniques can be used to determine an optimal value for the resistive component of the shunt impedance. One possibility is to set up an optimization problem to minimize the H_2 norm of a transfer function associated with the structure as a function of the parameter R. Such a procedure is employed in reference [16] and if performed properly, will result in the placement of poles so that structural damping is augmented. Alternatively, one could employ a pole placement method, such as explained in reference [82].

A parallel circuit variation where the resistance and inductance are placed in parallel rather than series was proposed by Wu [197]. Although the two circuits, series and parallel, achieve similar performance, it is claimed that the parallel structure is less sensitive to sub-optimal resistance values.

The foremost difficulty in implementing resonant shunt circuits is the need for very large inductive elements. As an example, to implement a single-mode RL shunt for control of a vibratory mode of 11 Hz using a piezoelectric transducer with capacitance of $C_p = 101\ nF$ requires an inductor of 2072 H. Such an inductor can only be synthesized using active electronic circuitry. Two possible implementations are demonstrated in Figure 4.4[1].

As illustrated in Figure 4.4 (a), a grounded inductor can be synthesized by two op-amps. The equivalent impedance observed at the terminals of this Gyrator is

$$Z_{in} = \frac{Z_1 Z_3 Z_5}{Z_2 Z_4}.$$

Allowing Z_4 to be a capacitor, C, and replacing other impedances with resistors, the resulting circuit will act as an inductor $L = kC$, where

$$k = \frac{R_1 R_3 R_5}{R_2}.$$

This circuit is all that is needed for implementation of single-mode RL shunts. To implement certain classes of multi-mode shunts, however, one would need to implement floating inductors. This can be done using four op-amps, shown in Figure 4.4 (b).

Replacing the inductor needed in a single-mode RL shunt by a Gyrator is practical. However, particular care must be taken in implementing such circuits. Voltages generated across piezoelectric transducers could be very large, particularly when disturbances acting on the structure have frequency content that is very close to resonance frequencies of the base structure. Subsequently, high-voltage op-amps are to be used for this purpose. The designer should also be careful to avoid saturation of internal nodes of the system. For a single-mode RL shunt this is straightforward, due to the simplicity of the required circuit. For multi-mode circuits such implementation becomes overly complex and impractical. This issue will be further discussed in Section 4.4.

4.2.4 Resonant Multi-mode Shunt Circuits

Extension of the single-mode shunt damping technique to allow for multiple-mode vibration suppression has been the subject of intense research in recent years. A trivial solution is to attach a number of piezoelectric transducers to a structure, each one shunted by an RL circuit tuned to a specific mode. This is clearly not a viable option as there is only a limited amount of surface area over which the transducers can be mounted. The main focus in this area, therefore, has been on devising multiple-mode vibration damping methods which use a single piezoelectric transducer. The remainder of this chapter is devoted to the study of this issue.

[1] It is possible to reduce the size of necessary inductors by adding an extra capacitance in parallel to the piezoelectric transducer. This issue will be further elaborated on in the next chapter.

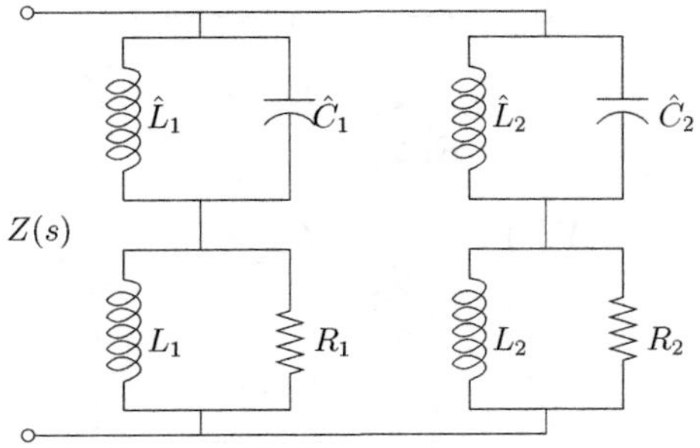

Figure 4.5. Two-mode shunt damping circuit [198]

Current-blocking

The first multi-mode techniques were proposed by Wu and his co-authors [199, 198, 203]. Their idea is centered around using an RL (either parallel, or series) shunt for each individual mode, and then inserting "current blocking" LC circuits into each branch. The electric shunt circuit for a two-mode system is depicted in Figure 4.5.

If vibration of the first two modes of the base structure are to be reduced, then L_1C_p is tuned to ω_1 while L_2C_p is tuned to ω_2. Furthermore, $\hat{L}_1\hat{C}_1$ is tuned to ω_2 while $\hat{L}_2\hat{C}_2$ is tuned to ω_1. Therefore, L_1R_1 and L_2R_2 are effectively separated at ω_1 and ω_2. For three modes, two current blocking circuits are inserted inside each branch, and so on.

The main difficulty with this method is that the circuit order increases very rapidly with the number of modes to be shunt damped simultaneously. This seriously complicates the task of implementing the required circuits.

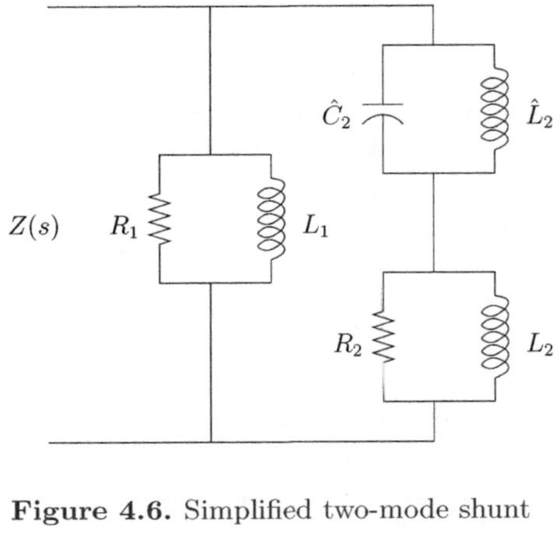

Figure 4.6. Simplified two-mode shunt

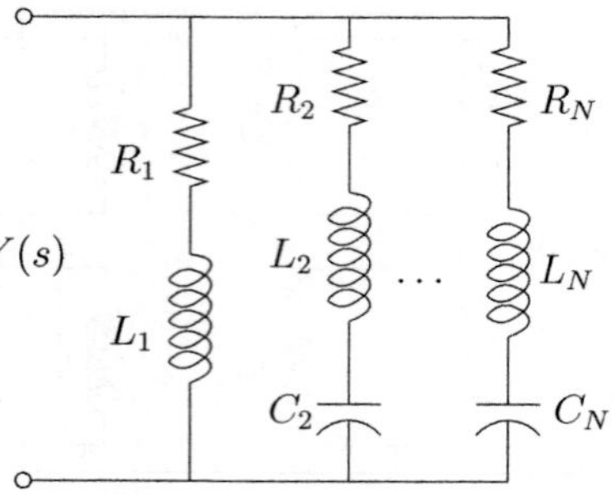

Figure 4.7. Hollkamp circuit

The number of components needed in this multi-mode shunt circuit can be almost halved by using straightforward circuit theory. For example, the two-mode shunt in Figure 4.5 can be reduced to that in Figure 4.6. For this circuit to work properly, its components must be chosen according to

$$L_1 = \frac{1}{\omega_1^2 C_p} \tag{4.3}$$

$$L_2 = \frac{\left(L_1\tilde{L}_2 + \tilde{L}_2\hat{L}_2 - L_1\hat{L}_2 - \omega_2^2 L_1\tilde{L}_2\hat{L}_2\hat{C}_2\right)}{\left(L_1 - \tilde{L}_2\right)\left(1 - \omega_2^2\hat{L}_2\hat{C}_2\right)} \tag{4.4}$$

and

$$\tilde{L}_2 = \frac{1}{\omega_2^2 C_p} \tag{4.5}$$

$$\hat{L}_2 = \frac{1}{\omega_1^2 \hat{C}_2}. \tag{4.6}$$

Here, $\hat{C}_2$ is an arbitrary capacitor used in the current blocking network.

This class of multi-mode impedances can be tuned to damp a small number of modes in an effective way. An example of this is contained in [16] where the second and third modes of a simply supported beam are reduced in magnitude by $18.9dB$ and $19.1dB$.

Hollkamp

Hollkamp [89] suggested a specific resonant shunt structure, depicted in Figure 4.7. The shunt circuit consists of a number of parallel RLC shunts, with the very first branch being an RL circuit. For one mode, Hollkamp's circuit reduces to the one proposed by Hagood and von Flotow. However, for each additional mode, an RLC branch has to be added. When an extra branch is added, the previous resistive and inductive elements must be re-tuned to

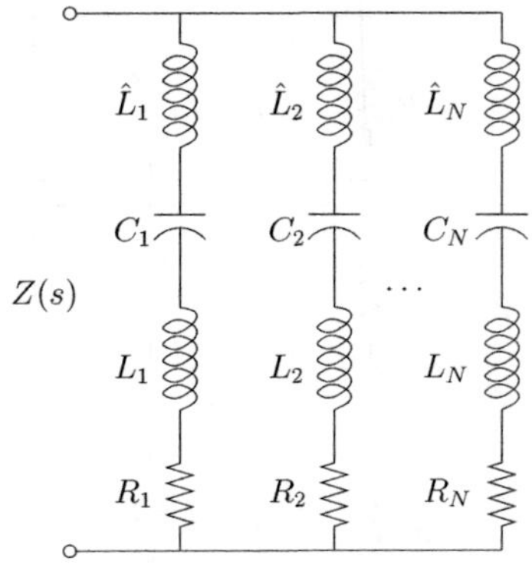

Figure 4.8. Current-flowing multi-mode shunt circuit [17]

ensure satisfactory performance. No closed form tuning solution has been proposed for this technique. However, in [89] values for the shunt circuit electrical elements are determined by numerical optimization of a non-linear objective function. Given that all circuit elements are to be determined numerically, for a large number of modes this procedure may result in a complicated optimization problem. Nevertheless, this method has been applied to a cantilevered beam in [89], in which vibration of the second and third modes were reduced by 19 and 12 dB respectively.

Current-flowing

The *current-flowing* shunt circuit [17] was introduced as a means for simplifying the implementation of high-order multi-mode shunt circuits. Shown in Figure 4.8, a current-flowing shunt circuit requires one parallel branch for each structural mode to be controlled. The current-flowing network $\hat{L}_i C_i$ in each branch is tuned to approximate a short circuit at the target resonance frequency whilst approximating an open circuit at the frequencies of adjacent branches *i.e.*

$$\hat{L}_1 = \frac{1}{\omega_1^2 C_1}, \quad \cdots \quad, \hat{L}_N = \frac{1}{\omega_N^2 C_N}. \tag{4.7}$$

The remaining inductor and resistor in each branch $L_i R_i$, are tuned to damp the ith target structural mode in a manner analogous to single-mode shunt damping *i.e.*,

$$L_1 = \frac{1}{\omega_1^2 C_p}, \quad \cdots \quad, L_N = \frac{1}{\omega_N^2 C_p}. \tag{4.8}$$

The two inductors in each branch can be combined, resulting in a series RLC circuit in each parallel arm of the shunt circuit. This further simplifies the task of implementing the shunt circuit. The total shunt admittance is then

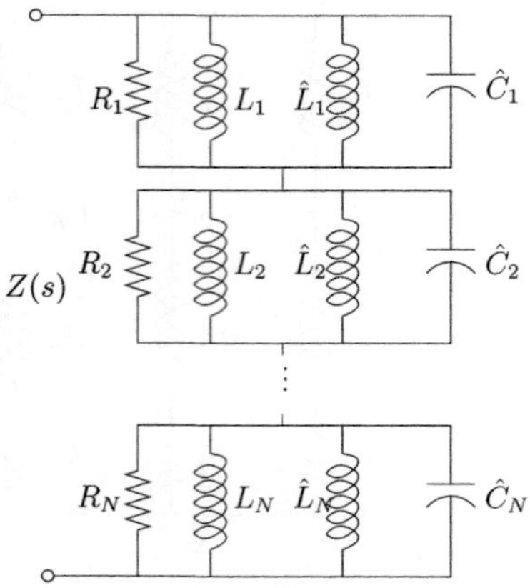

Figure 4.9. The series-parallel shunt damping circuit of [17].

$$
\begin{aligned}
Y(s) &= \sum_{i=1}^{N} Y_i(s) \\
&= \sum_{i=1}^{N} \frac{(1/\tilde{L}_i)s}{s^2 + (R_i/\tilde{L}_i)s + (1/\tilde{L}_i C_i)},
\end{aligned}
$$

where

$$\tilde{L}_i = L_i + \hat{L}_i.$$

Compared to the circuit proposed by Wu and his co-authors [199, 198, 203], the resulting shunt circuit is of a considerably lower order. Furthermore, in comparison with the technique proposed in [89], this is a more systematic way of designing a shunt impedance circuit. The only parameters left to be chosen are the resistors $R_1, \ldots, R_N$. These parameters can be determined via optimization or tuned experimentally.

Series-parallel

The electrical dual of a current-flowing circuit, the so-called $series-parallel$ circuit shown in Figure 4.9 was proposed as a method for reducing inductive component values in [67].

Each network in series $C_i\hat{L}_iL_iR_i$ contains two sub-networks: a current-blocking network $\hat{C}_i\hat{L}_i$, and a parallel single-mode damping network L_iR_i. Both the current-blocking and damping networks, $\hat{C}_i\hat{L}_i$ and L_iR_i, are tuned to the same target resonance frequency ω_i. At this frequency, the current-blocking network has an extremely large impedance. All of the remaining current-blocking networks, tuned to other structural resonance frequencies, have a low impedance at ω_i. A voltage applied at the terminals results in a current that flows freely through the detuned current-blocking networks but is forced to flow through the active damping network. In this way, the circuit is decoupled so that each damping network L_iR_i can be tuned individually to a target resonance frequency. As described above, the series-parallel structure offers no great advantage over comparable techniques. Benefits arise from a suitable choice in the arbitrary capacitances $\hat{C}_i$. The recommended capacitance value is $10-20$ times that of the piezoelectric capacitance. In this case, the current blocking inductors become significantly smaller than the damping inductors. When the circuit is simplified by combining the damping and current-blocking inductors in each series network, the resulting single inductor is a fraction of that required in other single- or multi-mode circuits.

Control Orientated

All of the multi-mode techniques discussed thus far are more or less direct extensions of the original single-mode circuits. A new approach to the design of passive piezoelectric shunt damping circuits was presented in [140]. By viewing the electrical impedance as parameterizing an equivalent collocated strain feedback controller, a shunt impedance can be found by working backwards from an effective feedback controller. Under certain conditions, the passivity, and hence stability of the shunted system can be guaranteed [140]. Present controller designs have benefits similar to that of current-flowing circuits, they are low in order, easy to tune, and suitable for modally dense systems. The design and analysis of these shunt circuits is discussed in Section 5.5.

4.2.5 Non-linear Techniques

In an attempt to eliminate the need for large inductors, literature has also developed on the so-called *switched shunt* or *switched stiffness* techniques [41]. Three major subclasses exist where the piezoelectric element is switched into a shunt circuit comprising either of: another capacitor [49], a resistor [37], or an inductor [163]. The required inductance is typically one-tenth of that required to implement a simple LR resonant shunt circuit designed to damp the same mode. Similar to virtual circuit implementation, an external power source is required for the gate drive and timing electronics.

In reference [148] a hybrid systems approach is taken to derive optimal switching rules for the aforementioned switching shunts. It is shown theoretically and experimentally that the empirical switching laws can be improved. Although the optimal solution is numerically intensive to compute in real-time, a simplified quasi-optimal solution is easily implemented.

The Strain Amplitude Minimization Patch (STAMP) damper was presented as a self-powered shunt vibration controller in references [107] and [108]. Achievable performance is less than switched inductor designs but without the requirement for a power supply.

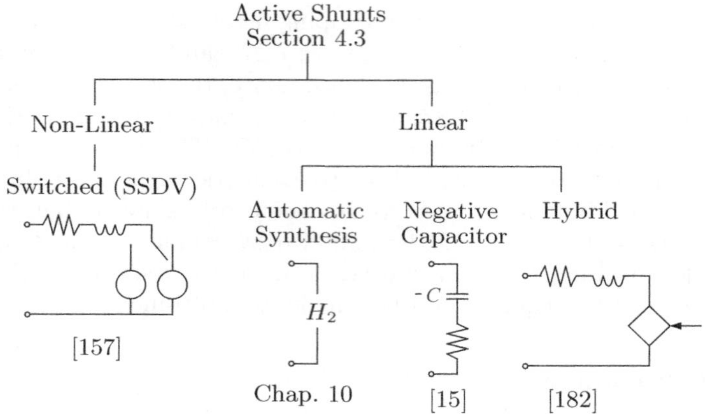

Figure 4.10. Active piezoelectric shunt damping techniques

A better performing autonomous shunt using two transducers was presented in [148]. While one transducer is connected to a switching circuit closely resembling a switched inductor network, the second transducer is utilized to provide energy for MOSFET gate drives and to correctly time the switching events.

Another non-linear technique, the time varying resistance, was suggested for maximizing the loss factor of a shunted system [193]. The foremost drawbacks are the complicated design process and complexity of implementation.

4.3 Active Shunt Damping

Active shunt impedances cannot be realized using passive physical components. Although passivity, and hence stability is not guaranteed, active shunts have been known to provide greater vibration suppression than passive circuits.

The negative capacitor shunt circuit [201, 15] is a simple technique for broad-band structural damping. By treating the internal piezoelectric voltage source as a supply, and the shunt impedance as a load, the traditional concept of maximum power transfer can be applied. The optimal impedance is equal in magnitude to the source impedance, but opposite in phase; hence the negative capacitor. Detrimentally, the optimal impedance applies large control voltages from DC to out-of-bandwidth frequencies. An effective technique for stabilizing and *toning down* the control effort of a negative capacitor is presented in [15]. Although negative capacitor shunts are somewhat immune to variation in the structural resonance frequencies, any variation in the transducer impedance can heavily degrade performance and lead to instability. Due to their simplicity, negative capacitors are discussed further and evaluated experimentally in Chapter 9.

Optimal shunts are discussed at length in Chapter 10. By viewing the problem of shunt impedance design as a standard feedback problem. Synthesis techniques such as LQG, H_2, and H_∞ are readily employed to design optimal impedances. Optimal shunts can provide excellent vibration suppression (greater than 30dB, see Chapter 10) and can be made less sensitive to variation in resonance frequencies. The main disadvantage associated with optimal shunts is the requirement for a dynamic system model and the dependency on transducer capacitance.

The state-switched active shunt (SSDV) [157] was proposed to increase the performance of switched inductor techniques. Rather than switching a series inductor-resistor network between open- and closed-circuit states, the network is switched between two voltage sources. Damping performance is increased but at the expense of additional power requirements.

The final vibration control technique falling loosely into the category of piezoelectric shunt damping is hybrid active-passive control [182]. This technique comprises a sensor and piezoelectric actuator driven by a voltage source in series with a resonant shunt circuit. There are a number of advantages in augmenting an active feedback system with a passive shunt: 1) Damping is still provided by the shunt circuit if the sensor fails or if the voltage source fails in short-circuit; 2) The system transfer function seen by the controller consists of the true structural system with damping augmented by the passive shunt. This can simplify control design and provide greater immunity to changes in resonance frequency.

4.4 Implementation of Resonant Shunt Circuits

For typical values of piezoelectric capacitance and structural resonance frequency, resonant shunt circuits require inductance values in the order of tens or hundreds of Henrys. Inductors of this size must be implemented by virtual circuits or otherwise.

4.4.1 Virtual Circuits

Virtual inductors and Riordan gyrators [164] have been employed to implement large inductance values. They are most useful for the implementation of single-mode shunt circuits. In reference [152], a practical single-mode circuit is described with the option for online analog tuning.

The foremost difficulties associated with virtual circuit implementation are discussed below.

- Virtual inductors or Riordan Gyrators are difficult to tune and are sensitive to component age, temperature, and non-ideal characteristics. The complexity of the required circuitry is illustrated in Figure 4.11 where a simplified dual-mode current-blocking network is implemented by virtual inductors and Riordan Gyrators.

- As piezoelectric patches are capable of developing hundreds of volts from moderate structural excitations, the entire circuit must be constructed from high-voltage components. In the likely case of voltages greater than $\pm 45V$, the components may be prohibitively expensive. Virtual circuits may also contain internal voltage gains. In this case, the range of operation is dictated by the greatest circuit node voltage.
- The minimum number of op-amps required to implement a piezoelectric shunt damping circuit increases rapidly with the number of modes to be damped. At least 30 op-amps are required to implement a three-mode series configuration multi-mode shunt damping circuit with current-blocking networks in every branch. In general for this configuration, $2n + 4n(n-1)$ op-amps are required to damp n modes. If 5 modes are to be controlled, 90 op-amps are required. The advent of current-flowing shunt circuits has significantly alleviated circuit complexity issues since they require only $2n$ op-amps to control n modes.

Despite the associated practical problems, the majority of work within the field has adopted virtual circuits for shunt impedance implementation. This may be a cause for the slow transfer of piezoelectric shunt damping technology into the industrial domain.

4.4.2 Synthetic Admittance

From the discussion in the previous section it should be clear that a key difficulty associated with passive piezoelectric shunt damping is the need for large inductive elements. As demonstrated in Figure 4.11, the necessary circuit can be implemented using virtual inductors and gyrators. The complexity of such implementation precludes the extension to high-voltage or high-order shunt circuits. Here, the *synthetic admittance* is introduced as an alternative that is not limited by shunt circuit complexity or voltage.

As originally proposed in [59] and [60], the concept is illustrated in Figure 4.12. The *synthetic admittance* is a two terminal device consisting of a voltage buffer, signal filter $Y(s)$, and a voltage controlled current source (VCCS). Referring to Figure 4.12 (b), the applied current i_z is determined by the measured voltage v_z. By fixing i_z as the output of a linear transfer function $Y(s)$, the controlled current source can be made to synthesize an arbitrary terminal impedance $Z(s)$. In the Laplace domain,

$$i_z(s) = Y(s)v_z(s), \tag{4.9}$$

where $Z(s) = \frac{1}{Y(s)}$ is the desired terminal impedance.

In some circumstances it is advantageous to utilize a voltage source rather than a current source. In this case, referred to as *impedance synthesis*, Figure 4.13 (b), the applied voltage v_z is determined by the measured current i_z.

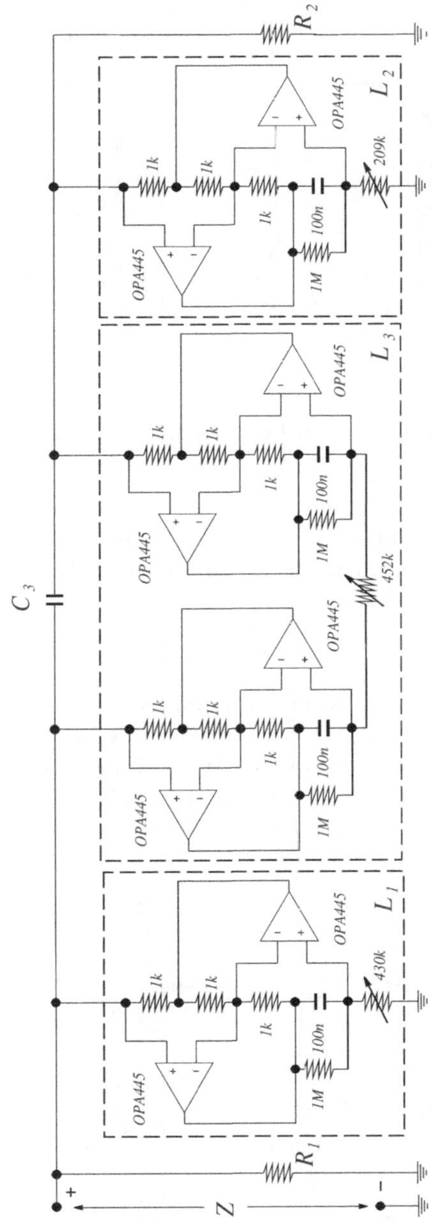

Figure 4.11. A simplified two-mode current-blocking shunt damping circuit implemented by a virtual inductors [16]

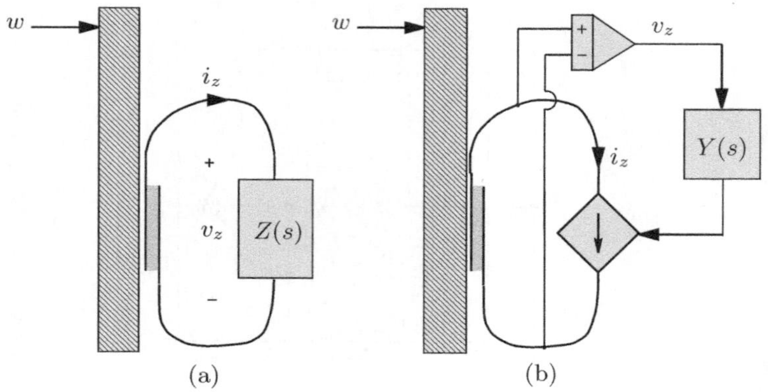

Figure 4.12. A synthetic admittance circuit (b) employed to implement a shunt impedance Z(s).

By fixing v_z as the output of a linear transfer function $Z(s)$, the controlled current source can be made to synthesize an arbitrary terminal impedance $Z(s)$. In the Laplace domain,

$$v_z(s) = Z(s)i_z(s), \tag{4.10}$$

where $Z(s)$ is the desired terminal admittance.

The choice of configuration, either synthetic impedance or admittance, will depend on the relative order of the desired impedance. As implementation of improper transfer functions is impractical [101], the choice should be made so that the required transfer function $Z(s)$ or $Y(s)$ is at least proper [101].

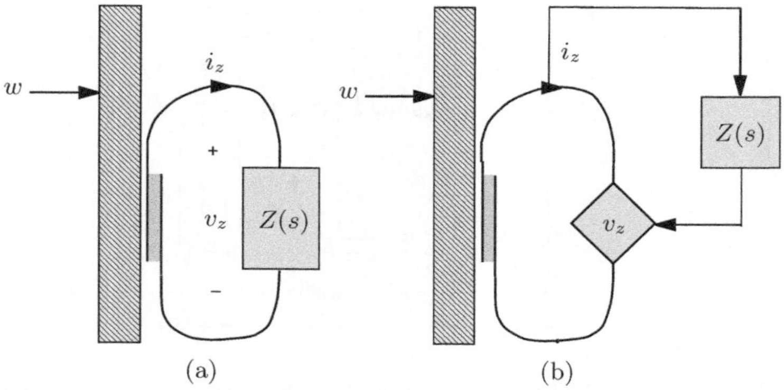

Figure 4.13. A synthetic impedance circuit (b) employed to implement a shunt impedance Z(s).

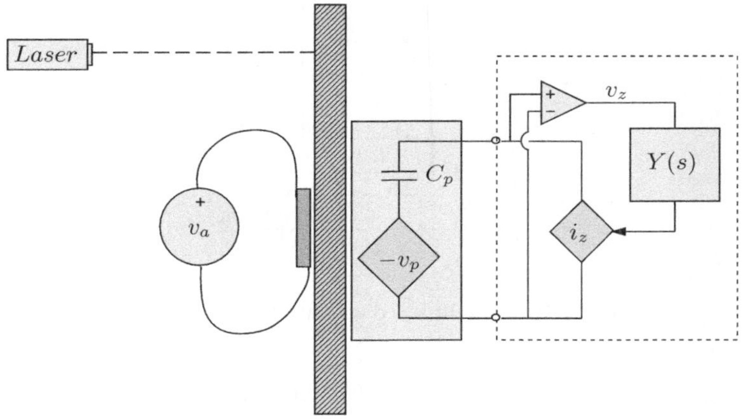

Figure 4.14. The experimental setup. An artificial disturbance is introduced by applying a voltage v_a to a disturbance transducer. The resulting vibration is suppressed by the presence of a shunt circuit implemented by a synthetic admittance

The former method, however, is believed to be more advantageous. It is often more straightforward to obtain high-precision voltage measurements from a piezoelectric transducer, while the current can be supplied with the required precision. Another justification for this observation is related to the hysteretic behavior of piezoelectric transducers at higher drives, when driven by a voltage source. It is known that a piezoelectric transducer displays negligible hysteretic nonlinearities if it is driven by a current source instead [78, 147, 39].

Further discussion on the implementation of synthetic admittance circuits, including current and charge amplifiers for piezoelectric loads can be found in Chapter 6.

4.5 Experimental Demonstration

In this section, a dual-mode current-blocking circuit is implemented by synthetic admittance to damp the second and third modes of a simply supported beam. The resonance frequencies of the experimental apparatus, shown in Figure 4.14, are 74.5 Hz and 171.3 Hz respectively. In this experiment two identical collocated transducers are utilized, one to introduce an artificial strain disturbance, and another to control resulting vibration. Both of the transducers are made from PIC151 piezoelectric ceramic and have a capacitance of 101 nF. Vibration of the beam at a point on its surface is measured using a PSV300 laser vibrometer.

The inductance values L_1, L_2 and $\hat{L}_2$ of the circuit shown in Figure 4.6 were determined from Equations (4.3) to (4.6). The capacitance $\hat{C}_2$, for the

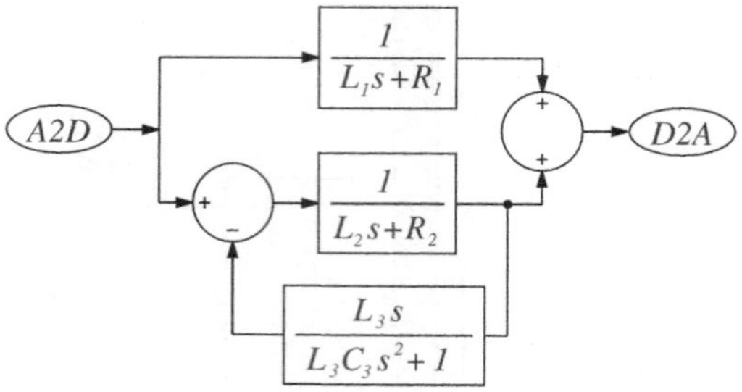

Figure 4.15. Simulink® diagram of the two-mode shunt

"blocking circuit" was chosen as $100nF$. The inductor parameters were found to be $L_1 = 43H$, $L_2 = 20.9H$, and $\hat{L}_2 = 45.2H$.

To determine optimal values for the resistive elements, a state-space model of the composite system was derived and parameterized in terms of R_1 and R_2. An optimization problem was then set up to minimize the H_2 norm of the transfer function from the applied disturbance voltage to the displacement on the surface of the beam. A minimum was found at $R_1^* = 262.75k\Omega$ and $R_2^* = 550.73k\Omega$.

The shunt was then implemented using a simple synthetic admittance circuit. The synthetic admittance circuit used in this experiment is depicted in Figure 6.7 and fully explained in Section 6.6. The admittance transfer function $Y(s)$ was implemented digitally on a dSPACE 1103 Digital Signal Processor. A block diagram of the required admittance transfer function is shown in Figure 4.15. C-code for the DSP was generated automatically using the Real Time Workshop for MATLAB®.

Frequency-domain results are demonstrated in Figure 4.16. The experimental resonant amplitudes for the $2nd$ and $3rd$ modes were successfully reduced by 22 and $18dB$ respectively. To demonstrate the effect of additional damping on the transient response of the structure, a second set of experimental data was collected. The experiment consisted of exciting the beam structure using a step voltage across the piezoelectric actuator and monitoring the velocity decay at a point on the beam structure. A Polytec Laser Scanning Vibrometer (PSV-300) was used to monitor the velocity decay at a point located on the surface of the beam. A dSPACE DS1103 system was used to generate the voltage step and capture the time response data from the vibrometer. The sampling frequency was set at $20kHz$. Within the dSPACE, a low-pass digital filter was used to remove high order structural modes and noise. The cut-off frequency for the digital filter was set at $230Hz$. After sampling the time response, the data was filtered with a high-pass digital filter to remove low frequency noise. The cut-off frequency was chosen to be $30Hz$ so as to remove the dynamics of the first structural mode, which was not

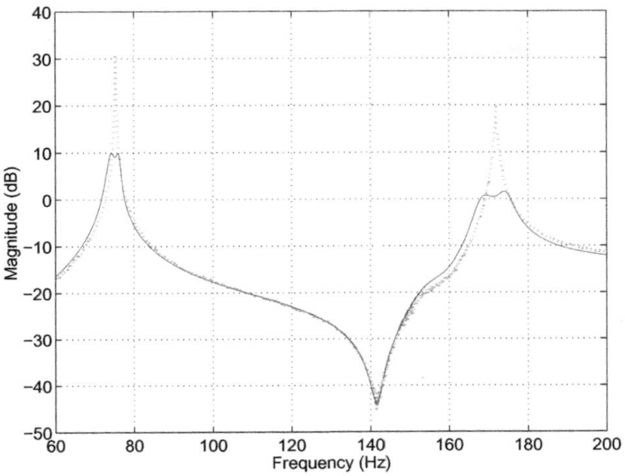

Figure 4.16. Magnitude frequency response of the beam measured from an applied disturbance voltage to the resulting displacement. (—) Open-loop, - - shunted

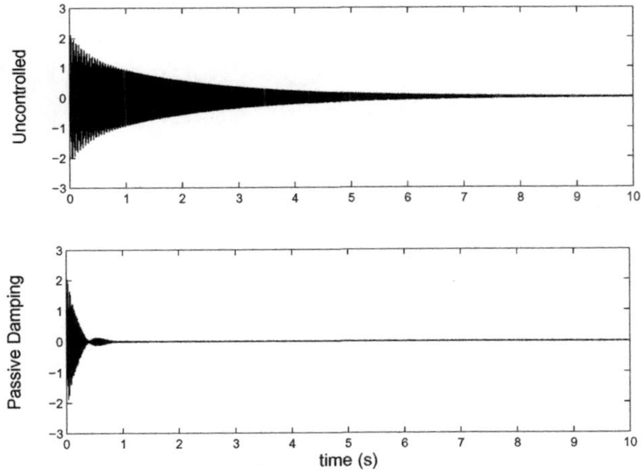

Figure 4.17. Transient step response of the beam

targeted for damping. Results from the transient time regime are shown in Figure 4.17.

5

Feedback Structure of Piezoelectric Shunt Damping Systems

5.1 Introduction

The previous chapter introduced the concept of piezoelectric shunt damping for vibration control of flexible structures. From the discussions, it should be clear that most of the methods proposed in the literature for the design of shunt circuits are based on *ad hoc* procedures. It turns out, however, that the problem of piezoelectric shunt damping can be interpreted as a feedback control problem. The feedback structure associated with shunted piezoelectric transducers was first reported in [140, 139].

Having made the observation that there is an underlying feedback structure associated with piezoelectric shunt damping we can propose new shunt impedances that are capable of damping structural vibrations in an efficient way. Furthermore, this can be done in a systematic way using the identified equivalence between the problems of piezoelectric shunt damping and collocated feedback control with an identical piezoelectric actuator and sensor.

Two classes of resonant shunts are introduced in this chapter. These shunts are closely related to the two resonant controller structures introduced in Chapter 3. In particular, we will illustrate that the closed-loop system consisting of a piezoelectric transducer shunted by a resonant impedance is equivalent to the closed-loop system associated with an identical structure consisting of a piezoelectric actuator/sensor pair and a resonant controller. A direct implication of this observation is that, in certain cases, it may be possible to replace a feedback controlled system by a shunted system, and thereby, reduce the number of necessary transducers by half.

In addition to a thorough analysis of shunt damping systems, we will demonstrate the effectiveness of the proposed shunt impedances by implementing them on two piezoelectric laminated structures.

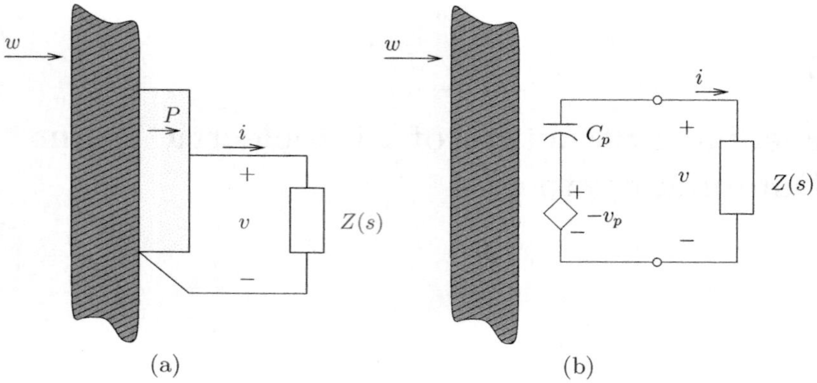

Figure 5.1. (a) A piezoelectric laminate shunted to an impedance $Z(s)$; (b) Electrical equivalent of the same system

5.2 Feedback Interpretations

Consider the shunted piezoelectric transducer in Figure 5.1 (a) and its electrical equivalent in Figure 5.1 (b). It should be clear that compared to active control methods shunting the piezoelectric transducer with an impedance Z removes the need for an additional sensor. It will be shown, however, that this is achieved at the expense of having to deal with a slightly more complicated feedback control problem.

To visualize the underlying feedback control structure, we need to identify a number of variables such as the control signal, the measurement, the disturbance and the physical variable that is to be regulated. Furthermore, we have to choose either $Z(s)$ or $Y(s)$ as the controller. The feedback structure can be identified by noticing that the current flowing out of the piezoelectric transducer may be written as:

$$i(s) = -\left(v_p(s) + v(s)\right) C_p s. \tag{5.1}$$

Furthermore,

$$v(s) = Z(s)i(s). \tag{5.2}$$

Also, given the linear behavior of the system, we may write:

$$v_p(s) = G_{vv}(s)v(s) + G_{vw}(s)w(s). \tag{5.3}$$

Here, G_{vw} represents the transfer function from the disturbance to the induced piezoelectric voltage v_p, when the piezoelectric transducer is short circuited. Also, G_{vv} is the collocated transfer function of the piezo. That is, the transfer function from a voltage applied across the conducting electrodes of the transducer to the induced piezoelectric voltage v_p. Furthermore, as explained in Chapter 2, typically

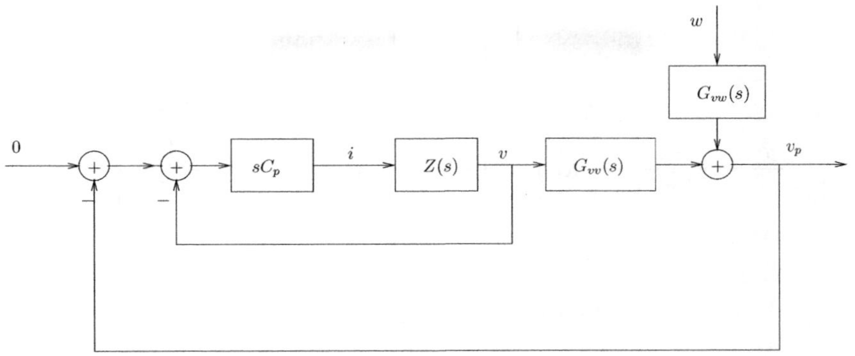

(a)

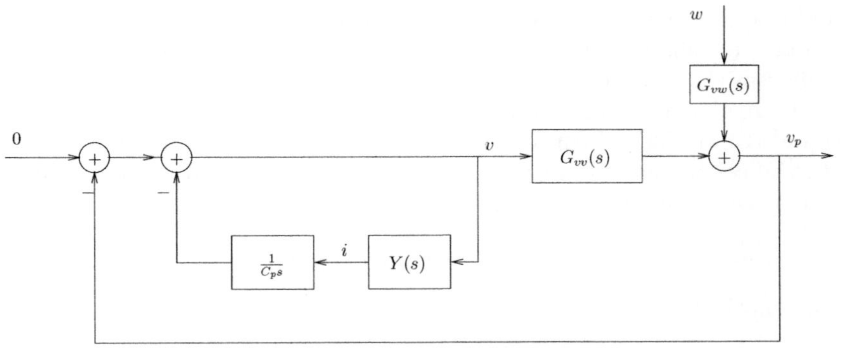

(b)

Figure 5.2. The feedback structure associated with the shunt damping problem in Figure 5.1: (a) $Z(s)$ functions as the controller; (b) $Y(s)$ serves as the controller

$$G_{vv}(s) = \sum_{i=1}^{M} \frac{\gamma_i}{s^2 + 2\zeta_i \omega_i s + \omega_i^2}, \tag{5.4}$$

and G_{vw} has a similar structure, *i.e.*

$$G_{vw}(s) = \sum_{i=1}^{M} \frac{\eta_i}{s^2 + 2\zeta_i \omega_i s + \omega_i^2}. \tag{5.5}$$

G_{vv} represents a collocated transfer function, hence $\gamma_i \geq 0$ for $i = 1, 2, \ldots, M$. However, due to the non-collocated nature of G_{vw}, a similar assertion cannot be made, *i.e.* in general η_i may be non-positive.

Equations (5.3), (5.1) and (5.2) amount to the feedback structure depicted in Figure 5.2(a). The block diagram suggests a slightly complicated feedback structure as the controller, $Z(s)$ is itself inside an inner feedback loop.

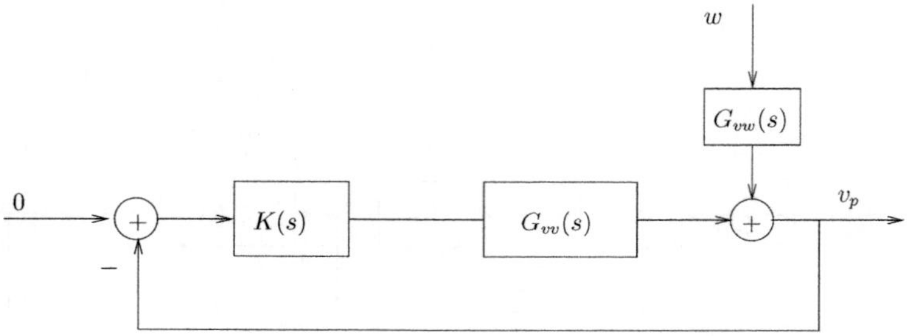

Figure 5.3. Equivalent feedback loop

If $Y(s)$ is to be viewed as the controller, the block diagram of Figure 5.2(a) should be redrawn as in Figure 5.2(b). There are specific reasons as to why $Z(s)$ or $Y(s)$ should be chosen as a controller. Some of these reasons will be clarified in this chapter.

The reader may notice that the feedback systems depicted in Figure 5.2 (a) and (b) are indeed equivalent to the feedback control problem associated with the collocated system $G_{vv}(s)$, and a feedback controller $K(s)$ depicted in Figure 5.3, where

$$K(s) = \frac{sC_pZ(s)}{1 + sC_pZ(s)} \tag{5.6}$$

or alternatively

$$K(s) = \frac{1}{1 + \frac{Y(s)}{sC_p}}. \tag{5.7}$$

An implication of this observation is that the shunted system in Figure 5.4(b) and the feedback system in Figure 5.4(a) are essentially equivalent as long as $K(s)$ and $Z(s)$ are related according to (5.6) or (5.7). This observation, however, could be misleading as it may lead the reader to the conclusion that having designed a controller for the former system, one may obtain an impedance for the latter. While this may be true in certain cases, such a procedure may result in an impedance, or an admittance transfer function that is not implementable digitally. Therefore, more practical impedance design methods are needed. A number of techniques are discussed in the following sections.

Identification of the underlying feedback structure associated with shunt damping is an important step in designing high-performance impedance shunts. In particular, the knowledge of this feedback structure enables us to address issues that would be very difficult to tackle otherwise. This includes problems such as: fundamental performance limitations in vibration damping; dealing with actuator saturation; multivariable shunt design; robustness issues, *etc.* Some of these issues will be discussed in the remainder of this book.

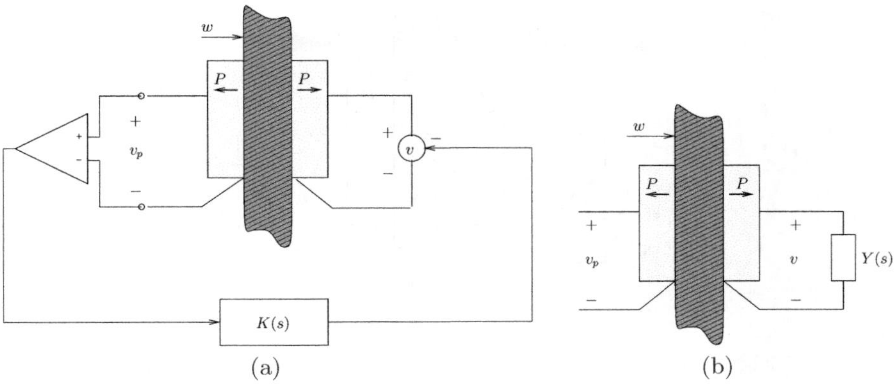

Figure 5.4. (a) Feedback control of a collocated system - notice the high-impedance buffer between the sensing patch and the controller. (b) A shunt damped system with a redundant open-circuited piezoelectric transducer.

To this end, it should be mentioned that if the disturbance signal is applied to an identical collocated piezoelectric transducer, such as shown in Figure 5.5, then $G_{vw} = G_{vv}$. This configuration is frequently used in laboratory experiments. In fact, some experimental results obtained from a similar setup will be demonstrated at the end of this chapter.

A point that needs to be clarified before this section is concluded is the effect of the redundant piezoelectric transducer in Figure 5.4(b). Existence of this patch is essential, if the systems depicted in two parts of this figure are to be equivalent. If the redundant transducer is removed, or short circuited, the transfer function $G_{vv}(s)$ associated with Figure 5.4(b) should be replaced with $\tilde{G}_{vv}(s)$, where

$$\tilde{G}_{vv}(s) = \frac{G_{vv}(s)}{1 + G_{vv}(s)}. \tag{5.8}$$

This should be of no particular concern since $\tilde{G}_{vv}(s)$ represents a collocated system. Hence, any methodology developed for one system can still be applied to the other. In most practical applications, of course, this redundant transducer should be removed. Indeed, a significant advantage of piezoelectric shunt damping, compared with using a collocated feedback control system, is that the need for a sensor is abolished.

5.3 Feedback Structure of Passive Shunts

In this section we attempt to apply the observation made in the previous section, on the feedback nature of piezoelectric shunt damping, to develop a deeper understanding of how efficient, or otherwise, certain passive shunts may be.

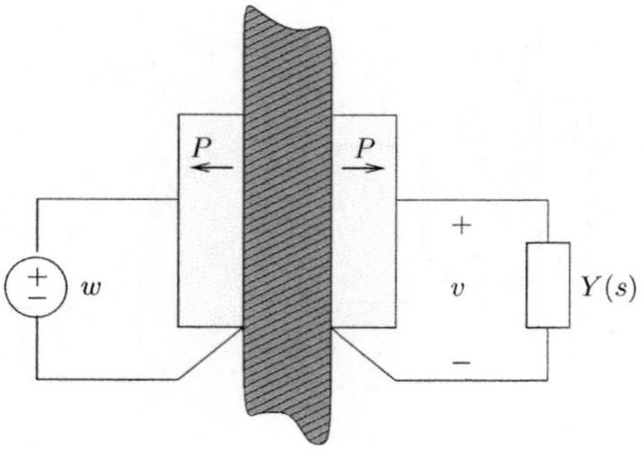

Figure 5.5. A structure with collocated piezoelectric transducers.

The first shunt to be studied is a purely resistive shunt, that is, $Z(s) = R$. This amounts to the feedback system consisting of $G_{vv}(s)$ and the feedback controller

$$K_R = \frac{s}{s + \omega_c}, \tag{5.9}$$

where

$$\omega_c = \frac{1}{RC_p}.$$

It can be observed that in this case the equivalent feedback controller is a first order high-pass filter. If the cut-off frequency of this filter, ω_c, is chosen to be high enough to include several modes of G_{vv}, the controller will approximate a differentiator within that bandwidth. In Section 3.4 it was demonstrated that a collocated system can be effectively damped using a differentiator. However, given the fixed nature of K_R, the gain that is needed to provide sufficient damping cannot be supplied. Consequently, a purely resistive shunt cannot act as an effective damper. Another possibility, of course, is to lower the cut-off frequency, by choosing a large enough resistor. In this case, the controller will, effectively, perform as a gain, which cannot provide sufficient damping either.

If the shunt is a purely capacitive load, *i.e.* $Z(s) = \frac{1}{Cs}$, the equivalent feedback controller will be

$$K(s) = \frac{C_p}{C_p + C}.$$

Therefore, the controller will be a fixed gain. This gain will always be less than one. Consequently, a significant level of damping cannot be expected from this shunt.

A purely inductive load, $Z(s) = Ls$, results in the feedback controller

$$K_L = \frac{s^2}{s^2 + \omega_\ell},$$

where

$$\omega_\ell = \frac{1}{\sqrt{C_p L}}.$$

This controller has both its poles on the $j\omega$ axis and is very ineffective in adding damping to the structure. Suppose the resonant system is made up of second order systems with no damping. That is, all poles of G_{vv} are on the $j\omega$ axis. Then, all the poles of the closed-loop system will be on the $j\omega$ axis too. Therefore, K_L adds no damping to the system. In reality, however, each mode of G_{vv} has a small amount of damping. Introduction of this controller "may" add very small damping to the system. The added damping, however, will not be significant.

Now, if the shunt consists of the series connection of a resistor R, and an inductor L, the equivalent feedback controller will be:

$$K^s_{RL}(s) = \frac{s(s + 2d_\ell \omega_\ell)}{s^2 + 2d_\ell \omega_\ell s + \omega_\ell^2}, \tag{5.10}$$

where

$$\omega_\ell = \frac{1}{\sqrt{C_p L}}$$

and

$$d_\ell = \frac{R}{2}\sqrt{\frac{C_p}{L}}.$$

It can be observed that the equivalent feedback control problem consists of $G_{vv}(s)$ and the controller $K^s_{RL}(s)$ which is a narrow band-pass filter centered around the ℓth resonance frequency of $G_{vv}(s)$. Hence, the controller would apply a high gain at that specific frequency, which would result in addition of extra damping to the structure, associated with that specific mode. The key difference between K_{RL} and K_L is d_ℓ. This parameter can be manipulated in a way that the damping added to the structure is maximized[1].

For a shunt which consists of the parallel connection of a resistor R with an inductor L, the equivalent feedback controller will be

$$K^p_{RL}(s) = \frac{s^2}{s^2 + 2d_\ell \omega_\ell s + \omega_\ell^2}$$

where ω_ℓ is as above, and

$$d_\ell = \frac{1}{2R}\sqrt{\frac{L}{C_p}}.$$

Given the structure of the equivalent controller, it is natural to expect this shunt to be equally effective in adding damping to the structure.

---

[1] This amounts to adjusting the resistor R

5.4 Reduction of Inductance Requirements by Adding a Parallel Capacitor

It was noted in Section 4.2.3 that one way of reducing the size of inductors needed in piezoelectric shunts is to add a capacitor across the shunted piezoelectric transducer. Although we have demonstrated that almost any passive circuit can be digitally implemented using either a synthetic impedance or a synthetic admittance arrangement, in certain applications it may be beneficial to reduce the size of the necessary inductors to the point that they are commercially available. The discussion in this section targets this particular set of applications.

Understanding the feedback structure associated with shunt damping systems enables us to explain how a capacitor should be chosen, and what compromises need to be made in order to make the system work properly.

Consider the shunted piezoelectric transducer in Figure 5.6 (a), where a capacitor, C_{add}, has been added in parallel to the shunt $Z(s)$. It can be proved that the shunted transfer function from w to v_p is [67]:

$$T_{v_p w}(s) = \frac{G_{vw}(s)}{1 + \tilde{K}(s) G_{vv}(s)}, \tag{5.11}$$

where

$$\tilde{K}(s) = \frac{\alpha C_T Z(s)}{1 + C_T Z(s)},$$

$$C_T = C_{add} + C_p,$$

and

$$\alpha = \frac{C_p}{C_T}.$$

Therefore, the system illustrated in Figure 5.6 (a) is equivalent to that in Figure 5.6 (b). According to the choice of C_T, the structure of the system is changed. Hence, this may adversely affect the performance obtainable from the system. However, it could have the advantageous property of reducing the size of necessary inductors. The following analysis of an RL shunt will clarify this point.

If the piezo is shunted by a series RL circuit, *i.e.* if $Z(s) = R + Ls$, then the equivalent feedback controller is K^s_{RL}, given in (5.10). Adding a capacitor C_{add} in parallel to this impedance, assuming that the resistive and inductive components are changed to $\hat{R}$ and $\hat{L}$ respectively, results in the equivalent controller

$$\tilde{K}^s_{RL}(s) = \frac{s(s + 2d_\ell \omega_\ell)}{s^2 + 2d_\ell \omega_\ell s + \omega_\ell^2}, \tag{5.12}$$

where

$$\omega_\ell = \frac{1}{\sqrt{C_T \hat{L}}}$$

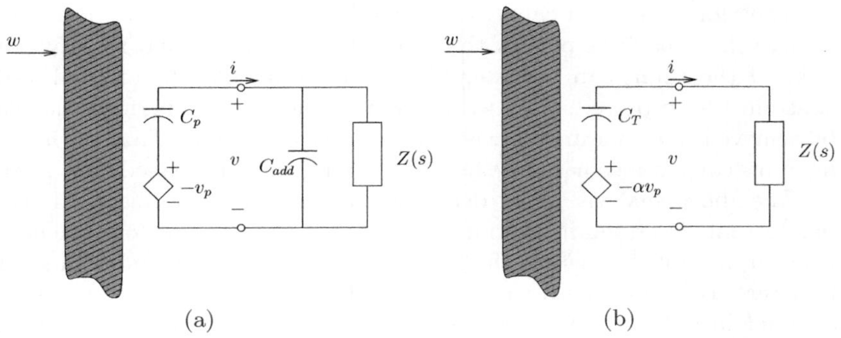

Figure 5.6. (a) A shunted piezoelectric transducer with a parallel capacitor and (b) its equivalent circuit

Table 5.1. Shunt circuit component values

	Natural PZT	**Modified Cap.**
C_p	104.8 nF	104.8 nF
C_{add}	0 nF	212.2 nF
$\hat{C}_2$	104.8 nF	317.0 nF
L_1	41.8 H	13.9 H
L_2	20.8 H	6.9 H
$\hat{L}_2$	41.8 H	13.9 H
R_1	1543 Ω	514 Ω
R_2	1145 Ω	381 Ω

and

$$d_\ell = \frac{\hat{R}}{2}\sqrt{\frac{C_T}{\hat{L}}}.$$

The original feedback controller can be retained by enforcing the following conditions:

$$\begin{aligned}\hat{L} &= L\frac{C_p}{C_T} \\ &= L\alpha\end{aligned}$$

and

$$\begin{aligned}\hat{R} &= R\frac{C_p}{C_T} \\ &= R\alpha.\end{aligned}$$

Thus, to ensure $\tilde{K}^s_{RL}(s) = K^s_{RL}(s)$, the inductive and resistive elements should be reduced by the same factor that the capacitance is increased.

For multi-mode shunt circuits, such as those described in Chapter 4, similar results can be obtained. By increasing the effective piezoelectric capacitance, the resistance and inductance of each branch should be reduced.

Therefore, adding a capacitor in parallel to a piezoelectric transducer increases the capacitive part of the transducer. Inductive and resistive parameters of the shunt can be manipulated in a way that the original feedback controller is recovered. This way, however, the original performance cannot be achieved since parameter α is always smaller than one. That is, the supplementary capacitance has the effect of reducing gain of the feedback controller.

The above analysis is best demonstrated on a two-mode series shunt, similar to that described in Section 4.2.4. A summary of the component values is given in Table 5.1, where the first column represents the original shunt parameters. By attaching a capacitance of $C_{add} = 212.2\ nF$ across the shunt, a threefold reduction of component values can be achieved, as demonstrated in Column 2 of Table 5.1. Figure 5.7 shows the frequency response of the system for three cases: open-loop, shunted and shunted with a supplementary capacitance. Addition of this supplementary capacitance has resulted in a decrease in the performance of the shunted system, as expected from the above analysis.

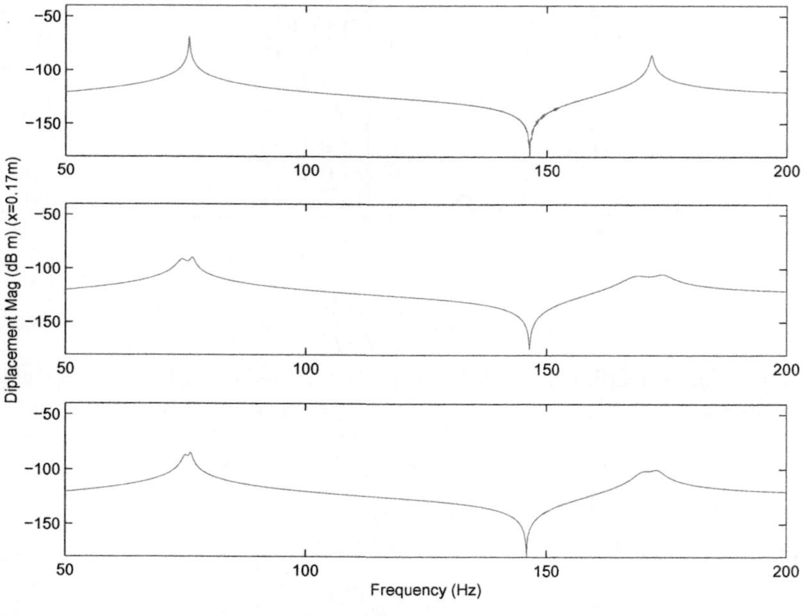

Figure 5.7. Displacement magnitude frequency response of the shunted beam when the piezo is open circuited, shunted and a supplementary capacitance is placed across the shunt

5.5 Resonant Shunts

In the Section 5.2 it was demonstrated that the problem of piezoelectric shunt damping is equivalent to a feedback control problem in which the impedance shunted to the piezoelectric transducer functions as the controller. The feedback structure resembles a cascade feedback system. This observation has interesting implications since it enables us to employ systems theoretic tools in designing effective impedances for piezoelectric shunt damping systems.

Let us consider the feedback problem associated with the system in Figure 5.2 (b). The system is equivalent to a feedback loop described by

$$T_{v_p w}(s) = \frac{G_{vw}(s)}{1 + K(s)G_{vv}(s)}, \tag{5.13}$$

where the real controller $Y(s)$ is itself inside an inner feedback loop (Figure 5.2 (b)); *i.e.*

$$K(s) = \frac{1}{1 + \frac{Y}{C_p s}}.$$

To this end, we wish to determine a parameterization of all admittance transfer functions, $Y(s)$ that renders the inner feedback loop internally stable. The Youla parameterization [190] of all stabilizing controllers for a stable plant can be used for this purpose. Once such a parameterization is determined, we may proceed to determine a specific admittance that guarantees closed-loop stability of the overall system. It should be pointed out that internal stability of the inner feedback loop is indeed important since it represents the interactions between the shunted admittance and the capacitive component of the piezoelectric transducer. By ensuring that this loop is internally stable we can guarantee that problems such as unstable pole zero cancellation will not occur.

The Youla parameterization of all stabilizing controllers $\tilde{K}(s)$ for a given stable plant $\tilde{G}(s)$ can be written as

$$\tilde{K}(s) = \frac{Q(s)}{1 - Q(s)\tilde{G}(s)},$$

where $Q(s)$ can be any stable transfer function (not necessarily proper). However, it should be noted that for the inner feedback loop in Figure 5.2 (b) the plant is an integrator, *i.e.* $\tilde{G}(s) = \frac{1}{C_p s}$ and hence unstable. If the plant $\tilde{G}(s)$ has unstable poles, instead of using Youla parameterization for an unstable plant [190], we may still use the above parameterization, as long as $Q(s)$ satisfies the following conditions (see Lemma 15.4 of [79]):

1. $Q(s)$ must be stable and proper,
2. $Q(s)$ must have zeros at the unstable poles of $\tilde{G}(s)$,
3. $1 - Q(s)\tilde{G}(s)$ must have zeros at the unstable poles of $\tilde{G}(s)$.

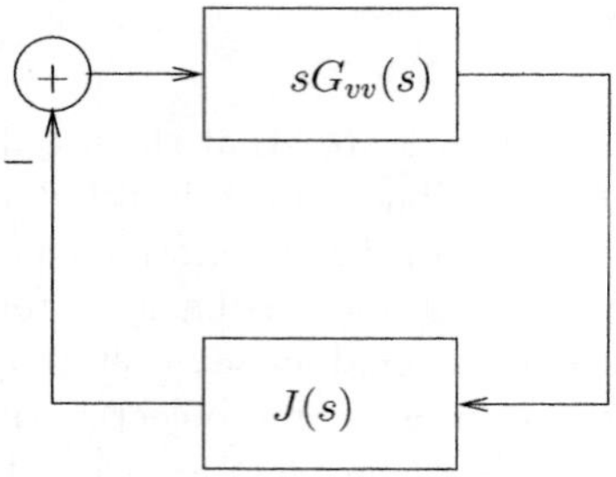

Figure 5.8. Equivalent system for study of closed-loop stability

Noting that $\tilde{G}(s) = \frac{1}{C_p s}$, for the inner feedback loop, all these conditions can be enforced as long as $Q(s)$ is chosen as

$$Q(s) = H(s)C_p s, \tag{5.14}$$

with $H(s)$ satisfying the following conditions:

1. $H(s)$ is stable and strictly proper, and
2. $1 - H(s)$ has a zero at the origin.

Therefore, any admittance chosen as

$$Y(s) = \frac{H(s)}{1 - H(s)} C_p s,$$

with $H(s)$ satisfying the above two conditions will guarantee internal stability of the inner feedback loop resulting in the equivalent controller

$$K(s) = 1 - H(s),$$

which is stable. Furthermore, if $H(s)$ is chosen to be

$$H(s) = 1 - sJ(s)$$

with $J(s)$ a strictly positive real system, the closed-loop stability of the overall system can be guaranteed. This can be verified by observing that stability of the overall system is equivalent to the stability of the negative feedback loop in Figure 5.8.

It can be verified that $sG_{vv}(s)$ is a positive real system [104]. Therefore, if $J(s)$ is chosen to be a strictly positive real system, closed-loop stability of the overall system will be guaranteed [99, 104].

The purpose of the shunting admittance is to add damping to the structure within a specific bandwidth. This amounts to moving closed-loop poles of the system deeper into the left half of the complex plane. Given the highly resonant nature of the base structure, as evident from (5.4), one way of achieving this goal would be to choose a highly resonant structure for the shunted impedance. An impedance structure that results in favorable closed-loop performance can be constructed by choosing the following form for $J(s)$,

$$J_\alpha(s) = \sum_{i=1}^{N} \frac{\alpha_i s}{s^2 + 2d_i\tilde{\omega}_i s + \tilde{\omega}_i{}^2}, \tag{5.15}$$

where $\alpha_i \geq 0$ for $i = 1, 2, \ldots, N$ and

$$\sum_{i=1}^{N} \alpha_i = 1. \tag{5.16}$$

This choice for $J_\alpha(s)$ results in the equivalent controller

$$K_\alpha(s) = \sum_{i=1}^{N} \frac{\alpha_i s^2}{s^2 + 2d_i\tilde{\omega}_i s + \tilde{\omega}_i{}^2}.$$

Also, the admittance transfer function that needs to be implemented is

$$Y_\alpha(s) = \frac{\sum_{i=1}^{N} \frac{\alpha_i(2d_i\tilde{\omega}_i s + \tilde{\omega}_i{}^2)}{s^2+2d_i\tilde{\omega}_i s+\tilde{\omega}_i{}^2}}{1 - \sum_{i=1}^{N} \frac{\alpha_i(2d_i\tilde{\omega}_i s+\tilde{\omega}_i{}^2)}{s^2+2d_i\tilde{\omega}_i s+\tilde{\omega}_i{}^2}} C_p s. \tag{5.17}$$

It needs to be pointed out that the transfer function in (5.17) is not a minimal system. A minimal realization of (5.17) is discussed in the next section when we attempt to study properties of the shunting admittance.

Another possible form is based on:

$$J_\beta(s) = \sum_{i=1}^{N} \frac{\beta_i(s + 2d_i\tilde{\omega}_i)}{s^2 + 2d_i\tilde{\omega}_i s + \tilde{\omega}_i{}^2}, \tag{5.18}$$

with

$$\sum_{i=1}^{N} \beta_i = 1.$$

This results in the admittance

$$Y_\beta(s) = \frac{\sum_{i=1}^{N} \frac{\beta_i\tilde{\omega}_i{}^2}{s^2+2d_i\tilde{\omega}_i s+\tilde{\omega}_i{}^2}}{1 - \sum_{i=1}^{N} \frac{\beta_i\tilde{\omega}_i{}^2}{s^2+2d_i\tilde{\omega}_i s+\tilde{\omega}_i{}^2}} C_p s\ , \tag{5.19}$$

which generates the resonant controller:

$$K_\beta(s) \sum_{i=1}^{N} \frac{\beta_i s(s + 2d_i\tilde{\omega}_i)}{s^2 + 2d_i\tilde{\omega}_i s + \tilde{\omega}_i{}^2}.$$

A distinct characteristic of (5.17) and (5.19) is their inherent stabilizing properties. The existence of out-of-bandwidth modes in (5.4) has the potential to cause closed-loop instabilities as a result of the spill-over effect [12, 13]. However, due to the specific structure of $Y(s)$, the closed-loop stability in the presence of out-of-bandwidth modes is guaranteed. A further interesting property of $Y(s)$ is its robustness with respect to an inaccurate in-bandwidth model of the base structure. If $Y_a(s)$ is designed based on a model (5.4) with inaccurate parameters, $\tilde{\omega_i}$, ζ_i and α_i, the closed-loop system will still remain stable, although the closed-loop performance may significantly deteriorate.

In general $Y_\alpha(s)$ is a biproper transfer function while $Y_\beta(s)$ is a strictly proper transfer function. Therefore the task of digital implementation of $Y_\beta(s)$ may be more straightforward than $Y_\alpha(s)$. However, despite its biproper nature, in many applications $Y_\alpha(s)$ can be implemented without difficulty. In Section 5.7 we implement both impedances on piezoelectric laminate structures and we demonstrate that both impedances function well.

5.6 Properties of Resonant Shunts

In this section we concentrate on the impedance of the resonant shunting element, *i.e.* $Z_\alpha(s) = \frac{1}{Y_\alpha(s)}$, and $Z_\beta(s) = \frac{1}{Y_\beta(s)}$. Although tedious, it can be verified that both $Z_\alpha(s)$ and $Z_\beta(s)$ are positive real transfer functions, *i.e.* they are stable and $\mathbf{Re}\{Z(j\omega)\} \geq 0$ for all $\omega \in \mathbf{R}$. Therefore, both impedances must be implementable using passive circuit elements, *i.e.* resistors, inductors and capacitors. To synthesize a network of passive elements for $Z_\alpha(s)$ or $Z_\beta(s)$, one has to start with the following representations

$$Z_\alpha(s) = \frac{\sum_{i=1}^{N} \alpha_i s \prod_{\ell=1,\ \ell\neq i}^{N} (s^2 + 2d_i\tilde{\omega_i}s + \tilde{\omega_i}^2)}{C_p \sum_{i=1}^{N} \alpha_i (2d_i\tilde{\omega_i}s + \tilde{\omega_i}^2) \prod_{\ell=1,\ \ell\neq i}^{N} (s^2 + 2d_i\tilde{\omega_i}s + \tilde{\omega_i}^2)} \tag{5.20}$$

and

$$Z_\beta(s) = \frac{\sum_{i=1}^{N} \beta_i (s + d_i\tilde{\omega_i}) \prod_{\ell=1,\ \ell\neq i}^{N} (s^2 + 2d_i\tilde{\omega_i}s + \tilde{\omega_i}^2)}{C_p \sum_{i=1}^{N} \beta_i \tilde{\omega_i}^2 \prod_{\ell=1,\ \ell\neq i}^{N} (s^2 + 2d_i\tilde{\omega_i}s + \tilde{\omega_i}^2)}. \tag{5.21}$$

A number of synthesis techniques can be used to implement such an impedance (refer to [179, 6]). However, the task of extracting a passive network from (5.20) or (5.21) may not be quite straightforward since standard synthesis techniques result in circuits that require elements such as Gyrators. Nevertheless, if such an implementation is found it is likely that it would be practically unrealizable due to the size of the necessary inductors. We have already established that in a typical vibration control problem, one is often interested in suppressing low-frequency modes. Hence, we should expect to need large inductors for this purpose. A practical approach to implementing such an impedance is to use the synthetic impedance arrangement discussed in Section 4.4.

Figure 5.9. Simply supported beam with a pair of collocated piezoelectric patches

It is instructive to consider the situation where only one mode of the structure is to be controlled. This can be achieved by setting the admittance $Y_\alpha(s)$ in (5.17) equal to the ith term, that is,

$$Z_\alpha(s) = \frac{s/C_p}{2d_i\tilde{\omega}_i s + \tilde{\omega}_i^2}.$$

This is equivalent to the parallel connection of a resistor

$$R = \frac{1}{2d_i\tilde{\omega}_i C_p},$$

with an inductor

$$L = \frac{1}{\tilde{\omega}_i^2 C_p}.$$

This is the same RL shunt which was proposed in reference [197], and discussed in Section 5.3, if $\tilde{\omega}_i = \omega_i$.

Now, if the same procedure is followed for $Z_\beta(s)$, the series RL shunt of Hagood and von Flotow can be recovered.

5.7 Experimental Implementation of Resonant Shunts

This section is devoted to the experimental validation of the two resonant shunts proposed in Section 5.5. The shunts were applied to two flexible structures: a piezoelectric laminate beam and a piezoelectric laminate plate. The beam apparatus consists of a uniform aluminum bar with rectangular cross

Figure 5.10. Piezoelectric laminated plate bounded structure

section and experimentally pinned boundary conditions at both ends. Likewise, the plate structure also consists of an aluminum rectangular plate with pinned boundary conditions at all external edges.

The two experimental structures are shown in Figures 5.9 and 5.10 respectively. For both structures, two piezoelectric patches were bonded to the surface in a collocated fashion using strong adhesive material, as shown in Figures 5.11 and 5.12. The patches used in the beam experiment are PIC151[2] piezoelectric patches while those used in the plate are T140-A4E-602[3] patches. On each structure, one piezoelectric patch is used as an actuator to generate a disturbance through the structure, and the other as a shunting transducer.

In the first part of the experiment, the resonant shunt Y_α was validated experimentally on the piezoelectric laminate plate structure. Close examination of the experimental setup reveals that the disturbance is the voltage applied to a piezoelectric actuator collocated with the shunted piezoelectric patch. Since the two piezoelectric transducers are identical, and considering the orientation of polarization vectors of the two transducers, we may write

$$G_{vw}(s) = -G_{vv}(s).$$

Therefore, the block diagram in Figure 5.2 (b) can be reduced to that shown in Figure 5.13. An advantage of using this configuration is that by measuring the unshunted G_{vw}, the collocated transfer function G_{vv} can be automatically determined. Otherwise, if a collocated piezoelectric transducer

---

[2] PIC151 piezoelectric patches are manufactured by *Physics Instruments Corp.*
[3] T140-A4E-602 piezoelectric patches are manufactured by *Piezoceramics Inc.*

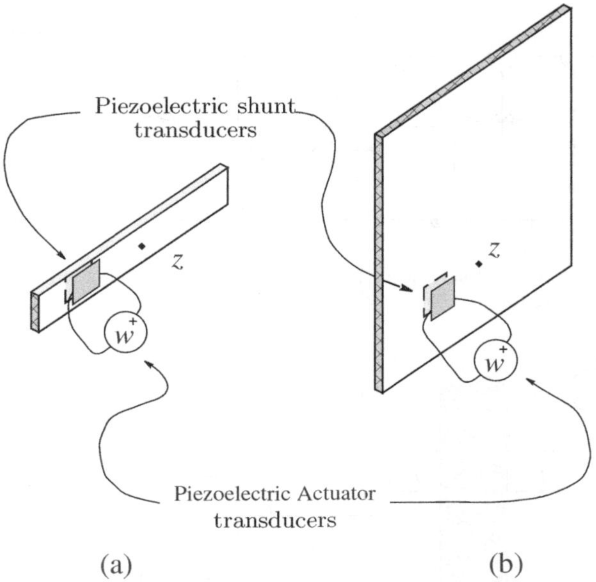

Figure 5.11. Experimental piezoelectric laminated structures: (a) beam and (b) plate. Note that the disturbance is the voltage applied to one of the collocate patches w and z is the displacement at some point on the structure.

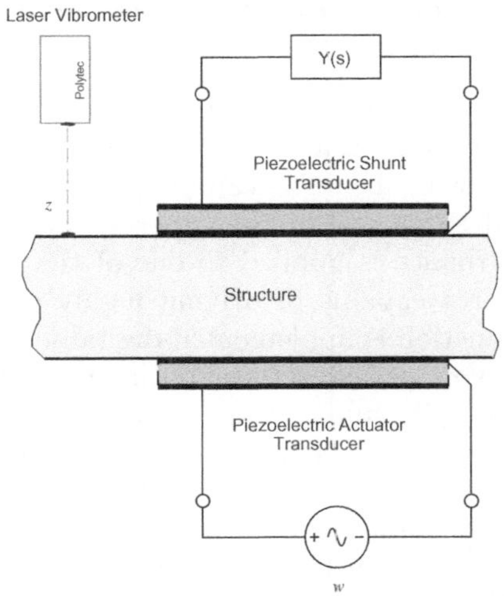

Figure 5.12. The experimental structure with collocated piezoelectric transducers

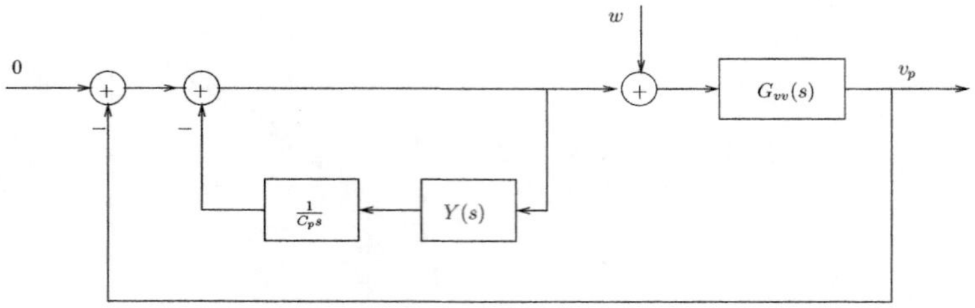

Figure 5.13. The feedback structure with the disturbance applied to the collocated piezoelectric transducer

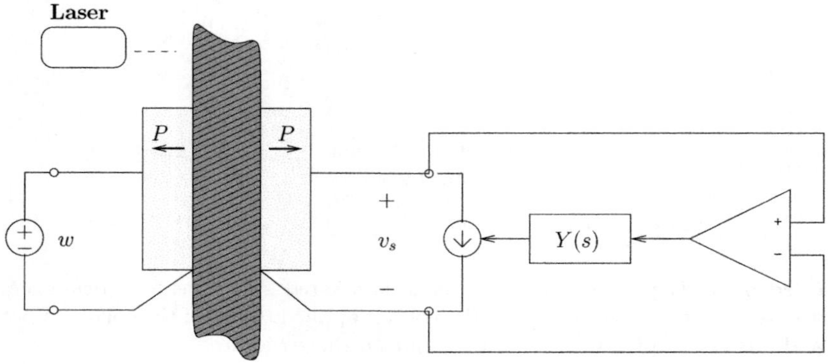

Figure 5.14. Experimental setup

is not available, G_{vv} can be determined by applying a voltage to the transducer and measuring the current that flows as a result of it. If the piezoelectric capacitance C_p is known, then G_{vv} can be determined immediately.

A more detailed schematic of the experimental setup is plotted in Figure 5.14. The disturbance is applied to one of the piezoelectric transducers using a voltage source capable of driving highly capacitive loads. The admittance transfer function is implemented digitally using a current controlled voltage source. The voltage across the shunting piezoelectric layer is measured and a current $i = Y_\alpha v_s$ is supplied to the same transducer. The admittance transfer function $Y_\alpha(s)$ is implemented digitally using a dSPACE DS1103 rapid prototyping system. A detailed description of the synthetic admittance system was given in Section 4.4.

In order to design an effective shunt it is necessary to obtain analytical models for G_{vv} and G_{zw}, the transfer function from the disturbance voltage to the displacement at a point on the surface of the beam, as measured by a laser vibrometer. For the majority of systems obtaining analytical models may be rather difficult, so again we opt to use a system identification method

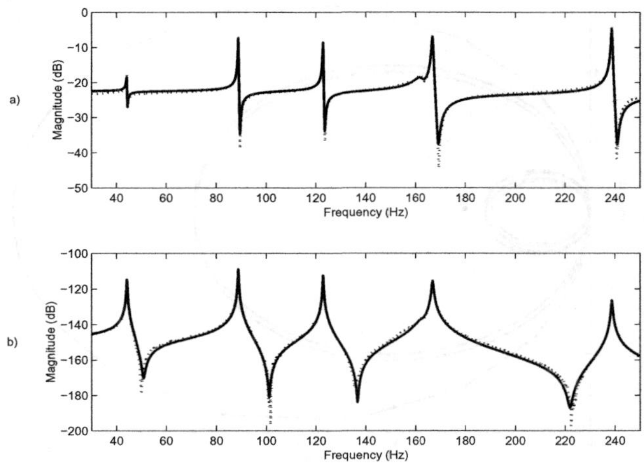

Figure 5.15. Frequency response a) $|G_{vv}(s)|$ (v/v) and b) $|G_{zw}(z, s)|$ (m/v), for the piezoelectric laminated plate structure. Experimental data ($\cdots$) and model obtained using subspace based system identification (—).

to develop a model of the system. In particular, we employ the subspace-based frequency domain system identification of reference [128, 129].

Using a Polytec laser scanning vibrometer (PSV-300) and a Hewlett Packard spectrum analyzer (35670A), the frequency responses for G_{vv} and G_{zw} were obtained. The measurements were carried out in open loop, *i.e.* the terminals of the shunting layer were left open when G_{zw} was being measured while they were connected to the spectrum analyzer when G_{vv} was being measured. Using the already mentioned system identification technique, G_{vv} and G_{zw} were identified over the frequency range of $30 - 250\ Hz$. The models include the first 6 modes of the plate. Magnitude frequency responses of the identified transfer functions are shown in Figure 5.15. The 1st, 2nd, 3rd, 5th and 6th structural modes were chosen for control due to their highly resonant nature.

An admittance of the form proposed in (5.17) was then designed. Resonance frequencies of the admittance were chosen to be identical to the resonance frequencies of the structure, *i.e.* $\tilde{\omega}_i = \omega_i$. In order to achieve good performance, appropriate values for the damping parameters $\alpha_1, \ldots, \alpha_6$ and $d_1, \ldots, d_6$ needed to be determined. To simplify the design, parameters $\alpha_1, \ldots, \alpha_3$ and α_5 and α_6 were all set equal to $1/5$ while α_4 was fixed at 0, since its associated mode was to be left uncontrolled. This ensured that Condition (5.16) was satisfied. The piezoelectric capacitance C_p was experimentally determined to be $67.9nF$. Furthermore, the first six resonance frequencies of

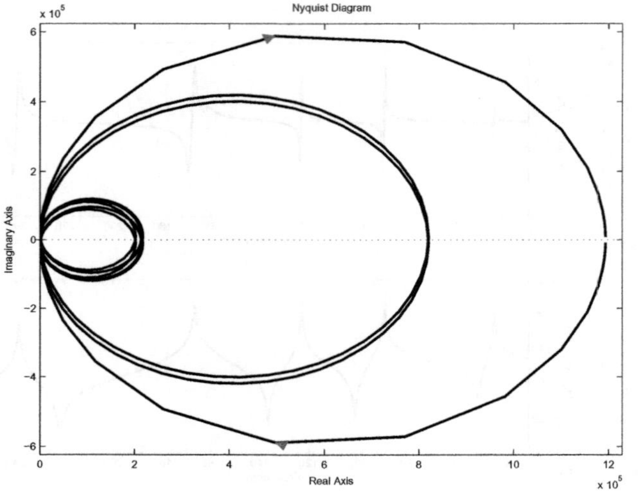

Figure 5.16. Nyquist plot for impedance Z(s)

the structure, $\Omega = \{\omega_1, \omega_2, \ldots, \omega_6\}$ were extracted from the identified models. The following optimization problem was then solved:

$$D^* = \arg\min_{D>0} \|T_{zw}\|_2 , \tag{5.22}$$

where $D^* = \{d_1^*, d_2^*, \ldots, d_6^*\}$, and T_{zw} represents the shunted transfer function from the disturbance voltage w to the displacement x as measured by a vibrometer. The purpose of the optimization is to minimize the H_2 norm of T_{zw}. This will subsequently result in adding extra damping to the structure by shifting closed-loop poles of the system deeper into the left half of the complex plane. The optimization problem was solved for a number of initial guesses, and a solution was found;

$$\begin{aligned} d_1^* &= 0.0049 \\ d_2^* &= 0.0076 \\ d_3^* &= 0.0063 \\ d_5^* &= 0.0087 \\ d_6^* &= 0.0080. \end{aligned}$$

Notice that the fourth mode is left uncontrolled. The reason for this will be clarified soon.

Using the values D^*, Ω and C_p we can sketch the Nyquist diagram for the impedance $Z(s)$. This is shown in Figure 5.16. From the Nyquist diagram we can see that the impedance is strictly positive real [179], that is, $Z(s)$ is stable and $Re[Z(j\omega)] > 0 : \forall\omega \in \mathbf{R}$, therefore guaranteeing that the impedance is

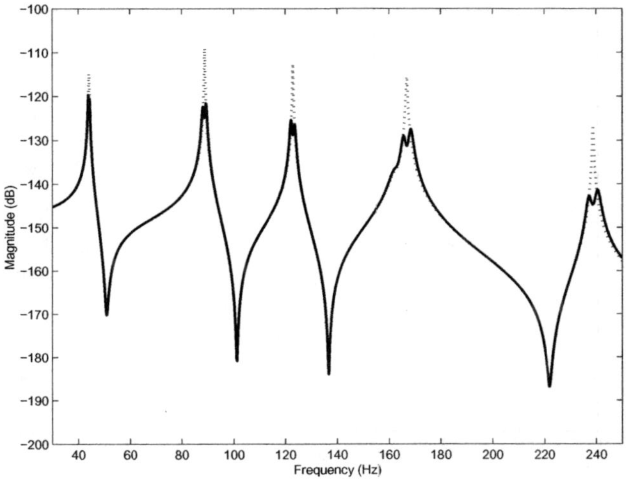

Figure 5.17. Simulated open-loop (···) vs. closed-loop response (–)

realizable using passive circuit elements. Such a realization using resistors, capacitors and inductors may be impossible as the size of the required inductive elements may be too large. Hence, the need for digital implementation of the impedance.

Simulations of the closed-loop response, as shown in Figure 5.17, demonstrate that the $1st$, $2nd$, $3rd$, $5th$ and $6th$ resonant amplitudes are expected to decrease by 5.0 dB, 12.6 dB, 12.7 dB, 11.8 dB and 14.7dB respectively. Figure 5.18 shows the locations of the closed-loop and open-loop poles. We can see from Figure 5.18, that the closed-loop poles have been pushed deeper into the left half of the complex plane. Notice that the 4th mode has an acceptable natural damping. Hence, there is no need for additional damping, *i.e.* $\alpha_4 = 0$. By shifting the poles to the left, effectively, more damping has been added to the structure.

A similar procedure was repeated for the beam. However, this time, a resonant shunt of the form Y_β was designed and implemented. In this experiment the first four modes of the beam were to be controlled. Transfer functions $G_{zw}(s)$ and $G_{vv}(s)$ were measured and identified. These are plotted in Figure 5.20. Within the bandwidth of interest, the identified models were found to be good representations of the piezoelectric laminate systems.

Following the optimization procedure explained above, a shunt impedance $Y_\beta(s)$ as described by Equation (5.19) was determined. The admittance was digitally implemented, using the synthetic admittance circuit, and then applied to the shunt transducer. A comparison of the experimental undamped and damped responses for $|G_{zw}(s)|$ is shown in Figure 5.21. The experimental resonance magnitudes were successfully reduced. In particular, it can be ob-

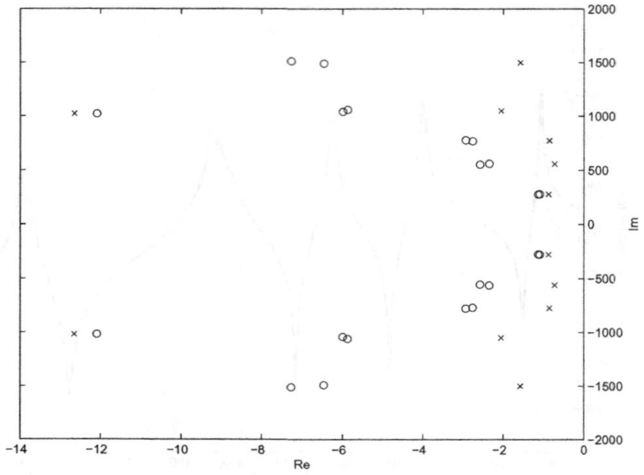

Figure 5.18. Simulated open-loop (×) and closed-loop (o) poles

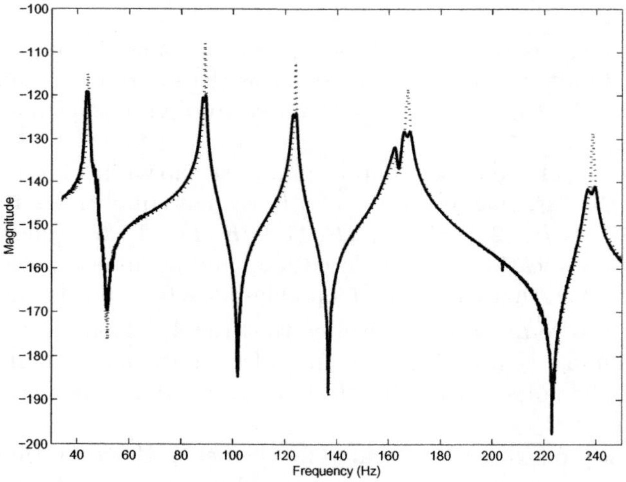

Figure 5.19. Experiential open-loop (···) vs. closed-loop response (–)

served that the first four modes of the beam were damped by 2 dB, 16.2 dB, 19.9 dB and 24.1 dB respectively.

Figure 5.22 shows the simulated closed-loop and open-loop poles of the beam. It can be observed, from Figure 5.22, that the closed-loop poles have been pushed deeper into the left half of the complex plane, hence adding to the natural damping associated with the first four modes of the structure.

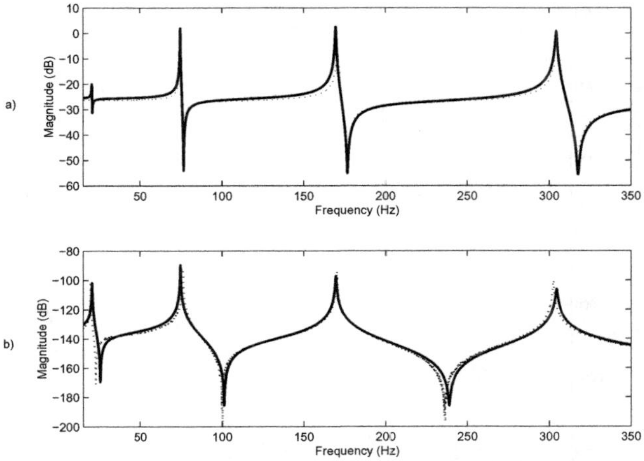

Figure 5.20. Frequency response (a) $|G_{vv}(s)|$ and (b) $|G_{zw}(s)|$, for the piezoelectric laminated beam structure. Experimental data $(\cdots)$ and model obtained using subspace based system identification (—).

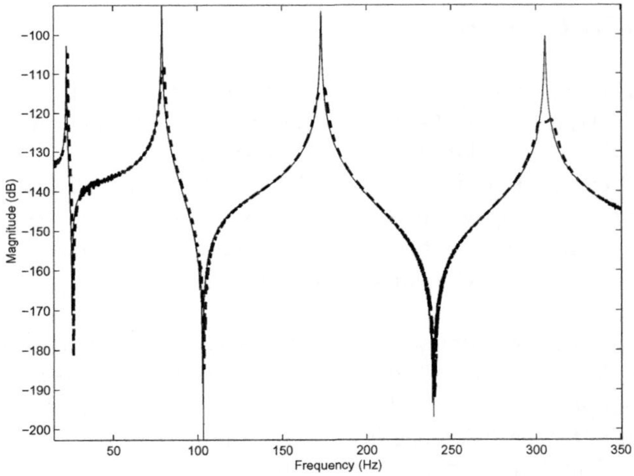

Figure 5.21. Experimental beam undamped (—) and damped (- - -) magnitude response of $G_{zw}(s)$

We observe that in both cases performance of the shunts in damping the first vibration mode of the structure is very limited. This can be attributed to the locations at which the piezoelectric shunt transducers are mounted on the beam and plate structure.

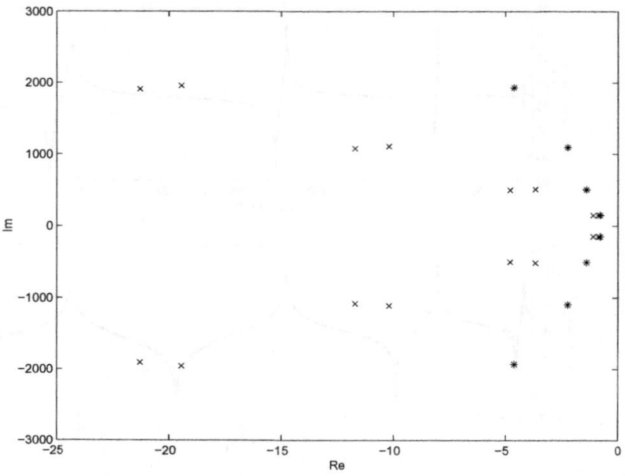

Figure 5.22. Simulated open-loop (*) and closed-loop (x) poles of the piezoelectric laminated beam

5.8 Hybrid Control

So far in this chapter we have dealt with flexible structures with shunted piezoelectric transducers. Our approach has been to identify the feedback control structure asscociated with these systems, and hence make it possible to use feedback control design methodologies to devise effective shunts with good damping properties. In this section we take this idea one step further and we illustrate that a hybrid system consisting of a shunted piezoelectric transducer and a feedback controller can also be viewed as a feedback control system.

There could be a number of advantages associated with a hybrid structure of this kind. We have indicated several times that feedback control of highly resonant systems is indeed a difficult problem. This difficulty is often attributed to the existence of a large number of lightly damped poles in the transfer functions of flexible structures. In a hybrid system, the shunted piezoelectric transducer can be used to augment the natural damping of in-bandwidth structural modes. This will make the design of the feedback controller a less complicated task since the damped system is now better conditioned.

The hybrid system that we consider in this section is depicted in Figure 5.23 (a). Here, a piezoelectric transducer is shunted by an impedance and a voltage v_k is being applied to the same transducer. We assume that v_k is related to the voltage induced in the collocated piezoelectric sensor, v_p, via

$$v_k(s) = -K(s)v_p(s). \tag{5.23}$$

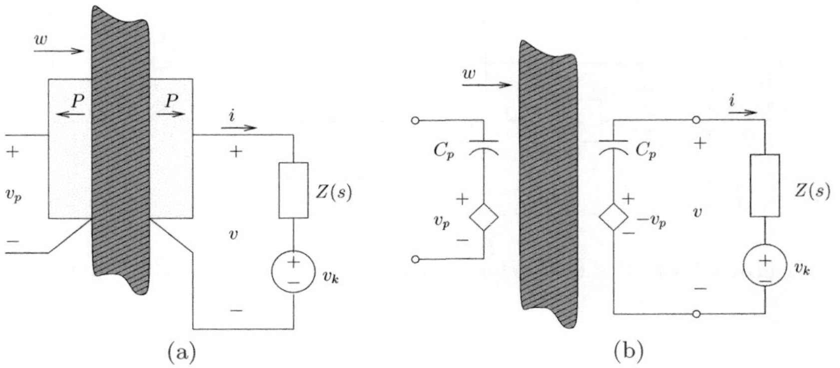

Figure 5.23. (a) A piezoelectric laminate shunted to an impedance $Z(s)$ incorporating a negative feedback controller; (b) Electrical equivalent of the same system.

That is, a negative feedback control system in addition to the shunt damped structure. Now, we can determine the feedback structure associated with this system from the following equations

$$i(s) = -\left(v_p(s) + v(s)\right) c_p s \tag{5.24}$$

$$v(s) = z(s)i(s) + v_k(s) \tag{5.25}$$

$$v_k(s) = -K(s)v_p(s) \tag{5.26}$$

$$v_p(s) = G_{vv}(s)v(s) + G_{vw}(s)w(s) \tag{5.27}$$

which are obtained by inspection from the electrical equivalent of the hybrid system depticted in Figure 5.23 (b). Equation 5.24 follows from the KVL, Equation 5.25 expresses v in terms of the current flowing out of the piezoelectric transducer and the feedback control signal, Equation 5.26 asserts the negative feedback structure, and Equation 5.27 is due to the linearity of the system.

The feedback structure of the hybrid system, as obtained from (5.24)-(5.24), is depicted in Figure 5.24.

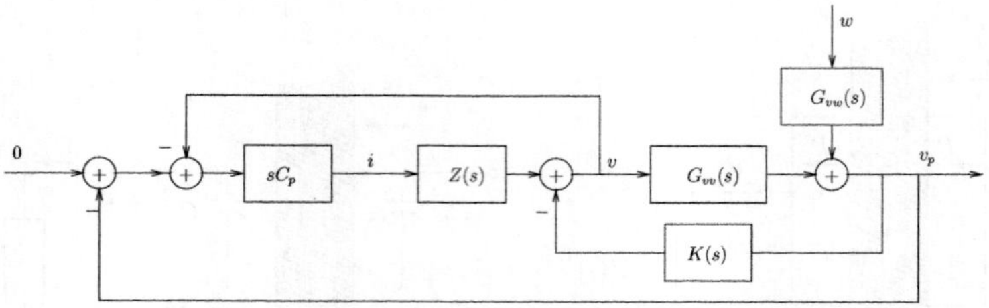

Figure 5.24. The feedback structure associated with a hybrid system

6

Instrumentation

6.1 Introduction

In addition to modeling and control design, the performance of piezoelectric laminate structures and positioning devices is also heavily dependent on the quality of instrumentation. Poorly designed instrumentation can result in high electronic noise levels, poor bandwidth, and when in feedback, instability due to unmodeled amplifier dynamics. The following seeks to provide an introduction to the instrumentation and techniques required for active and passive piezoelectric vibration control, and positioning.

Active vibration control by piezoelectric strain voltage feedback is the most straightforward technique in implementation. As discussed in Chapter 3, at least two piezoelectric transducers are employed, one as an actuator, and another as a sensor. The actuator is driven directly with either a voltage or charge source while the sensor transducer is open circuited to allow direct measurement of the strain voltage. A discussion of the issues involved with strain voltage measurement, voltage amplifiers, current amplifiers, and charge amplifiers follows in Sections 6.2, 6.3, 6.4 and 6.5.

Although shunt control systems do not require a feedback sensor, practical implementation requires a more sophisticated driver than feedback systems. While applying voltage, the driver is also required to provide an accurate measurement of the corresponding current or charge. Analogously, when applying a current or charge, a load voltage measurement is required.

6.2 Strain Voltage Measurement

In active control systems, it is often convenient to use the piezoelectric strain voltage measured from a sensor transducer v_p as the feedback variable. Some consideration must be given to the electrical nature of the transducer in order to design a suitable instrumentation circuit. In the frequency range of interest for vibration control, typically $1Hz$ - $1kHz$, the equivalent model of a piezoelectric transducer is shown in Figure 6.1. As the dielectric resistance R_p is usually extremely large, the open-circuit voltage v_z is effectively equal to the strain voltage v_p. For the same reason, the presence of R_p is neglected in the analysis of feedback and shunt damping systems in Chapters 3, 4 and 5. Problems arise when the high impedance source is connected to the input of an instrumentation or differential amplifier. As shown on the right side of Figure 6.1, the input of an instrumentation amplifier can be modeled as an ideal differential amplifier in combination with a parallel resistance R_{in}, voltage noise v_n and current noise i_n. The parallel resistance R_{in} models the amplifier's differential impedance and an additional resistance placed across the terminals to allow for bias currents. Although the differential impedance of an instrumentation amplifier can generally be neglected, the impedance of single or dual op-amp differential amplifier is often low and must be considered.

The most basic design objective is to minimize the input voltage offset due to the input bias current i_n. This is easily achieved by specifying R_{in} as low as possible. Unfortunately, decreasing R_{in} has other undesirable effects.

Neglecting the dielectric resistance R_p, the transfer function from the strain voltage v_p to the instrumented signal v_{out} is a first-order high-pass filter with corner frequency $\frac{1}{2\pi R_{in} C_p}$,

$$\frac{v_{out}(s)}{v_p(s)} = \frac{s}{s + \frac{1}{R_{in} C_p}}. \tag{6.1}$$

At low frequencies, R_{in} results in a phase-lead, and heavy attenuation of the strain signal. As the capacitance C_p cannot be changed, the only option for increasing low-frequency bandwidth is to increase R_{in}. Thus, there is a trade-off between low-frequency dynamic performance and bias-current-generated input offset voltages.

The next consideration is noise performance. In precision sensing applications, designers often select instrumentation amplifiers with ultra-low input voltage noise v_n, for example, the INA103 from Texas Instruments with an input noise voltage of $\frac{1nV}{\sqrt{Hz}}$. Unfortunately, amplifiers that achieve such noise performance are invariably based on Bipolar Junction Transistors (BJTs)

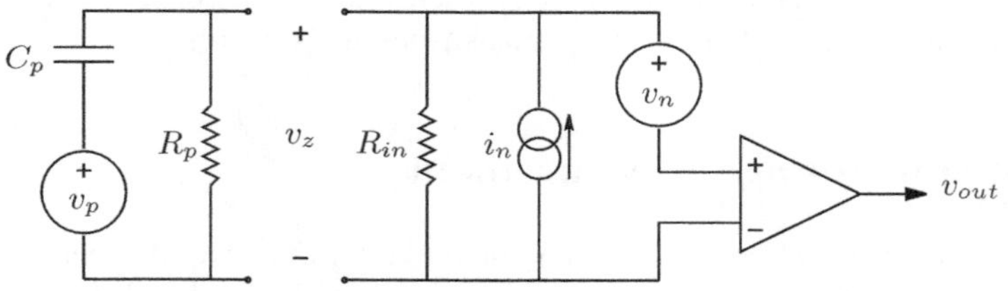

Figure 6.1. The electrical model of a piezoelectric transducer (left) and instrumentation amplifier (right)

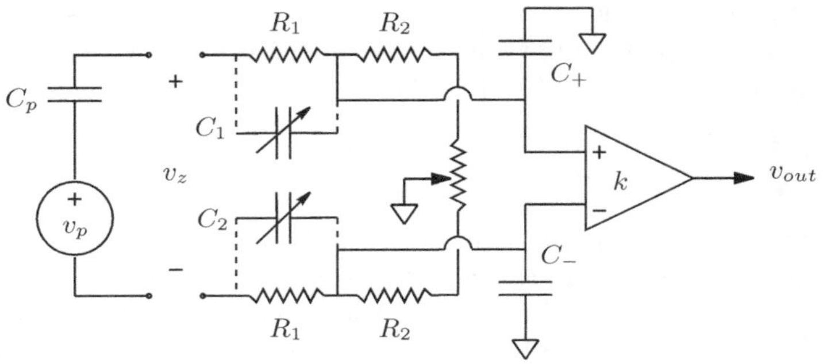

Figure 6.2. A balanced high-impedance instrumentation circuit

with high input-stage bias currents and high current noise i_n. As the source impedance of a piezoelectric transducer is extremely high at low-frequencies, even small amounts of current noise can result in large input noise voltages. Thus, low current noise is preferable to low input voltage noise.

In summary, to select a suitable instrumentation amplifier, and design the circuit, the following points should be considered:

- A high quality instrumentation amplifier should be selected with the least possible input bias and noise current specifications. A high differential impedance is also desirable. Instrumentation amplifiers with a Junction Field Effect Transistor (JFET) input stage are more suitable.
- What is the lowest structural frequency f_{low} to be controlled? A high-pass corner frequency two decades below will provide excellent signal integrity over the bandwidth of interest, *i.e.*

$$R_{in} = \frac{1}{0.01 f_{low} 2\pi C_p}. \tag{6.2}$$

  If this results in unacceptable input voltage offsets or current-induced noise, R_{in} will have to be reduced until satisfactory performance is obtained. For values of R_{in} larger than that specified in Equation (6.2), the high-pass filter dynamics cannot be neglected and should be included in the system model.
- If satisfactory noise performance and low-frequency bandwidth cannot be achieved, a change in transducer may be required. Multi-layer transducers, thinner transducers or transducers with greater surface area have a greater capacitance and relax the requirements on R_{in}. Composite fiber or other transducers with interdigitated electrodes have extremely low capacitance and are not suitable as low-frequency (1 - 1 kHz) strain sensors.

In applications involving high sensor voltages, a balanced voltage divider such as that illustrated in Figure 6.2 is required. The variable resistor is used to tune the division ratio of each branch and should be chosen as small as possible. The capacitors C_1 and C_2 are used to negate the effects of the parasitic input capacitances C_+ and C_-. The overall gain of the circuit is

$$\frac{v_{out}}{v_z} = k \frac{R_2}{R_1 + R_2}. \tag{6.3}$$

In addition to the considerations discussed previously, circuits of this configuration also suffer from problems due to parasitic input capacitance. The input capacitances C_+ and C_- effectively form a low-pass filter with the resistances R_1 and R_2. The high-frequency cut-off is increased by decreasing either or both of R_1 and R_2. As R_1 must be relatively large to maintain good low-frequency performance, significant phase delays in the 1-10 kHz frequency range can result if R_2 is too large.

If the instrumentation amplifier has a good noise performance and Gain-Bandwidth Product, R_2 can be chosen small, which implies a large k. Depending on the instrumentation amplifier, k values of up to 10 or 100 are possible without significant bandwidth reduction or poor noise performance. A more complicated alternative involves the inclusion of the capacitances C_1 and C_2 to maintain a constant voltage division even at high frequencies. With the presence of C_1 and C_2, an additional capacitive voltage divider is formed at each input. By tuning these capacitors so that $R_1C_1 = R_2C_2$, the parasitic capacitors are actually 'tricked' into increasing the bandwidth. Although this approach is time-consuming to calibrate, bandwidths increases of 10 - 100 times can be realized.

6.3 Voltage Amplifiers

6.3.1 Linear Amplifiers

Linear voltage amplifiers are used extensively to drive piezoelectric transducers in both open and closed loop. An extensive variety of commercial devices are available to suit most applications in vibration control and positioning. The major parameters to consider when selecting a voltage amplifier are: Output current and voltage, maximum output power, bandwidth, and slew-rate. These are discussed below.

The most obvious amplifier parameter is output voltage range. Although the amplifier should be capable of utilizing the transducers full rated voltage, care should be taken not to overspecify the amplifier. In fault, or during periods of high disturbance, a voltage greater than the maximum tolerated by the transducer will cause depolarization, that is, begin to realign the piezoelectric dipoles within the material. An extended period of reverse over-voltage will completely depolarize the transducer. In some applications involving periodic high temperatures greater the Curie temperature, the transducers are unavoidably depolarized. In such cases, a high voltage is applied to the transducer after cooling to repolarize it before service. As the depolarization voltage is proportional to material thickness, thicker transducers will tolerate higher voltages. Piezoelectric transducers, particularly stacks, often have a much higher depolarization voltage in the same axis in which they are poled, *e.g.* $-200V$ and $+1000V$. If this is the case, an amplifier with non-symmetric output is beneficial. Although typical active noise and vibration control applications require a symmetric actuation range, the transducers can be electrically prestressed about the middle of their operational voltage range. This requires no additional power as no DC current flows through a capacitor.

The required output current can be approximated by modeling the transducer as a purely capacitive load, and assuming conditions of maximum voltage and frequency, *i.e.*

$$i_{\max} = v_{\max} 2\pi C_p f_{\max}. \tag{6.4}$$

For a purely capacitive load driven by a sinusoid, the worst case power dissipation can be calculated from the following formula,

$$P = 4V_s^2 C_p f_{\max}, \tag{6.5}$$

where $\pm V_s$ is the supply voltage, P is the worst case power dissipation in $Watts$, and $f_{\max}$ is the maximum frequency of operation. Note that this is the worst case dissipated power, not the output power. Worst case dissipated power is of the most interest as it determines the internal heat generation. Sustained operation within the voltage, current, and output power limits, while violating the maximum dissipated power, will overheat the amplifier and trigger a thermal shutdown.

Another important parameter, especially when enclosing the amplifier in a feedback loop, is the amplifier bandwidth or dynamic transfer function. The majority of commercial amplifiers are voltage-feedback in construction and are internally stabilized by dominant pole compensation. This results in a second order input-output response:

$$\frac{v_{out}(s)}{v_{in}(s)} = \frac{K\omega_{amp}^2}{s^2 + 2\zeta\omega_{amp}s + \omega_{amp}^2}, \tag{6.6}$$

where K is the DC gain, ω_{amp}^2 is the resonance or corner frequency, and ζ is the damping ratio. Unless the amplifier is significantly greater in bandwidth than that of the controller, it is generally essential to include the amplifier dynamics in the model, or at the very least, take the phase response into consideration when assessing the closed-loop phase margin.

The final parameter of interest, slew-rate, is of little interest in vibration control, but important to some positioning applications. In dynamic operation, the worst case slew-rate occurs at maximum voltage and frequency. Assuming the amplifier is within its other operating limits, a slew-rate constraint limits the output voltage rate-of-change. The maximum slew-rate in $V/\mu S$ of a sine wave is $SR = 2\pi f_{\max} V_{\max}$.

6.3.2 Switched-mode Implementation

During the evaluation of a smart materials system, the weight and power requirements of the drive and processing electronics cannot be neglected. Although piezoelectric transducers are themselves quite efficient, the highly reactive load impedance presented to a linear amplifier results in extremely poor efficiency. As an example, consider a $1\mu F$ capacitance driven at $1kHz$ by a $200V$ linear amplifier, although no real power is dissipated by the load, the worst case power dissipated by the amplifier is $160W$. Reactive power supplied to a capacitor is simply burnt by the output transistors.

An analysis of power flow to a piezoelectric transducer reveals similarly low levels of efficiency [192], a phase angle of 80-85 degrees results in large reactive power flows dissipated as heat by the amplifier. The efficiency is even worse when a piezoelectric transducer, integrated into a smart structure, is used in feedback to damp vibration. It has been shown that the impedance at the terminals has a negative real component [29]. This implies that while the transducer is damping structural vibration, it is effectively supplying power back to the amplifier. In addition to the wasted reactive power flow, such generated power is simply lost as heat.

Switched-mode power amplifiers [132] have recently emerged as an alternative to traditional linear devices. The main advantage of switched-mode amplifiers is that they are capable of absorbing or recycling reactive power, and as a result, require only a fraction of the supply power and heatsinking of linear devices [117]. A number of switched-mode power amplifiers, mainly for driving piezoelectric stack actuators, have appeared on the commercial market.

The main drawbacks associated with switched-mode implementation are the poor dynamic performance and increased high-frequency noise. The bandwidth of a typical switched-mode amplifier is more likely to be in the hundreds of Hertz range rather than the Kilohertz. Output signal noise is generally dependent on the switching frequency and the quality of the output filter. A good quality output filter and high switching frequency provides the best noise performance albeit at the expense of efficiency.

Switched-mode amplifiers dedicated to driving piezoelectric loads, can generally be grouped into three categories: H-bridges, boost converters and hybrid amplifiers. These are characterized below.

- H-bridge amplifiers are the simplest in conception, design, and implementation. They also provide good bandwidth, and can be controlled using conventional feedback theory. The optimized design and construction of an H-bridge amplifier dedicated to piezoelectric loads can be found in reference [30].
- Boost or flyback converters [116] have the advantage of not requiring a high-voltage supply bus. A Low-voltage supply is dynamically 'boosted' to track a reference voltage. Unfortunately boost or flyback converters are

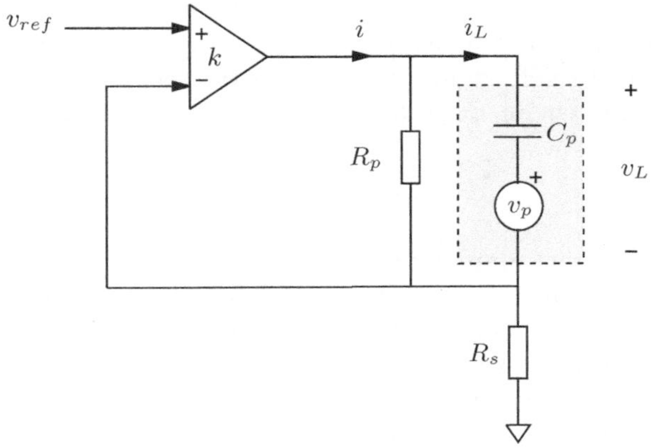

Figure 6.3. The simplified schematic of a generic current source. The piezoelectric transducer is shown in grey.

highly nonlinear and consequently difficult to control. Bandwidths in the tens of Hertz are typical.

- Two classes of hybrid switched-mode amplifier have shown promise in literature. In order to reduce the output voltage variation due to switching, the hybrid-linear amplifier [178] contains an efficient switched-mode stage in series with a floating linear stage. Since the linear stage does not drop large voltages across its output transistors, little power is dissipated. Significant improvements to the output waveform fidelity can be achieved with this approach. The hybrid-charge switching amplifier [125] utilizes a charge reference capacitor to deposit known electrical charges onto the load. Similar to its linear counterpart presented in [147], this method is both efficient and improves the strain linearity. A full discussion of charge control, and the reduction of hysteresis in piezoelectric actuators can be found in Chapter 11.

6.4 Current Amplifiers

The primary application of current amplifiers is in implementation of passive and semi-active piezoelectric shunt damping circuits. Charge amplifiers are a better choice for all positioning and active structural control systems. When connected to a capacitive load, the compliance voltage (or output voltage) of a current amplifier is inversely proportional to frequency. This can result in an impractically large dynamic range if the frequency range of operation is equal to, or greater than 2 decades.

The simplified schematic of a current feedback amplifier is shown in Figure 6.3 [91]. The op-amp feedback loop works to equate the applied reference voltage v_{ref} to the voltage measured across the sensing resistance R_s. As no current flows into the op-amp terminals, the output current i is equal to

$$i = \frac{1}{R_s} v_{ref} \ , \tag{6.7}$$

i.e. we have a current amplifier with gain $1/R_s \ A/V$.

The foremost difficulty in employing such devices to drive highly capacitive loads is that of DC current offsets. Inevitably, the voltage measured across the sensing impedance will contain a non-zero voltage offset, this and other sources of voltage or current offset within the circuit result in a net uncontrolled output current. As a capacitor integrates DC current, the load voltage will ramp upward and saturate at the power supply rail. Any offset in the load voltage v_L limits the compliance range of the current source and may eventually cause saturation. To alleviate this difficulty a parallel resistance R_p is required to provide a path for DC output currents. In this configuration, the actual current $i_L(s)$ applied to the transducer is as follows. Note that the load is assumed to be purely capacitive, *i.e.* we are neglecting the internal source v_p. This is a valid assumption if v_p is small in comparison to the load voltage and this occurs if there are no structural resonances below the electrical cutoff frequency.

$$i_L(s) = i(s) \frac{s}{s + \frac{1}{R_p C_p}} . \tag{6.8}$$

Additional dynamics have been added to the current source. The transfer function now contains a high-pass filter with cutoff $\omega_c = \frac{1}{R_p C_p}$. That is,

$$\frac{i_L(s)}{v_{ref}(s)} = \frac{1}{R_s} \frac{s}{s + \frac{1}{R_p C_p}} . \tag{6.9}$$

In contrast to the infinite DC impedance of a purely capacitive load, the load impedance now approaches R_p below $\omega_c = \frac{1}{R_p C_p}$. Thus, a DC offset current of i_{dc}, results in a load voltage offset of $v_L = i_{dc} R_p$. In a typical piezoelectric driving scenario, with $C_p = 100 \, nF$, and $i_{dc} = 1 \mu A$, a $1 \, M\Omega$ parallel resistance is required to limit the DC voltage offset to $1 \, V$. The frequency response from an applied reference voltage to the actual capacitive load current $i_L(s)$ is shown in Figure 6.4. Phase lead exceeds 5 degrees below 18 Hz. Such poor low frequency performance precludes the use of current amplifiers in such applications as scanners and low frequency structural control. At the expense of amplifier complexity, an additional voltage feedback loop can be added to improve low frequency performance [64].

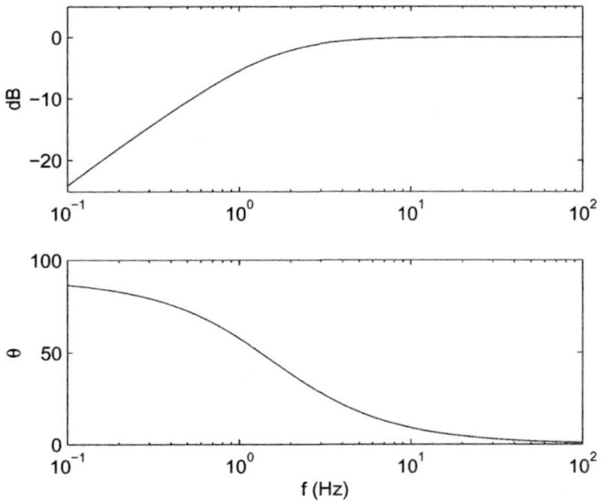

Figure 6.4. Typical frequency response from an applied reference voltage to the actual capacitive load current $i_L(s)$

6.5 Charge Amplifiers

As discussed in Chapter 11, charge amplifiers provide the benefit of increased actuator linearity without the dynamic range problems associated with current amplifiers. Charge amplifiers are the preferred driver for open- and closed-loop positioning applications, active structural control, and shunt circuit implementation.

Consider the simplified diagram of a generic charge source shown in Figure 6.5. The piezoelectric load, modeled as a capacitor and voltage source v_p, is shown in grey. The high gain feedback loop works to equate the applied reference voltage v_{ref}, to the voltage across a sensing capacitor C_s. Neglecting the resistances R_p and R_s, at frequencies well within the bandwidth of the control loop, the load charge q_L is equal to

$$q_L = V_{ref} C_s, \tag{6.10}$$

i.e. we have a charge amplifier with gain C_s *Columbs/V*.

As with current amplifiers, there are practical considerations associated with DC and low-frequency performance. In Figure 6.5, the resistances R_p and R_s model the op-amp input impedance, capacitor dielectric leakage, and possibly an additional resistance associated with load voltage measurement. In practice, these parasitic resistances are often swamped by physical resistances included to reduce bias current induced voltage drift.

For vibration control applications where DC accuracy may not be required, the sensing resistance R_s can be neglected. With a parallel load resistance R_p, the actual charge $q_L(s)$ flowing through the load transducer becomes

$$q_L(s) = q(s) \frac{s}{s + \frac{1}{R_p C_p}}. \tag{6.11}$$

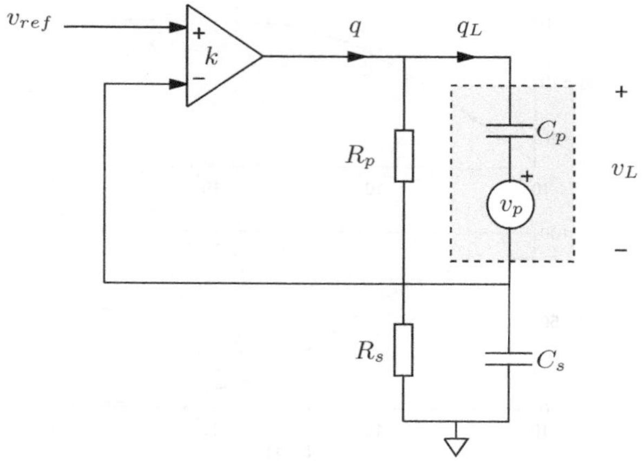

Figure 6.5. The simplified schematic of a charge source. The piezoelectric transducer is shown in grey.

The amplifier now contains a high-pass filter with cutoff frequency $\omega_c = \frac{1}{R_p C_p}$, that is,

$$\frac{q_L(s)}{V_{ref}(s)} = C_s \frac{s}{s + \frac{1}{R_p C_p}}. \tag{6.12}$$

Similar to the current source scenario discussed in Section 6.4, with a capacitance of $C_p = 100\ nF$, and a DC output offset current of $i_{dc} = 1\mu A$, a $1\ M\Omega$ parallel resistance is required to limit the DC voltage offset to $1\ V$. The frequency response from an applied reference voltage to the actual load charge $q_L(s)$ is exactly the same as that plotted for the current amplifier in Figure 6.4.

An improved solution to the poor low-frequency performance and current-induced voltage drift was presented in [63]. An auxiliary voltage feedback loop is included to correct low-frequency behavior and allow for constant charge offsets. Although still simpler than a current source controller, the additional circuitry requires the design of separate voltage and charge feedback controllers. A simplified design utilizing the intrinsic voltage feedback offered by the parasitic resistances was later presented in [204].

If both of the resistances R_p and R_s are considered, it can be shown that the transfer function between the reference signal and the load charge is

$$\begin{aligned} \frac{q_{LC}(s)}{v_{ref}(s)} &= \frac{q_L(s)}{v_{ref}(s)} \frac{q_{LC}(s)}{q_L(s)} \\ &= C_s \frac{s + \frac{1}{C_s R_s}}{s} \frac{s}{s + \frac{1}{R_L C_L}}. \end{aligned} \tag{6.13}$$

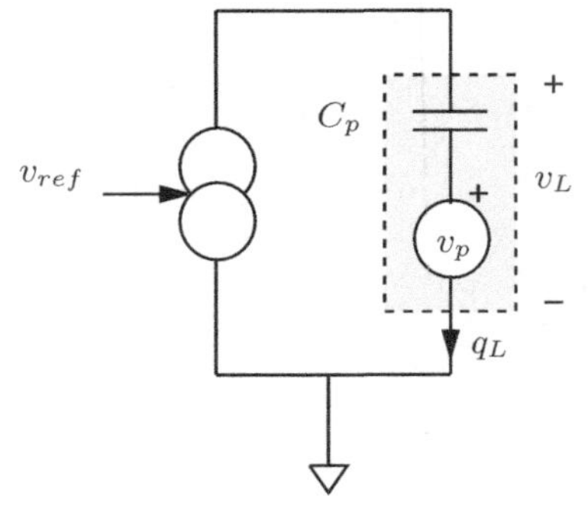

Figure 6.6. Test for voltage / charge dominance

By setting $C_L R_L = C_s R_s$, *i.e.*

$$\frac{R_L}{R_s} = \frac{C_s}{C_L}, \tag{6.14}$$

the low-pass voltage dynamics cancel the high-pass charge dynamics to yield a constant gain of C_s *Columbs/Volt*. Effectively the voltage amplifier, comprised of the two resistances R_p and R_s, synthesizes the operation of an ideal charge amplifier at low frequencies.

If the amplifier can be viewed as the concatenation of a voltage and charge amplifier, it is important to identify the region in which it operates as a pure charge amplifier. Consider the schematic shown in Figure 6.6 with $v_{ref} = 0$. During perfect charge operation, *i.e.* when q_L is correctly regulated to zero, the voltage v_L will be equal to v_p. During voltage dominant behavior, v_L will be regulated to zero. These characteristics can easily be measured experimentally. Although the voltage dynamics have been designed to perfectly synthesize the operation of an ideal charge amplifier, during voltage dominant operation, if the load is not purely capacitive, errors in q_L will occur.

When $v_{ref} = 0$, which implies $q_L = 0$, the transfer function from v_p to v_L reveals the voltage or charge dominance of the amplifier. At frequencies where $v_L \approx v_p$, the amplifier is charge dominant, and voltage dominant when $v_L \approx 0$. For the hybrid amplifier shown in Figure 6.5, when $v_{ref} = 0$,

$$\frac{v_L(s)}{v_p(s)} = \frac{s}{s + \frac{1}{R_p C_p}}, \tag{6.15}$$

i.e. at frequencies above $\frac{1}{2\pi R_p C_p}$ Hz the amplifier is charge dominant, and voltage dominant below. Obviously, given Equation (6.15), the objective will be to select a load resistor R_L as large as possible. This may be limited by other factors such as op-amp current noise attenuation, bias-current induced offset voltages, and the common-mode and differential leakage of the op-amp.

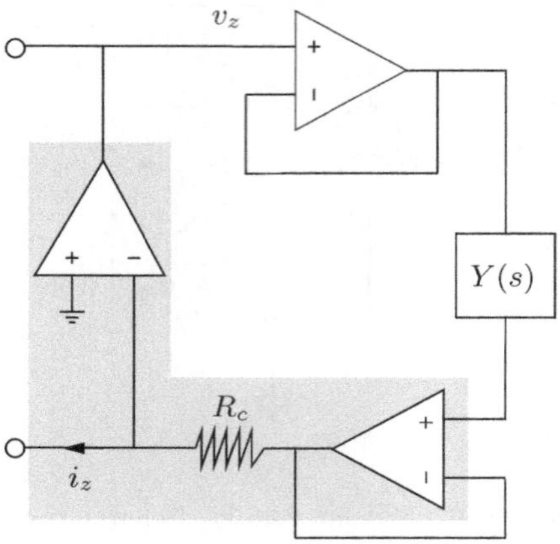

Figure 6.7. A simple synthetic admittance circuit requiring only a single high-voltage op-amp

In practice $\frac{v_L(s)}{v_p(s)}$ is best measured by applying a voltage to a collocated transducer and using that as a reference. At frequencies lower than the first mechanical mode, the voltage applied to a collocated transducer will be proportional to $v_p(s)$.

A description of a modified charge amplifier suitable for driving grounded loads can be found in Section 12.5.1. A switched-mode implementation can be found in [125].

6.6 Synthetic Admittance

Resonant piezoelectric shunt damping circuits, as discussed in Section 4.2, are a preferred method for vibration reduction. They provide good damping performance, guarantee the stability of a structural system, are simple to design, and can be adapted online to maintain optimal performance (see Chapter 8). In Section 4.4 the synthetic admittance was introduced as a simplified technique for the implementation of multi-mode shunt damping circuits. The synthetic admittance is superior to virtual circuit technologies as it requires only a single high-voltage op-amp; is capable of implementing high quality-factor inductor networks; and can easily be extended to facilitate multiple ports, circuits of high-orders, and non-linear or adaptive circuits.

The schematic diagram of a simple synthetic admittance is shown in Figure 6.7. The shaded area represents a voltage controlled current source (VCCS) with gain $\frac{1}{R_c}$ A/V. The filter $Y(s)$ can be realized either as an analog or digital transfer function.

Synthetic implementation is a unifying replacement for various virtual circuits including: virtual capacitors and inductors, negative impedance converters, transformers and gyrators. Only a single high-voltage op-amp is required to provide near ideal implementation of any arbitrary terminal impedance.

An example of a high-voltage amplifier customized for use as a synthetic admittance is pictured in Figure 6.8. This device is constructed from discrete bipolar junction transistors and can supply a peak current of 32 amps from a ± 250 Volt supply. In order to provide maximum flexibility in implementation, the amplifier is reconfigurable to drive either current, charge, voltage, or current rate-of-change. A high-bandwidth measurement of terminal voltage, compliance voltage, current, charge, and current rate-of-change is also available. At the time of writing, the total component cost was under US $50 retail.

6.7 Switched-mode Synthetic Admittance

As discussed in Section 6.3.2, linear electronic devices are not capable of efficient operation when driving highly reactive loads such as piezoelectric transducers. Therefore, they may not be suitable for applications where power is limited, or where the large requirement for heat dissipation is either prohibitively expensive or impractical. The *switched-mode synthetic admittance*

Figure 6.8. A high-voltage current, charge, or voltage amplifier used as part of a synthetic admittance or impedance

requires no high voltage linear components, is small in size, and is ideal for high-power or industrial scale applications.

6.7.1 Device Operation

A simplified circuit diagram of the switched-mode synthetic admittance is shown in Figure 6.9. The basic operation is much the same as that discussed previously, the device maintains an arbitrary relationship between the measured terminal voltage and applied current, *i.e.* between i_z and v_z.

We begin with some preliminary circuit analysis. In the Laplace domain

$$i_z(s) = \frac{v_z(s) - v_{pwm}(s)}{Z_c(s)}. \tag{6.16}$$

In this case, the desired relationship between terminal voltage and current is the impedance $Z(s)$,

$$i_z(s) = \frac{1}{Z(s)} v_z(s). \tag{6.17}$$

Combining (6.16) and (6.17) yields the relationship required to maintain (6.17) at the terminals,

$$v_{pwm}(s) = v_z(s) \left(1 - \frac{Z_c(s)}{Z(s)}\right). \tag{6.18}$$

The reader may recognize the similarity between the circuit on the right-hand side of Figure 6.9 and a controlled single phase switched-mode rectifier, or a four quadrant switched-mode amplifier [132]. Indeed the only difference between such devices is the selection of control impedance and bridge control algorithm. Although v_{pwm} can only take the values of $\pm v_{dc}$, an arbitrary average voltage between these limits can be generated by controlling the switching duty cycle. The relationship between the desired v_{pwm} and the control duty cycle is [132]

$$D = \frac{1}{2} \left(\frac{v_{pwm}}{v_{dc}} + 1\right). \tag{6.19}$$

6.7.2 Boost Configuration

Here we consider a specific choice of control impedance $Z_c(s)$ - a series inductor and resistor $Z_c(s) = R_c(s) + L_c(s)$. In this configuration, the structure of the circuit resembles that of a single phase boost rectifier. The primary motivation is to allow the flow of real and reactive power back to the source.

Assuming that the inductance is large enough to maintain an approximately constant current over the switching interval, when the applied potential v_{pwm} opposes the current i_z, the inductor overcomes the source potential and forces current through the anti-parallel diodes back to v_{dc}.

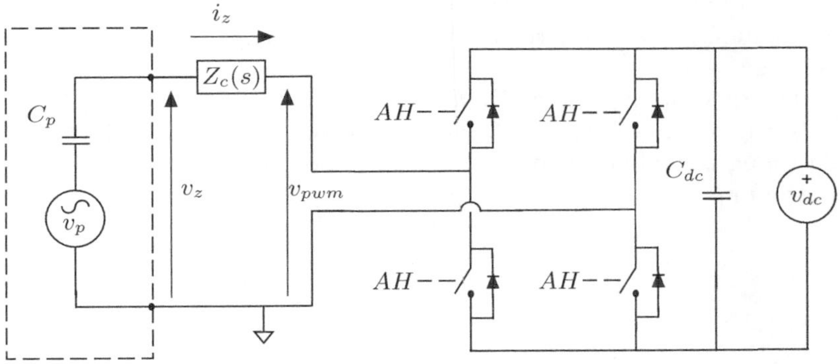

Figure 6.9. The switched-mode synthetic admittance

This configuration also has the advantage of greatly reducing high frequency content applied to the piezoelectric transducer. The inherent transducer capacitance together with the control impedance creates a second order low pass filter,

$$\frac{v_z(s)}{v_{pwm}(s)} = \frac{\frac{1}{L_c C_p}}{s^2 + \frac{R_c}{L_c}s + \frac{1}{L_c C_p}}. \tag{6.20}$$

For reasonable values of R_c, L_c, and C_p (300 Ω, 0.1 H, 400 ηF), the filter has a cutoff frequency of around 800 Hz. If we consider a system with a switching frequency of 8 kHz, such a filter would attenuate the fundamental switching component by 40 dB. Taking into account the additional low pass dynamics of the plant, the actual realized disturbance due to switching is negligible.

6.7.3 Efficiency

If we consider a sinusoidal voltage source V_s connected to an impedance $Z(s)$, the real dissipated power is

$$P_T = \frac{1}{2}|V_s|^2 \mathbf{Re}\left\{\frac{1}{Z(s)}\right\} = \frac{1}{2}\left|\frac{V_s}{Z(s)}\right|^2 \mathbf{Re}\{Z(s)\}. \tag{6.21}$$

We define the efficiency of a switched-mode synthetic admittance as the ratio of power absorbed by v_{dc}, to the power that would normally be dissipated if the impedance $Z(s)$ was implemented using ideal physical components, *i.e.*

$$\eta(Z_c(s), Z(s), \omega) = 100\% \times \frac{P_{v_{dc}}}{P_T}. \tag{6.22}$$

By this definition, virtual or linear synthetic implementations will always result in a negative efficiency, *i.e.* they absorb no real power. In fact, the situation is worse - such implementations must actually supply power to synthesize the flow of apparent power. For our application, *i.e.* synthesizing inductors to form a highly resonant circuit, the realized efficiency is extremely poor (large and negative).

The quantity $P_{v_{dc}}$ is easily found by performing a power balance. Obviously, the real power as seen from the terminals will be equal to P_T. The only remaining contribution to the net real power flow is that dissipated by the control impedance,

$$P_c = \frac{1}{2}\left|\frac{V_s}{Z(s)}\right|^2 \mathbf{Re}\{Z_c\} \tag{6.23}$$

$$\eta(Z_c, Z_T, j\omega) = 100\% \times \frac{P_T - P_c}{P_T} \tag{6.24}$$

$$= 100\% \times \left[1 - \frac{\mathbf{Re}\{Z_c\}}{\mathbf{Re}\{Z_T\}}\right]. \tag{6.25}$$

The best efficiency (100%) is achieved if the control impedance contains no real component. If the control impedance has a larger real component than the terminal impedance, the efficiency is negative, *i.e.* the source V_{dc} must supply real power to the system.

6.7.4 Practical Advantages and Considerations

The switched-mode synthetic admittance has a number of advantages over its linear counterpart. Some difficulties also arise that are not present in the linear case.

- **Cost.** Discrete power switches can be obtained for a fraction of the cost of HV linear components.
- **Size / Density.** Besides the small FET transition losses, the circuit shown in Figure 6.9 does not dissipate any real or reactive power flowing between the source and the controlling impedance. There is also no requirement for quiescent or bias current. Coupled with the small physical size of power switches, a low heat dissipation allows the circuit to be manufactured in an extremely small enclosure. Another significant factor is the size of the power supply. In the linear case, a large supply is required to power the components and to supply reactive power to the structure. As we have seen, in the switching case, not only is the power supply small, but if the synthesized terminal impedance has a larger real component than the controlling impedance, no power supply is required at all.
- **Control Conditioning.** The switched-mode synthetic admittance manipulates the terminal current by controlling the average voltage across a control impedance $Z_c(s)$. In practice, the circuit must be conditioned so that the expected current range results in realizable voltage differences across the control impedance. At a specific frequency, this is easily achieved by ensuring $|Z_c(s)| >> |Z(s)|$, *i.e.* by choosing a control impedance much

Figure 6.10. The experimental beam

greater in magnitude than $Z(s)$. Another simple technique is to design $Z_c(s)$ having an opposite or significantly different phase angle with respect to $Z(s)$. In the boost configuration, we are limited in choice to an inductor and resistor.

The impedance of passive shunt damping circuits is typically comprised of inductive resistive branches. In the active frequency range, the reactance of each branch is heavily dominated by the inductor. This is expected as resonant circuits operate at very low power factors (implying small real impedance).

We must consider a number of factors: For efficiency we wish to keep the control resistance R_c small. If R_c is small, the only way to increase the control impedance, is to increase the size of the inductance L_c. As both the control and terminal impedance have a similar impedance angle (approximately $+\pi$), we cannot improve the control conditioning by relying on a phase difference. Thus, to obtain a well conditioned voltage drop across the control impedance $Z_c(s)$, the control inductance must be a reasonable fraction of the terminal inductance. *e.g.* $L_c = \frac{L}{10}$. Multi-mode shunt circuits include at least one inductance per branch, in this case, we must consider the lowest frequency branch, (the branch with the greatest inductance).

- **Common Mode Instrumentation Performance.** The return terminals of the load PZT and supply v_{dc} must be electrically isolated. Ideally, the acquisition of v_z should be performed using a circuit completely isolated from both references. As this is impossible in practice, the instrumentation amplifier must have a high common mode rejection ratio to attenuate components resulting from the varying potential between the two references.

Table 6.1. Experimental Beam Parameters

Length, L	$0.6\ m$
Width, w_b	$0.05\ m$
Thickness, h_b	$0.003\ m$
Youngs Modulus, E_b	$65 \times 10^9\ N/m^2$
Density, ρ	$2650\ kg/m^2$

Table 6.2. Piezoelectric Transducer Properties

Length	$0.070\ m$
Charge Constant, d_{31}	$-210 \times 10^{-12}\ m/V$
Voltage Constant, g_{31}	$-11.5 \times 10^{-3}\ Vm/N$
Coupling Coefficient, k_{31}	0.340
Capacitance, C_p	$0.105\ \mu F$
Width, w_s w_a	$0.025\ m$
Thickness, h_s h_a	$0.25 \times 10^{-3}\ m$
Youngs Modulus, E_s E_a	$63 \times 10^9\ N/m^2$

6.7.5 Experimental Application

The switched-mode synthetic admittance was employed to implement a two-mode current-blocking piezoelectric shunt damping circuit. The experimental beam apparatus, pictured in Figure 6.10, is a uniform aluminum bar with rectangular cross section and experimentally pinned boundary conditions at both ends. A pair of piezoelectric ceramic patches (PIC151) are attached symmetrically to either side of the beam surface. One patch is used as an actuator and the other as a shunting layer. Physical parameters of the experimental beam and piezoelectric transducers are summarized in Tables 6.1 and 6.2. Note that the transducer location offers little control authority over the first mode. In this work, the structure's second and third modes are targeted for reduction.

A prototype switched-mode admittance is shown in Figure 6.11. The board contains two isolated sub-circuits, the gate-drive and MOSFET bridge on the left, and the ground referenced instrumentation on the right. The MOSFET devices used in this circuit permit a peak-to-peak output voltage of 120 V at 2 Amps. The displacement and voltage frequency responses are measured using a Polytec scanning laser vibrometer (PSV-300) and HP spectrum analyzer (35670A).

As discussed in Chapter 4, a current-blocking piezoelectric shunt damping circuit was designed to minimize the H_2 norm of the beam displacement transfer function. The circuit schematic and component values can be found in Figure 6.20 and Table 6.3. The control impedance was selected as $Z_c(s) = 67\ mH + 33\ k\Omega$.

The experimental open- and closed-loop transfer functions from an applied actuator voltage to the resulting displacement at a point $0.17m$ from the beam

Figure 6.11. A switched-mode synthetic admittance connected to the experimental beam and dSPACE signal processor

Table 6.3. Component values

R_1	1543 Ω	L_1	43 H
R_2	1145 Ω	L_2	20.9 H
C_3	100 nF	L_3	45.2 H

end are shown in Figure 6.12. The amplitudes of the second and third modes are reduced by 21.6 and 21.3 dB respectively.

To assess the linearity of switched-mode implementation, a sine wave was applied at the second mode resonance frequency, the power spectral density of the resulting voltage applied to the piezoelectric transducer is shown in Figure 6.13. The harmonic content and switching noise applied to the piezoelectric transducer is negligible ($<$ 60 dB). The component at 50 Hz is due to the floating voltage measurement and common-mode mains interference.

6.8 Signal Processing

In order to synthesize an electrical network by either admittance or impedance synthesis, a filter is required that implements the electrical transfer function of the system. Although simple transfer functions are easily implemented by analog filters, greater flexibility, accuracy and ease of configuration/trouble-shooting are obtained when using a Digital Signal Processor (DSP) or micro-controller. As an example, consider the implementation of a three-mode 'current blocking' piezoelectric shunt circuit [200]. In this case, the electrical network could contain up to 18 individual passive components, the corresponding admittance transfer function would contain up to 15 states and be parameterized by up to 18 variables. Implementation and manual tuning of an analog

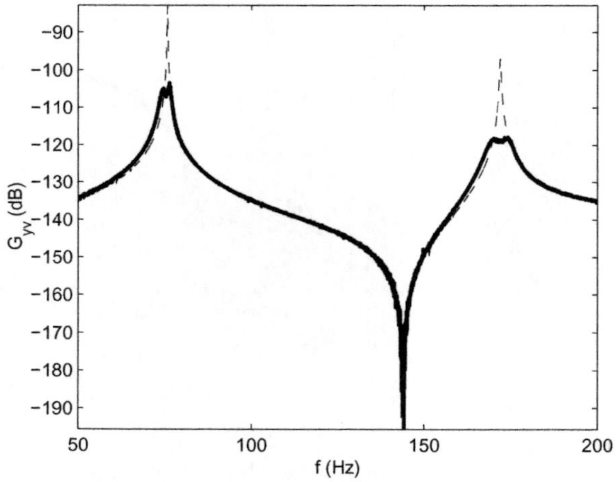

Figure 6.12. Experimental open-loop (- -) and damped system transfer (—) functions

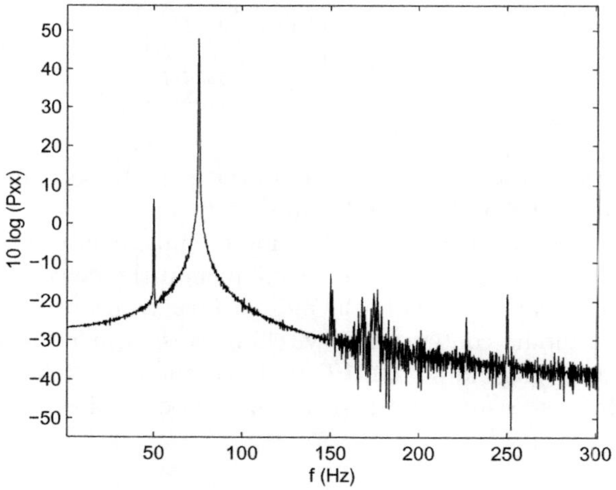

Figure 6.13. Power spectral density of the terminal voltage v_z applied to the piezoelectric transducer

filter this size would be prohibitively tedious. A more robust approach would utilize a basic DSP and include an automatic tuning algorithm such as that presented in Chapter 8. A low-cost 16 bit DSP such as that found in the Texas Instruments C2000 family would be well suited. In addition to all of the necessary processing and memory resources, these chips also include analog to digital converters and PWM generators.

Further difficulties are also experienced with analog implementation. Traditional filter synthesis techniques [184] typically require a partial fraction

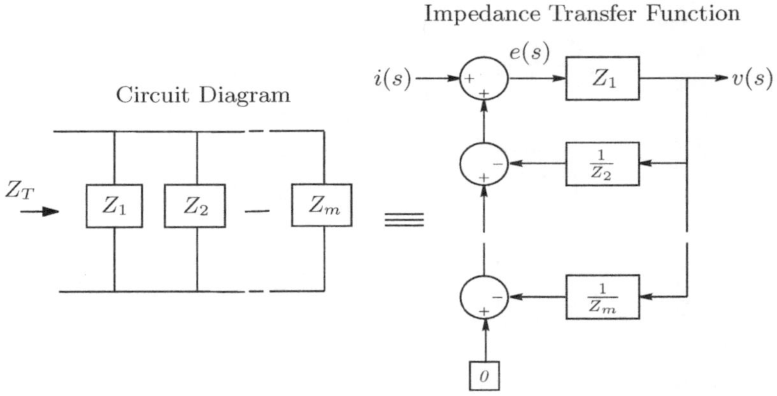

Figure 6.14. Parallel circuit equivalence for impedance block diagrams

decomposition, followed by the implementation of each second order section. To simplify the process of impedance or admittance transfer function implementation, this section introduces a link between system block diagrams and circuit schematics. In the digital case, if a graphical compilation package such as the real time workshop for MATLAB® or similar is available, no electrical transfer function calculations are required. The resulting block diagram bears a natural resemblance to the corresponding circuit, is clearly parameterized, and is consequently easy to tune. In the analog case, the circuit can be broken down into a number of simple op-amp integrators and amplifiers whose gains correspond directly to component values. The resulting filter circuit is practical, easy to implement, expandable and simple to tune.

Following are the transformations of interest for both the impedance and admittance synthesis cases. In Sections 6.8.3 and 6.8.4, two examples are presented to clarify the application.

6.8.1 Impedance Transformations

Parallel equivalence. Consider the parallel network components Z_1, Z_2, …, Z_m as shown in Figure 6.14. The terminal impedance and admittance corresponding to this network are:

$$Z_T(s) = \frac{1}{\frac{1}{Z_1}+\frac{1}{Z_2}+\ldots+\frac{1}{Z_m}} \quad Y_T(s) = \frac{1}{Z_1} + \frac{1}{Z_2} + \ldots + \frac{1}{Z_m} \; . \tag{6.26}$$

Now consider the transfer function block diagram, also shown in Figure 6.14,

$$\begin{aligned} G(s) = \frac{v(s)}{i(s)} &= \frac{Z_1}{1+Z_1\frac{1}{Z_2}+\ldots+Z_1\frac{1}{Z_m}} \\ &= \frac{1}{\frac{1}{Z_1}+\frac{1}{Z_2}+\ldots+\frac{1}{Z_m}} . \end{aligned} \tag{6.27}$$

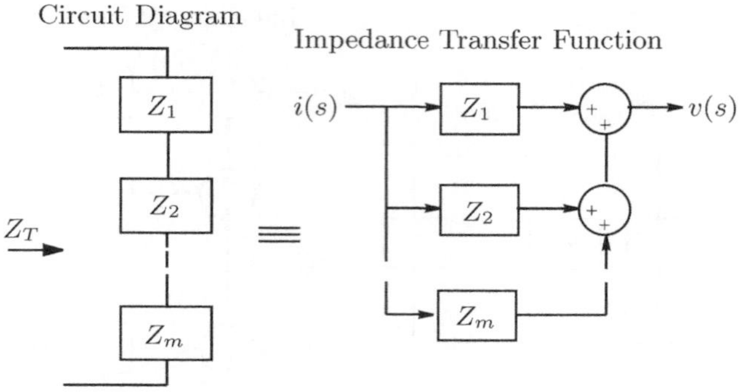

Figure 6.15. Series circuit equivalence for impedance block diagrams

Observe that $Z_T(s)$ and $G(s)$ as described in Equations (6.26) and (6.27) are identical. Therefore, this transformation can be used to generate impedance block diagrams from electrical sub-circuits in parallel.

Series equivalence. Consider the series network components $Z_1, Z_2, \ldots, Z_m$ as shown in Figure 6.15. The terminal impedance and admittance of this network are:

$$Z_T(s) = Z_1 + Z_2 + \ldots + Z_m \; Y_T(s) = \frac{1}{Z_1+Z_2+\ldots+Z_m}. \tag{6.28}$$

Now consider the transfer function block diagram, also shown in Figure 6.15,

$$G(s) = \frac{v(s)}{i(s)} = Z_1 + Z_2 + \ldots + Z_m. \tag{6.29}$$

Observe that $Z_T(s)$ and $G(s)$ as described in Equations (6.28) and (6.29) are identical. Therefore, this transformation can be used to generate impedance block diagrams from electrical sub-circuits in series.

6.8.2 Admittance Transformations

Parallel equivalence. Consider the parallel network components Z_1, Z_2, ..., Z_m as shown in Figure 6.16. The terminal impedance and admittance of this network is

$$Z_T(s) = (\tfrac{1}{Z_1} + \tfrac{1}{Z_2} + \ldots + \tfrac{1}{Z_m})^{-1} \; Y_T(s) = \tfrac{1}{Z_1} + \tfrac{1}{Z_2} + \ldots + \tfrac{1}{Z_m}. \tag{6.30}$$

Now consider the transfer function block diagram, also shown in Figure 6.16.

$$G(s) = \frac{i(s)}{v(s)} = \frac{1}{Z_1} + \frac{1}{Z_2} + \ldots + \frac{1}{Z_m}. \tag{6.31}$$

Observe that $Y_T(s)$ and $G(s)$ as described in Equations (6.30) and (6.31) are identical. Therefore, this transformation can be used to generate admittance block diagrams from electrical sub-circuits in parallel.

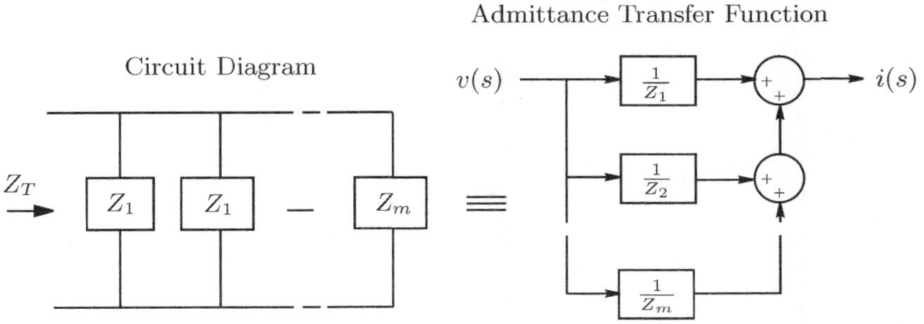

Figure 6.16. Parallel circuit equivalence for admittance block diagrams

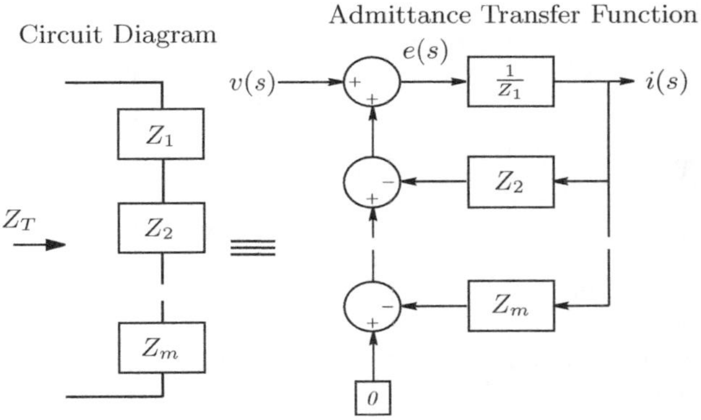

Figure 6.17. Series equivalence for admittance block diagrams

Series equivalence. Consider the series network components $Z_1, Z_2, \ldots, Z_m$ as shown in Figure 6.17. The terminal impedance and admittance of this network is:

$$Z_T(s) = Z_1 + Z_2 + \ldots + Z_m \; Y_T(s) = \frac{1}{Z_1+Z_2+\ldots+Z_m}. \tag{6.32}$$

Now consider the transfer function block diagram, also shown in Figure 6.17,

$$\begin{aligned} G(s) = \frac{i(s)}{v(s)} &= \frac{\frac{1}{Z_1}}{1+\frac{1}{Z_1}Z_2+\ldots+\frac{1}{Z_1}Z_m} \\ &= \frac{1}{Z_1+Z_2+\ldots+Z_m}. \end{aligned} \tag{6.33}$$

Observe that $Y_T(s)$ and $G(s)$ as described in Equations (6.32) and (6.33) are identical. Therefore, this transformation can be used to generate admittance block diagrams from electrical sub-circuits in series.

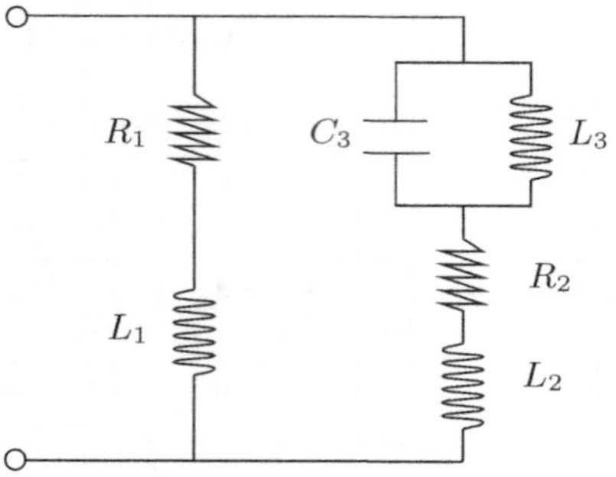

Figure 6.18. Current-blocking shunt circuit

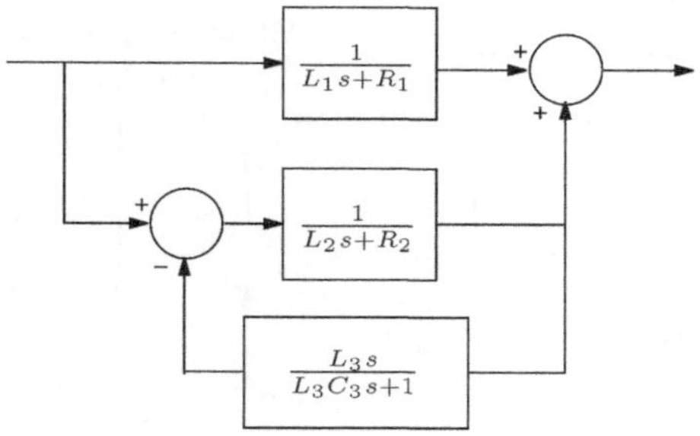

Figure 6.19. Admittance transfer function block diagram of a current-blocking shunt circuit

6.8.3 Example: Digital Synthesis

Consider the two-mode current-blocking circuit shown in Figure 6.18. The corresponding admittance block diagram is shown in Figure 6.19. Each subsystem can be further decomposed or implemented by parameterized state-space system, both methods facilitate simplified online tuning.

6.8.4 Example: Analog Synthesis

To implement the admittance of a current-flowing shunt circuit, a filter that represents a single circuit branch is required. The output of each branch filter can then be summed to produce a filter representing the entire multi-mode circuit.

One may first consider the traditional filter synthesis techniques of state-variable or Sallen-Key [184]. Such techniques result in a circuit whose component values are a complicated function of the original shunt components severely impeding any attempt at online tuning. Alternatively, using the transformations, each admittance branch can be implemented first as a system

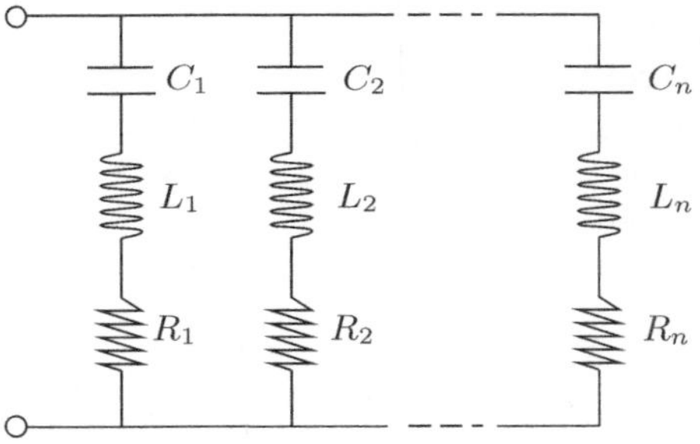

Figure 6.20. Current-flowing shunt circuit

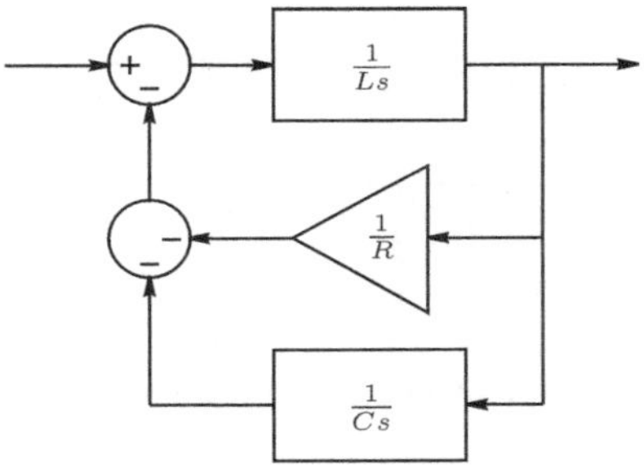

Figure 6.21. A single branch of a current-flowing admittance block diagram

block diagram, then as an analog circuit containing only summers, integrators and gains.

The admittance block diagram of a single-mode current-flowing shunt circuit is shown in Figure 6.21. A simple but effective analog implementation is shown in Figure 6.22. The transfer function is easily found to be

$$\frac{V_{out}(s)}{V_{in}(s)} = \frac{1}{R_1 C_1 s + \frac{R_2}{R_3} + \frac{1}{R_4 C_4 s}}. \tag{6.34}$$

The filter components are related to the original shunt circuit branch by

$$\begin{aligned} L &= R_1 C_1 \\ R &= \frac{R_2}{R_3} \\ C &= R_4 C_4. \end{aligned} \tag{6.35}$$

Although there are more op-amps than would normally be required, the transfer function is explicitly parameterized in terms of the parent circuit. The resistors R_1, R_2 and R_4 can be varied independently to tune the shunt circuit inductance, resistance and capacitance.

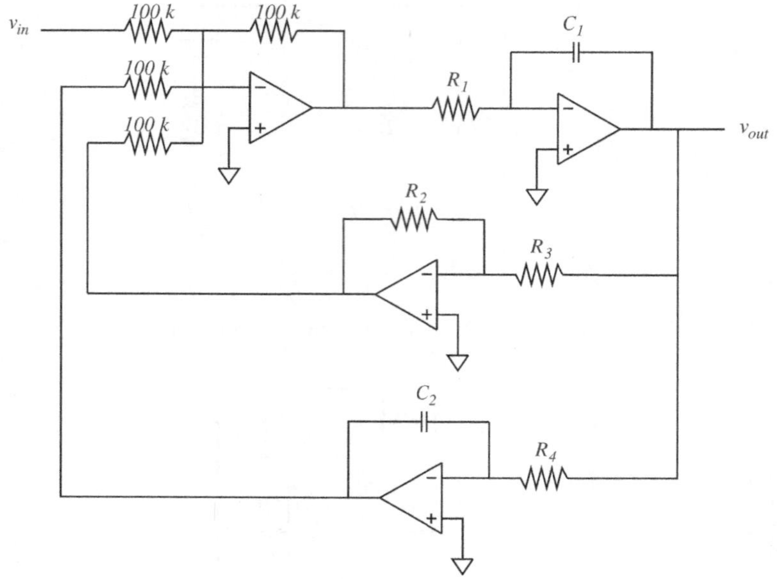

Figure 6.22. Analog implementation of the block diagram shown in Figure 6.21

Figure 6.23. An op-amp based current source with impedance card mounted vertically

A practical implementation is shown in Figure 6.23. For flexibility, the filter is manufactured as a small board that can be installed or removed as necessary. The pictured current source has a maximum supply voltage of ±45 Volts, includes an on-board low voltage supply and can hold up to two impedance cards.

7

Multi-port Shunts

7.1 Introduction

In Chapter 3 we showed that a piezoelectric shunt damping system is equivalent to a feedback control system with a well-defined feedback structure. Furthermore, we used this observation to develop two resonant shunt structures that are effective in terms of adding damping to the structure. All this, however, was done only for the case of a single shunted piezoelectric transducer.

One of the important properties of resonant shunts is that they guarantee closed-loop stability of the system in presence of out-of-bandwidth dynamics. This property is due to the passive nature of these shunts. Passivity restricts the level of performance one can obtain from the shunted system.

In many real-world vibration control applications we may require a level of damping that is not achievable from a single shunted piezoelectric transducer. In such a case we would typically use several piezoelectric transducers and shunt each one to an impedance. A more prudent approach, however, is to shunt the piezoelectric transducers to a multi-port impedance. This will make it possible to take full advantage of cross-couplings between piezoelectric transducers.

Another reason for using a multivariable piezoelectric shunt is due to the spatially distributed nature of flexible structures. If a large number of modes are to be shunt damped, it may not be practical to place one piezoelectric transducer on the surface the structure at a location where all the requisite modes can be damped efficiently. Therefore, in certain applications, a multivariable shunt may be the only possible option.

In this chapter we will look at the problem of multivariable piezoelectric shunt damping. In particular, we will show that a multivariable shunt damping system is equivalent to a feedback control problem associated with a multivariable system with identical piezoelectric actuator/sensor pairs. Furthermore, we will propose a number of multivariable piezoelectric shunts and will illustrate their effectiveness through a number of experiments. These multivariable piezoelectric shunts were first proposed in [141].

Another topic that will be revisited in this chapter is that of implementation of piezoelectric shunts. A new structure for charge/current amplifiers will be introduced and employed for implementation of the requisite shunts in our experiments. The coverage of this topic, however, is rather brief. A thorough analysis is presented in [64, 62].

7.2 Multi-port Piezoelectric Shunt

This section is concerned with developing models of flexible structures with several embedded piezoelectric sensors and actuators.

Consider the system depicted in Figure 7.1 where a number of piezoelectric transducers are bonded to a flexible structure, a beam in this case, and all piezos are electrically connected to a multi-port impedance $Z(s)$. The system is subject to a number of disturbances, $w_1, \ldots, w_\ell$, and the purpose of the shunt is to reduce the effect of these disturbances on the structure by adding damping to the system. In this section we demonstrate that this setting is equivalent to a multivariable feedback control problem. In that sense this is a generalization of the SISO structure investigated in Section 5.2.

A schematic of the system is depicted in Figure 7.1 (a) while the equivalent electrical circuit of the shunted piezoelectric transducers is drawn in Figure 7.1 (b). Notice that, as in previous chapters, each shunted piezoelectric transducer is modeled as a dependent voltage source in series with a capacitor. The piezoelectric voltage associated with the kth transducer, v_{p_k} is proportional to mechanical strain inside the device, which is in turn related to the disturbances acting on the structure. Furthermore, C_{p_k} represents the capacitance of the kth piezoelectric transducer.

Let

$$V(s) = \begin{bmatrix} v_1(s) \\ v_2(s) \\ \vdots \\ v_m(s) \end{bmatrix}; \quad V_p(s) = \begin{bmatrix} v_{p_1}(s) \\ v_{p_2}(s) \\ \vdots \\ v_{p_m}(s) \end{bmatrix};$$

$$W(s) = \begin{bmatrix} w_1(s) \\ w_2(s) \\ \vdots \\ w_m(s) \end{bmatrix}; \quad I(s) = \begin{bmatrix} i_1(s) \\ i_2(s) \\ \vdots \\ i_m(s) \end{bmatrix}. \tag{7.1}$$

Using the above notation we may write

$$V(s) = Z(s)I(s). \tag{7.2}$$

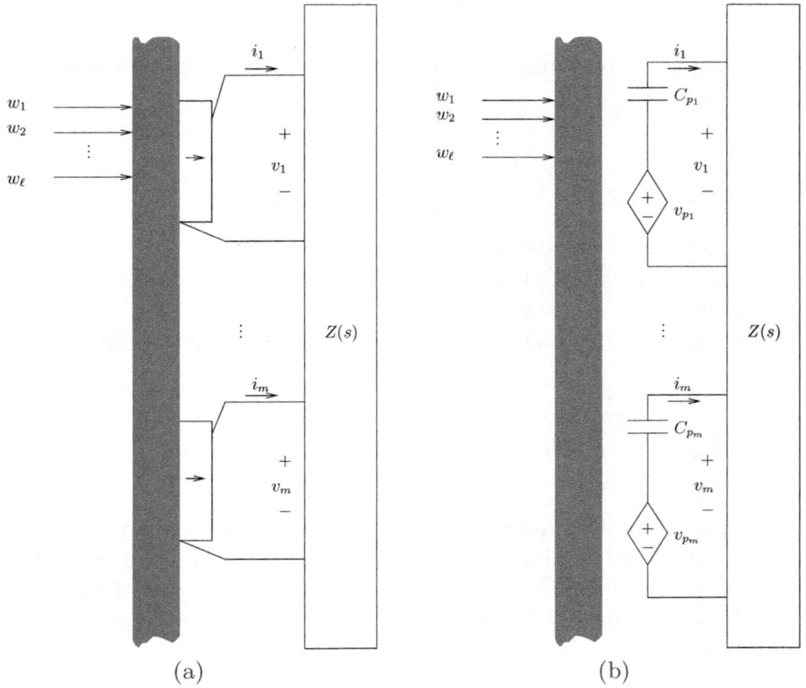

Figure 7.1. A piezoelectric laminated structure shunted to a multi-port impedance: (a) physical configuration; (b) electrical equivalence

Furthermore, writing the KVL around the kth loop we obtain

$$v_k(s) + v_{p_k}(s) + \frac{1}{C_{p_k} s} i_k(s) = 0$$

which implies

$$V(s) + V_p(s) + \frac{1}{s} \Lambda I_z(s) = 0, \tag{7.3}$$

where Λ is the following diagonal matrix

$$\Lambda = \begin{bmatrix} \frac{1}{C_{p1}} & & & \\ & \frac{1}{C_{p2}} & 0 & \\ & 0 & \ddots & \\ & & & \frac{1}{C_{pm}} \end{bmatrix}. \tag{7.4}$$

To capture the total effect of the disturbances as well as the effect of the electric shunt on the structure, we may write

$$V_p(s) = G_{vv}(s)V(s) + G_{vw}(s)W(s). \tag{7.5}$$

Here, $G_{vv}(s)$ is the multivariable collocated transfer function matrix of the system; *i.e.*

$$G_{vv}(s) = \sum_{k=1}^{M} \frac{\psi_k \psi_k'}{s^2 + 2\zeta_k \omega_k s + \omega_k^2}, \tag{7.6}$$

where resonance frequencies are ordered such that $\omega_1 \leq \omega_2 \leq \ldots \leq \omega_M$ and M can be an arbitrarily large number. In Section 2.9 we illustrated how such a model could be derived.

Next, Equations (7.2), (7.3) and (7.5) are combined to obtain,

$$V_p(s) = \left[I + G_{vv}(s)Z(s) \left(Z(s) + \frac{1}{s}\Lambda \right)^{-1} \right]^{-1} G_{vw}(s)W(s). \tag{7.7}$$

Transfer function matrix (7.7) amounts to the multivariable feedback structure depicted in Figure 7.2. In this system $G_{vv}(s)$ is a collocated transfer function matrix, hence it is a square system. Furthermore, the cascade structure of the system can be simplified by viewing the feedback connection of the impedance transfer function matrix and the integrator as a multivariable controller, *i.e.*

$$K(s) = Z(s) \left(Z(s) + \frac{1}{s}\Lambda \right)^{-1}. \tag{7.8}$$

This can also be directly inferred from equation (7.7). That is the transfer function matrix relating $W(s)$ to $V_p(s)$ is the feedback connection of $G_{vv}(s)$ with $K(s)$ above.

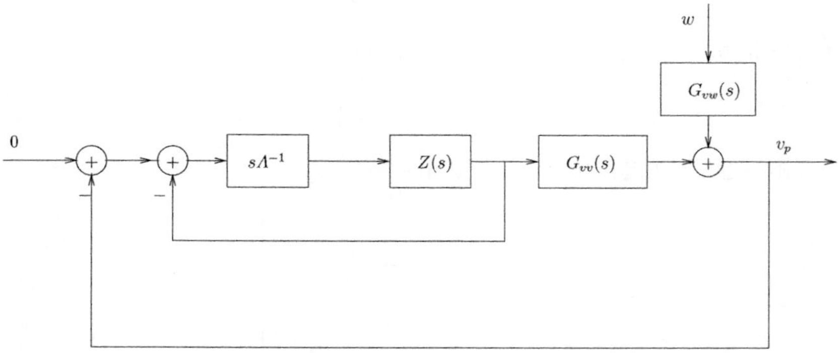

Figure 7.2. The feedback structure associated with the modified shunt damping problem

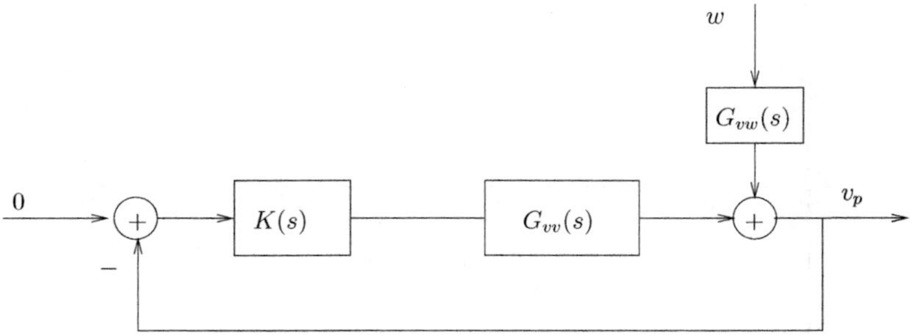

Figure 7.3. The equivalent of feedback system in Figure 7.2

The equivalent feedback control system is illustrated in Figure 7.3. Also the purpose of the system is to regulate V_p, in presence of disturbance W. The signal V_p, however, is not directly measurable. The reader may notice that, as in the SISO case, this is a very specific form of cascade feedback control structure (see Section 6.4 of [145]).

7.3 Stability of the Shunted System

A set of conditions under which stability of the closed-loop system depicted in Figure 7.2 is guaranteed, is derived in this section. Instead of considering the shunting impedance, $Z(s)$ as the controller, the closed-loop stability of the system is studied in terms of the shunted admittance, $Y(s) = Z(s)^{-1}$ noting that the closed-loop transfer function matrix in (7.7) can be re-written as:

$$V_p(s) = \left[I + G_{vv}(s) \left(I + \frac{1}{s} \Lambda Y(s) \right)^{-1} \right]^{-1} G_{vw}(s)\, W(s). \qquad (7.9)$$

The regulator problem associated with this system is depicted in Figure 7.4. A parameterization of stabilizing controllers for the system in (7.9) is introduced next.

Considering the structure of the feedback system, the Youla parameterization [121] of all stabilizing controllers for the inner feedback loop can be written as

$$Y(s) = (I - Q(s)\Lambda/s)^{-1}\, Q(s).$$

Although the inner loop contains integrators, the parameterization for a stable plant can be used as long as $Q(s)$ satisfies a number of conditions. Namely, $Q(s)$ must be stable, proper and have a transmission zero at the origin. Furthermore, $I - Q(s)\Lambda/s$ must have a transmission zero at $s = 0$. These conditions can be enforced by choosing

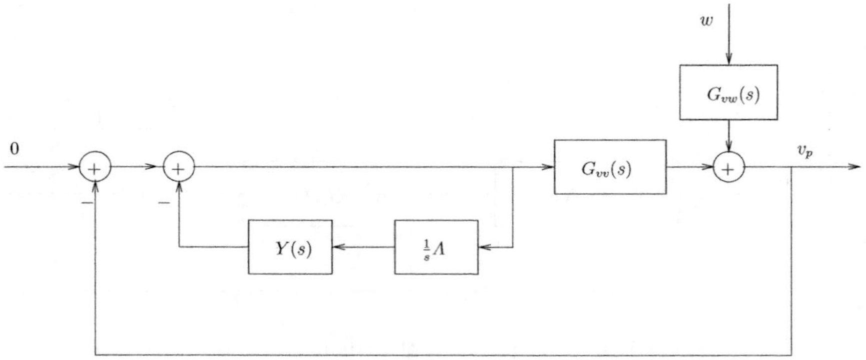

Figure 7.4. The feedback structure associated with the shunt damping problem with admittance as the control variable

$$Q(s) = H(s)\Lambda^{-1}s,$$

where $H(s)$ is stable, strictly proper and $I - H(s)$ has a transmission zero at the origin, *i.e.*

$$I - H(s) = sJ(s).$$

This choice for $Q(s)$ results in a closed-loop system with the transfer function matrix

$$\left[I + s\, G_{vv}(s)J(s)\right]^{-1} G_{vw}(s). \tag{7.10}$$

It is now possible to find closed-loop stability conditions in terms of $J(s)$ as the stability of (7.10) is equivalent to that of the system depicted in Figure 7.5.

It turns out that the closed-loop system will be stable as long as $J(s)$ is a strictly positive real transfer function matrix. The following two definitions and the subsequent theorem due to reference [99] are needed in the proofs.

Definition 7.1. *An $m \times m$ rational matrix $G(s)$ is said to be positive real (PR) if*

1. *All elements of $G(s)$ are analytic in $\mathrm{Re}(s) > 0$;*
2. *$G(s) + G^*(s) \geq 0$ in $\mathrm{Re}(s) > 0$ or equivalently*
   a) *Poles on the imaginary axis are simple and have nonnegative residues, and*
   b) *$G(j\omega) + G^*(j\omega) \geq 0$ for $\omega \in (-\infty, \infty)$.*

Definition 7.2. *An $m \times m$ stable rational matrix $G(s)$ is said to be strictly positive real in the weak sense (WSPR) if*

$$G(j\omega) + G^*(j\omega) > 0 \quad \textit{for} \quad \omega \in (-\infty, \infty).$$

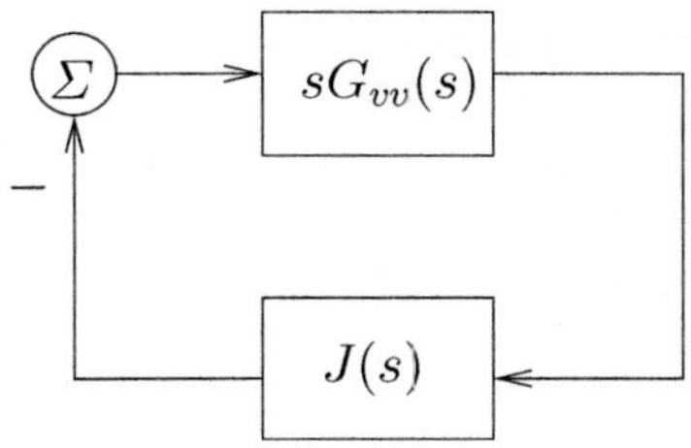

Figure 7.5. Feedback connection of $sG_{vv}(s)$ with $J(s)$

The following theorem is Corollary 1.1 of [99].

Theorem 7.3. *The negative feedback connection of a PR system with a WSPR controller is stable.*

It should be pointed out that there are a number of definitions in existing literature for strictly positive real systems. For a comprehensive review of these, the reader is referred to [194, 99]. For almost all such definitions, one would expect a similar result to that of Theorem 7.3, *i.e.* the negative feedback connection of a PR system with a SPR controller is stable. It turns out that for the problem at hand, Definition 7.2 is the most relevant.

Now we prove that

$$\tilde{G}_{vv}(s) = s\ G_{vv}(s) \tag{7.11}$$

is a positive real transfer function matrix. It can be noticed from (7.11) and (7.6) that all of the poles of $\tilde{G}_{vv}(s)$ are in the left half of the complex plane. Hence, the system is stable. Furthermore, the system has no poles on the $j\omega$ axis. To prove positive realness of $\tilde{G}_{vv}(s)$, we need to establish that $\tilde{G}_{vv}(j\omega) + \tilde{G}^*_{vv}(j\omega) \geq 0$ for all $\omega \in (-\infty, \infty)$.

$$\begin{aligned}
&\tilde{G}_{vv}(j\omega) + \tilde{G}^*_{vv}(j\omega) = \\
&\quad \sum_{k=1}^{N} \left\{ \frac{j\omega\psi_k\psi_k'}{\omega_k^2 - \omega^2 + j2\zeta_k\omega_k\omega} + \frac{-j\omega\psi_k\psi_k'}{\omega_k^2 - \omega^2 + -j2\zeta_k\omega_k\omega} \right\} \\
&= \sum_{k=1}^{N} \frac{4\zeta_k\omega_k\omega^2\psi_k\psi_k'}{(\omega_k^2-\omega^2)^2 + (2\zeta_k\omega_k\omega)^2} \\
&\geq 0 \qquad\qquad\qquad\qquad \text{for all} \ \ \omega \in (-\infty, \infty).
\end{aligned}$$

An implication of the above analysis is that to guarantee the closed-loop stability of the system it would suffice to choose an admittance

$$Y(s) = J(s)^{-1}\,(I - s\ J(s))\,\Lambda^{-1}, \tag{7.12}$$

with $J(s)$ a WSPR transfer function matrix.

7.4 Multivariable Shunts

The observation made in the previous section enables us to design impedance structures that guarantee closed-loop stability of the shunted system. This section introduces two specific multivariable structures that enforce the above conditions.

The two impedances are constructed using

$$J_\alpha = \sum_{i=1}^{\tilde{N}} \frac{s\alpha_i\alpha_i'}{s^2 + 2d_i\tilde{\omega}_i s + \tilde{\omega}_i^2} \tag{7.13}$$

and

$$J_\beta = \sum_{i=1}^{\tilde{N}} \frac{(s + 2d_i\tilde{\omega}_i)\beta_i\beta_i'}{s^2 + 2d_i\tilde{\omega}_i s + \tilde{\omega}_i^2}, \tag{7.14}$$

where

$$\alpha_i\alpha_i' = I \tag{7.15}$$

and

$$\beta_i\beta_i' = I. \tag{7.16}$$

It can be verified that both J_α and J_β are strictly proper and WSPR, that is, $J_\alpha(j\omega) + J_\alpha(j\omega)^* > 0$ and $J_\beta(j\omega) + J_\beta(j\omega)^* > 0$ for all $\omega \in (-\infty, \infty)$. Therefore, the resulting shunted system will be stable.

Now, using (7.13)-(7.16) and (7.12), we can obtain the corresponding shunts, $Y_\alpha(s)$ and $Y_\beta(s)$:

$$Y_\alpha(s) = \left(I - \sum_{i=1}^{N} \frac{(2d_i\tilde{\omega}_i s + \tilde{\omega}_i^2)\alpha_i\alpha_i'}{s^2 + 2d_i\tilde{\omega}_i s + \tilde{\omega}_i^2}\right)^{-1} \left(\sum_{i=1}^{N} \frac{(2d_i\tilde{\omega}_i s + \tilde{\omega}_i^2)\alpha_i\alpha_i'}{s^2 + 2d_i\tilde{\omega}_i s + \tilde{\omega}_i^2}\right) \Lambda^{-1}s \tag{7.17}$$

and

$$Y_\beta(s) = \left(I - \sum_{i=1}^{N} \frac{\tilde{\omega}_i^2\beta_i\beta_i'}{s^2 + 2d_i\tilde{\omega}_i s + \tilde{\omega}_i^2}\right)^{-1} \left(\sum_{i=1}^{N} \frac{\tilde{\omega}_i^2\beta_i\beta_i'}{s^2 + 2d_i\tilde{\omega}_i s + \tilde{\omega}_i^2}\right) \Lambda^{-1}s \tag{7.18}$$

Note that if only one piezoelectric transducer is shunted by an impedance, then the SISO shunts of Section 5.5 can be recovered. There is a clear advantage in using multi-port shunts. Due to the spatially distributed nature of flexible structures, it may not be possible to optimally place a piezoelectric transducer in a way that all necessary vibration modes can be controlled in an acceptable manner. Distributing a number of transducers on the structure, however, may enable one to position transducers such that all important modes are acceptably controllable.

One of the interesting properties of the above admittance transfer functions is that over a specific bandwidth, one has the option of choosing to control

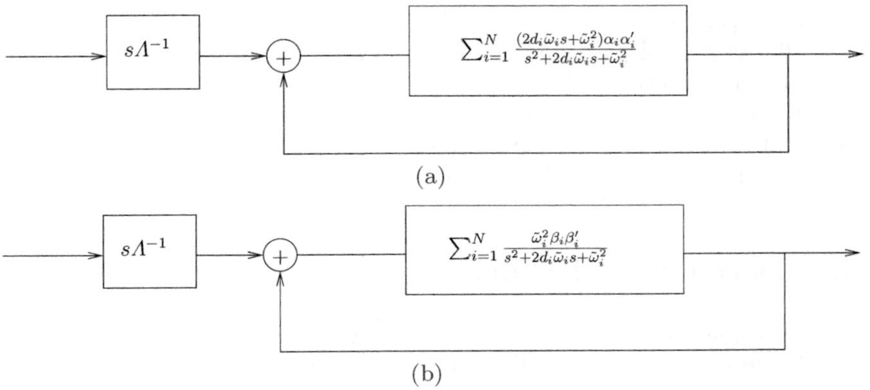

Figure 7.6. Feedback implementation of: (a) Y_α; and (b) Y_β

only those modes that are of importance. This is reflected in the constraint on parameters α_i and β_i in (7.15) and in (7.16), *e.g.* if mode ℓ is not to be controlled, then $\alpha_\ell \equiv 0$, or $\beta_\ell \equiv 0$, while Conditions (7.15) and (7.16) are to be satisfied for the remainder of modes. This is in contrast to control design methodologies such as LQG and H_∞ where the controller, at least, tends to have equal dimension to that of the system that is being controlled.

A further property of the controllers Y_α and Y_β is that in presence of out-of-bandwidth modes of the base structure they do not cause instabilities. As already explained, the spill-over effect [12, 13] is a serious cause of concern in control design for flexible structures. Often a feedback controller is designed using a model of the structure that contains a limited number of modes. Once the controller is implemented on the full order system, the presence of uncontrolled high frequency modes may destabilize the closed-loop system, or severely deteriorate the performance. Considering the discussion in Section 7.3, it should be clear that such a problem cannot happen if the above procedure is followed. Stability of the shunted system with the above class of shunts is guaranteed. Furthermore, due to their highly localized nature, these shunts have the additional property that their effect on the out-of-bandwidth modes of the system is minimal, hence minimizing the spill-over effect.

Now, it is straightforward but tedious to verify that both $Y_\alpha(s)$ and $Y_\beta(s)$ are strictly positive real transfer functions. Therefore, they can be realized by multi-port passive circuit components, *i.e.* resistors, inductors and capacitors. However, it is not clear how such a network may be obtained as standard synthesis techniques result in realizations that require Gyrators and op-amps. To this end, it should be pointed out that even if passive realizations for (7.17) and (7.18) are found, in practice, such an implementation is likely to be impractical. Given that often low frequency modes of a structure are targeted for shunt damping, the required inductors may be excessively large, in the order of several hundred to several thousand Henries. A practical way of implementing Y_α and Y_β is to use the synthetic admittance circuit as described in [59, 60] (and in Section 4.4) or the alternative and more effective method explained in Section 7.6.3.

If a multivariable impedance such as (7.17) or (7.18) is to be implemented digitally, it would be enormous help to identify the feedback structure of each shunt. These are shown in Figures 7.6 (a) and (b).

7.5 Decentralized Shunts

It is possible to modify the structure of the multivariable shunts, as proposed in the previous section, to allow for a decentralized structure. In certain applications, it may be easier and more straightforward to shunt each piezoelectric transducer by an impedance, rather than shunting all of them to a multi-port impedance. In particular, if online tuning of the impedance is needed, a decentralized impedance structure could be quite appealing. Two possibilities are explored below:

$$J_a(s) = \sum_{i=1}^{N} \text{diag}\left(\frac{\alpha_{1i} s}{s^2 + 2d_{1i}\tilde{\omega}_i s + \tilde{\omega}_i^2}, \frac{\alpha_{2i} s}{s^2 + 2d_{2i}\tilde{\omega}_i s + \tilde{\omega}_i^2} \right. \\ \left. , \dots, \frac{\alpha_{mi} s}{s^2 + 2d_{mi}\tilde{\omega}_i s + \tilde{\omega}_i^2} \right) \tag{7.19}$$

and

$$J_b(s) = \sum_{i=1}^{N} \text{diag}\left(\frac{\alpha_{1i}(s + 2d_{1i}\tilde{\omega}_i)}{s^2 + 2d_{1i}\tilde{\omega}_i s + \tilde{\omega}_i^2}, \frac{\alpha_{2i}(s + 2d_{2i}\tilde{\omega}_i)}{s^2 + 2d_{2i}\tilde{\omega}_i s + \tilde{\omega}_i^2} \right. \\ \left. , \dots, \frac{\alpha_{mi}(s + 2d_{mi}\tilde{\omega}_i)}{s^2 + 2d_{mi}\tilde{\omega}_i s + \tilde{\omega}_i^2} \right), \tag{7.20}$$

where, in both cases,

$$\alpha_{qi} \geq 0, \quad i = 1, 2 \dots, N \quad q = 1, 2, \dots, m \tag{7.21}$$

and

$$\sum_{i=1}^{N} \alpha_{qi} = 1, \quad q = 1, 2, \dots, m. \tag{7.22}$$

It can be verified that both $J_a(s)$ and $J_b(s)$ are strictly proper WSPR systems. Hence, the resulting admittances will guarantee closed-loop stability of the system.

Corresponding to $J_a(s)$ and $J_b(s)$, the expressions for $Y_a(s)$ and $Y_b(s)$ can be determined as:

$$Y_a(s) = \mathrm{diag}\left(\frac{\sum_{i=1}^{N} \frac{\alpha_{1i}(2d_{1i}\tilde{\omega}_i s+\tilde{\omega}_i^2)}{s^2+2d_{1i}\tilde{\omega}_i s+\tilde{\omega}_i^2}}{1-\sum_{i=1}^{N} \frac{\alpha_{1i}(2d_{1i}\tilde{\omega}_i s+\tilde{\omega}_i^2)}{s^2+2d_{1i}\tilde{\omega}_i s+\tilde{\omega}_i^2}}, \frac{\sum_{i=1}^{N} \frac{\alpha_{2i}(2d_{2i}\tilde{\omega}_i s+\tilde{\omega}_i^2)}{s^2+2d_{2i}\tilde{\omega}_i s+\tilde{\omega}_i^2}}{1-\sum_{i=1}^{N} \frac{\alpha_{2i}(2d_{2i}\tilde{\omega}_i s+\tilde{\omega}_i^2)}{s^2+2d_{2i}\tilde{\omega}_i s+\tilde{\omega}_i^2}}, \ldots, \frac{\sum_{i=1}^{N} \frac{\alpha_{mi}(2d_{mi}\tilde{\omega}_i s+\tilde{\omega}_i^2)}{s^2+2d_{mi}\tilde{\omega}_i s+\tilde{\omega}_i^2}}{1-\sum_{i=1}^{N} \frac{\alpha_{mi}(2d_{mi}\tilde{\omega}_i s+\tilde{\omega}_i^2)}{s^2+2d_{mi}\tilde{\omega}_i s+\tilde{\omega}_i^2}} \right) \Lambda^{-1} s \tag{7.23}$$

and

$$Y_b(s) = \mathrm{diag}\left(\frac{\sum_{i=1}^{N} \frac{\alpha_{1i}\tilde{\omega}_i^2}{s^2+2d_{1i}\tilde{\omega}_i s+\tilde{\omega}_i^2}}{1-\sum_{i=1}^{N} \frac{\alpha_{1i}\tilde{\omega}_i^2}{s^2+2d_{1i}\tilde{\omega}_i s+\tilde{\omega}_i^2}}, \frac{\sum_{i=1}^{N} \frac{\alpha_{2i}\tilde{\omega}_i^2}{s^2+2d_{2i}\tilde{\omega}_i s+\tilde{\omega}_i^2}}{1-\sum_{i=1}^{N} \frac{\alpha_{2i}\tilde{\omega}_i^2}{s^2+2d_{2i}\tilde{\omega}_i s+\tilde{\omega}_i^2}}, \ldots, \frac{\sum_{i=1}^{N} \frac{\alpha_{mi}\tilde{\omega}_i^2}{s^2+2d_{mi}\tilde{\omega}_i s+\tilde{\omega}_i^2}}{1-\sum_{i=1}^{N} \frac{\alpha_{mi}\tilde{\omega}_i^2}{s^2+2d_{mi}\tilde{\omega}_i s+\tilde{\omega}_i^2}} \right) \Lambda^{-1} s. \tag{7.24}$$

One of the advantages of decentralized shunts is that the damping parameter associated with each mode can be controlled according to the piezoelectric transducer to which the impedance is shunted. The full multivariable shunts of the previous section do not possess this property. Although, it may be possible to compensate for this using the cross-diagonal terms in (7.17) and (7.18).

7.6 Experimental Results

To validate the proposed concepts, experiments were carried out on a piezoelectric laminated beam. This section contains details of these experiments.

7.6.1 Experimental Setup

The test structure is a uniform aluminum beam with rectangular cross section and experimentally pinned boundary conditions. Two pairs of collocated piezoelectric patches are attached symmetrically to either side of the structure as shown in Figures 7.7 and 7.8. Piezoelectric transducers used in our experiments are PIC151[1] piezoelectric patches. Details of the beam and PIC151 piezoelectric patches are listed in Tables 7.1 and 7.2.

[1] These patches are manufactured by *Polytec PI Ceramics*.

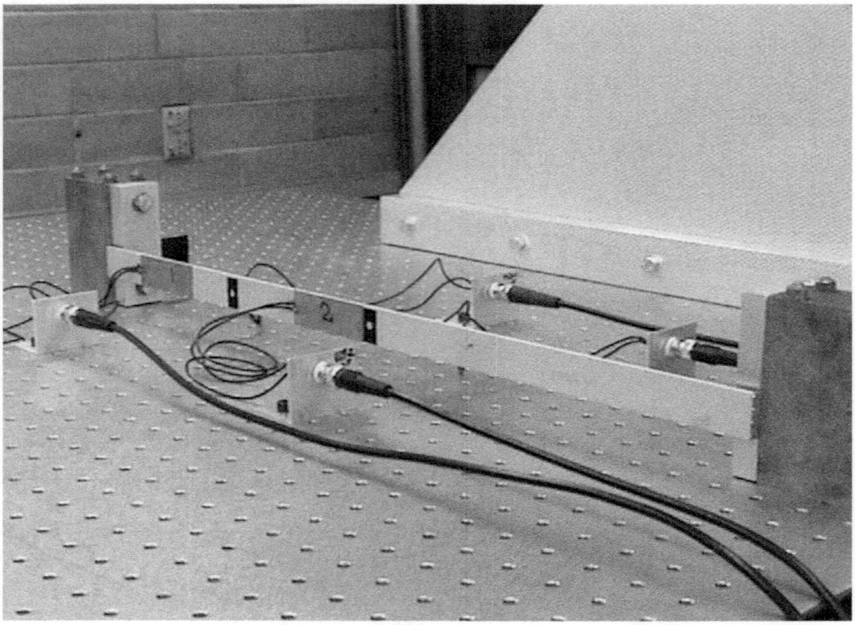

Figure 7.7. Experimental beam apparatus

The first pair of piezoelectric patches is placed close to one of the pinned ends of the beam, while the other pair is closer to the beam's center. Mode shapes of a simply-supported beam are sinusoidal functions [130]. Consequently, the first pair will be more effective in controlling the first vibration mode of the beam, while the second mode is better controlled by the second pair. Both transducers are effective in controlling the third mode. One of the main advantages of multivariable shunt damping, as evident from this application, is that if one transducer does not offer enough authority over a specific mode, another transducer may enable the designer to control that mode more effectively.

Table 7.1. Parameters of the simply-supported beam

Name	Symbol	Unit
Length	L	0.6 m
Width	w	0.025 m
Thickness	h	0.004 m
Young's modulus	E	65×10^9 N/m^2
Poisson's ratio	ν	0.3
Mass / unit area	ρ	10.6 kg/m^2

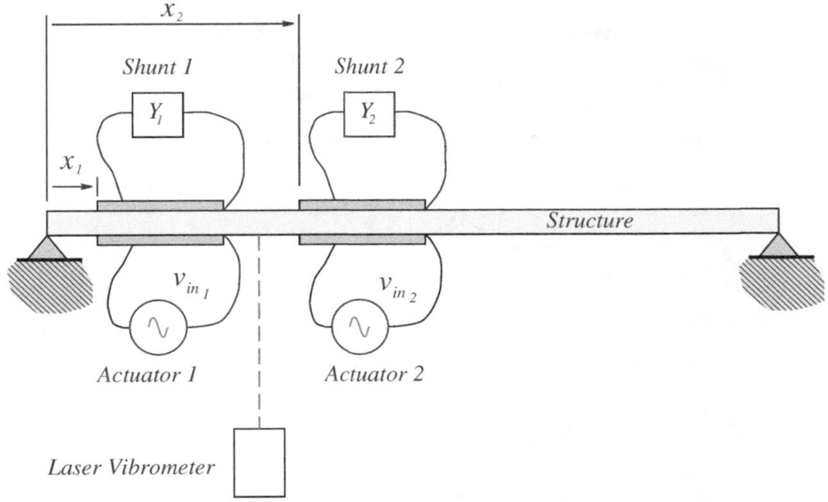

Figure 7.8. Beam apparatus

Table 7.2. PIC151 piezoelectric patch parameters

Name	Symbol	Unit
Location x-direction	x_1	$0.050\ m$
Location x-direction	x_2	$0.240\ m$
Length	L_p	$0.0724\ m$
Thickness	h_p	$0.00191\ m$
Width	w_p	$0.025\ m$
Capacitance	C_p	$471 \times 10^{-9}\ F$
Young's modulus	E_p	$62 \times 10^{9}\ N/m^2$
Poisson's ratio	ν_p	0.3
Strain Constant	d_{31}	$-320 \times 10^{-12}\ m/V$
Electromechanical coupling factor	k_{31}	0.44
Stress constant / voltage coefficient	g_{31}	$-9.5 \times 10^{-3}\ V\ m/N$

7.6.2 System Identification

The first step in the analysis involves procuring a model for the transfer function matrix $G_{vv}(s)$. This will enable us to simulate the effect of an attached piezoelectric shunt on the transfer function from the applied actuator voltages, $V_{in}(s)$, to the generated piezoelectric shunt layer voltages $V_p(s)$. These variables are internal and cannot be measured directly whilst an impedance is attached to the shunting layer. We will also consider the transfer function from the applied actuator voltages, $V_{in}(s)$, to the structural deflection at a point $D(x, s)$. This point is chosen such that the first three modes of the beam

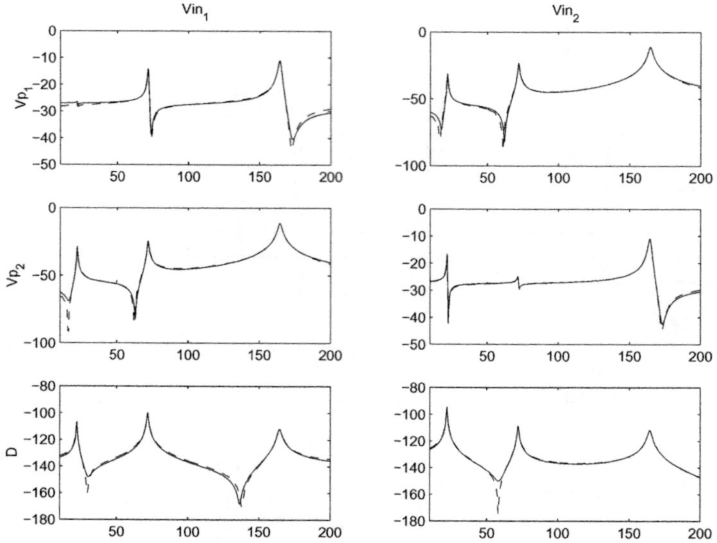

Figure 7.9. Experimental data (- -) and identified model (—). Horizontal axes are in Hz, vertical axes in dB. All the voltages are measured in volts. The displacement is measured in meters.

are observable at that location. In the case where there are two actuators and two sensors, we require a model with two inputs and three outputs

$$\begin{bmatrix} V_p(s) \\ D(x,s) \end{bmatrix} = G_p(s)\ V_{in}(s), \tag{7.25}$$

where $G_p(j\omega) \in \mathbf{C}^{3\times 2}$ is the open-loop plant transfer function matrix, V_p is the 2×1 vector of sensor voltages, V_{in} is the 2×1 vector of actuator voltages, and $D(x, s)$ represents the displacement measured at a point between the two piezos, as illustrated in Figure 7.8.

We have already indicated that modeling of piezoelectric laminate structures is generally accomplished in the literature by means of either analytic modeling, finite element analysis or system identification. Although, in Chapter 2 we demonstrated that for a simple structure such as a beam, an analytical model can be determined with relative ease, such models often need to be fine tuned using experimental data.

In this experiment, the continuous-time frequency domain system identification algorithm proposed of Van Overschee and De Moor [186] is employed. A full review and discussion of the various options for system identification can be found in Section 3.3.

We can now discuss the experiment performed to obtain a state-space representation of $G_p(s)$. Referring to Figure 7.8, a pre-filtered periodic chirp was applied to each actuating layer in succession, while the voltage applied to the

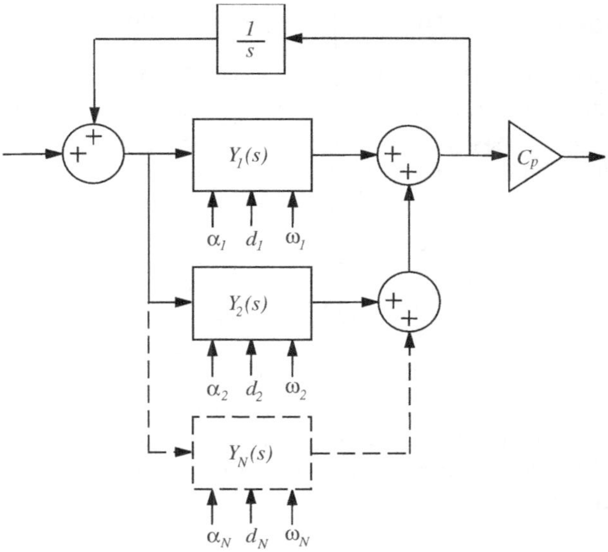

Figure 7.10. System diagram of (7.23) or (7.24)

other patch was set to zero. The resulting open circuit piezoelectric voltages and displacements were recorded using a dSPACE DS1103 rapid prototyping system. The chirp pre-filtering was performed by an FIR filter designed to reduce the power of the excitation in bands enclosing the resonance frequencies of the structure. This reduces the dynamic range and 'flattens out' the power spectral density and signal to noise ratio versus frequency at the outputs. An estimate for $G_p(j\omega)$ was then obtained using the empirical transfer function estimate [120]. The magnitude frequency response of $G_p(j\omega)$ is plotted in Figure 7.9. 340 frequency samples from 0 to 200 Hz were used to identify a 6th order state model for $G_p(s)$. The magnitude frequency response of the model is overlain on the experimental data in Figure 7.9.

7.6.3 Implementation of a Multi-port Synthetic Admittance

The synthetic admittance was introduced in Section 4.4.2 for the implementation of piezoelectric shunt damping circuits. One of the benefits of this technique is that the implemented admittance is not restricted to a network of physical components. Thus, admittance transfer functions such as (7.23) and (7.24), that do not have direct circuit interpretations, can be implemented in a straightforward manner. Further discussion on synthetic admittance circuits, and details on constructing a voltage controlled current source can be found in Sections 6.6 and 6.4 respectively.

7.6.4 Implementing the Admittance Transfer Function

On first inspection, the admittance transfer functions (7.23) and (7.24) may appear difficult to implement by means of either analog or digital signal processing. In fact the reverse is true, as the transfer function can be represented as a simple block diagram composed of second order subsystems. Consider the admittance required for a single piezoelectric transducer using the controller

$$Y_a(s) = \frac{\sum_{i=1}^{N} \frac{\alpha_i \omega_i^2}{s^2+2d_i\omega_i s+\omega_i^2}}{1-\sum_{i=1}^{N} \frac{\alpha_i \omega_i^2}{s^2+2d_i\omega_i s+\omega_i^2}} C_p s. \tag{7.26}$$

The structure (7.26) is shown diagrammatically in Figure 7.10. Each subsystem $Y_i(s, \alpha_i, d_i, \omega_i)$, parameterized for ease of online tuning, can be implemented by an analog state variable filter [91], or internally in a DSP algorithm. For digital implementation, each subsystem is most easily parameterized in state space form. For example,

$$Y_i(s, \alpha_i, d_i, \omega_i) = \frac{y_i}{u} = \frac{\alpha_i \omega_i^2 s}{s^2 + 2d_i\omega_i s + \omega_i^2}, \tag{7.27}$$

where

$$\begin{aligned} \dot{x} &= \begin{bmatrix} 0 & 1 \\ -\omega_i^2 & -2d_i\omega_i^2 \end{bmatrix} x + \begin{bmatrix} 0 \\ 1 \end{bmatrix} u \\ y_i &= \begin{bmatrix} 0 & \alpha_i \omega_i^2 \end{bmatrix} x. \end{aligned} \tag{7.28}$$

7.6.5 Results

In the experiments, one of the actuating piezoelectric transducers was used to disturb the structure. The two transducers on the other side of the beam were shunted with resonant impedances to attenuate the vibrations generated in the beam. Using the structure $Y_a(s)$ (Equation 7.23), shunt circuits were applied to both of the piezoelectric transducers. Specifically, $Y_1(s)$ that was shunted to the first piezoelectric transducer and tuned to control the $2nd$ and $3rd$ modes, and $Y_2(s)$ that was shunted to the second piezoelectric transducer and tuned to control the $1st$ and $3rd$ modes. Admittance parameters are shown in Table 7.3. Parameters $\alpha_{1,2}$, $\alpha_{1,3}$, $\alpha_{2,1}$ and $\alpha_{2,3}$ are set equal to 0.5 to satisfy Condition (7.22), and to put equal emphasis on every mode. The rest of the parameters are determined via the solution to an optimization problem.

The admittance has a diagonal structure

$$Y_a(s) = \mathrm{diag}\left(Y_1(s)\ ,\ Y_2(s)\right), \tag{7.29}$$

where

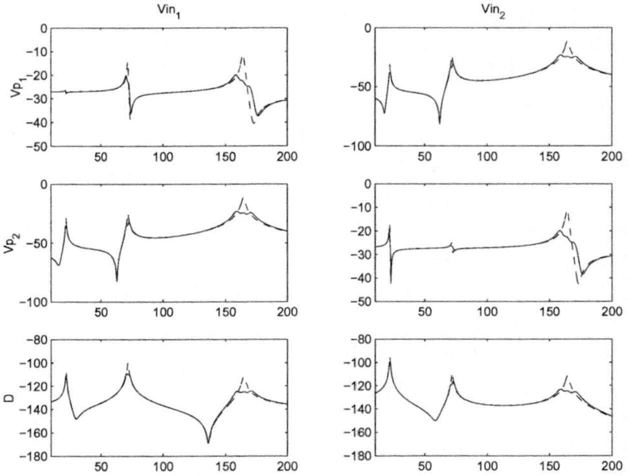

Figure 7.11. Simulated open- (- -) and closed-loop(—) magnitude frequency response. Horizontal axes are in Hz, vertical axes in dB. All the voltages are measured in volts. The displacement is measured in meters.

$$Y_1(s) = \frac{\left(\frac{\alpha_{1,2}\omega_{1,2}^2}{s^2+2d_{1,2}\omega_{1,2}s+\omega_{1,2}^2} + \frac{\alpha_{1,3}\omega_{1,3}^2}{s^2+2d_{1,3}\omega_{1,3}s+\omega_{1,3}^2}\right)}{1-\left(\frac{\alpha_{1,2}\omega_{1,2}^2}{s^2+2d_{1,2}\omega_{1,2}s+\omega_{1,2}^2} + \frac{\alpha_{1,3}\omega_{1,3}^2}{s^2+2d_{1,3}\omega_{1,3}s+\omega_{1,3}^2}\right)} C_p s,$$

and

$$Y_2(s) = \frac{\left(\frac{\alpha_{2,1}\omega_{1,1}^2}{s^2+2d_{2,1}\omega_{1,1}s+\omega_{1,1}^2} + \frac{\alpha_{2,3}\omega_{1,3}^2}{s^2+2d_{2,3}\omega_{1,3}s+\omega_{1,3}^2}\right)}{1-\left(\frac{\alpha_{2,1}\omega_{1,1}^2}{s^2+2d_{2,1}\omega_{1,1}s+\omega_{1,1}^2} + \frac{\alpha_{2,3}\omega_{1,3}^2}{s^2+2d_{2,3}\omega_{1,3}s+\omega_{1,3}^2}\right)} C_p s.$$

Table 7.3. Admittance parameters

Parameter	**Value (Hz)**
$\omega_{1,2}$	71.7
$\omega_{1,3}$	161.6
$\omega_{2,1}$	22.14
$\omega_{2,3}$	167.9

Parameter	Value
$d_{1,2}$	0.021
$d_{1,3}$	0.024
$d_{2,1}$	0.025
$d_{2,3}$	0.023

Parameter	Value
$\alpha_{1,2}$	0.5
$\alpha_{1,3}$	0.5
$\alpha_{2,1}$	0.5
$\alpha_{2,3}$	0.5

Figure 7.11 compares the simulated frequency response of the unshunted system with that of the shunted system. The figure is associated with a 2×3 system whose inputs are the voltages applied to the two actuating piezoelectric patches and whose outputs are the induced piezoelectric voltages and the displacement measurement at $x = 0.17\ m$. The displacement measurements were obtained using a PSV300 laser scanning vibrometer.

Figure 7.12 demonstrates the effect of the proposed admittance structure. It can be observed that by shunting the two piezoelectric patches with the proposed admittances the closed-loop poles of the system have been pushed further into the left half of the complex plane. We notice that both transducers are used to dampen the third mode, while the first two modes are damped using the second and first transducers respectively. This is due to the location at which the two patches are mounted on the beam. The first patch offers little authority over the first mode of the beam, while the second patch displays a similar lack of authority over the second mode. Both transducers, however, are effectively reducing vibration corresponding to the third mode of the structure.

In experiments, variables internal to the piezoelectric transducers are not directly measurable. Therefore, it was not possible to generate experimental results corresponding to all entries of the transfer function matrix displayed in Figure 7.11. However, as the displacement could be measured, results were obtained by applying a disturbance voltage to the first piezoelectric transducer and measuring the resulting displacement. The corresponding transfer functions are plotted in Figure 7.13. It can be observed that the experimental results closely match the simulations.

Experimental results show a considerable attenuation of the resonant peaks; 5 dB for the $1st$ mode, 10.5 dB for the $2nd$ mode and 14.4 dB for the $3rd$ mode.

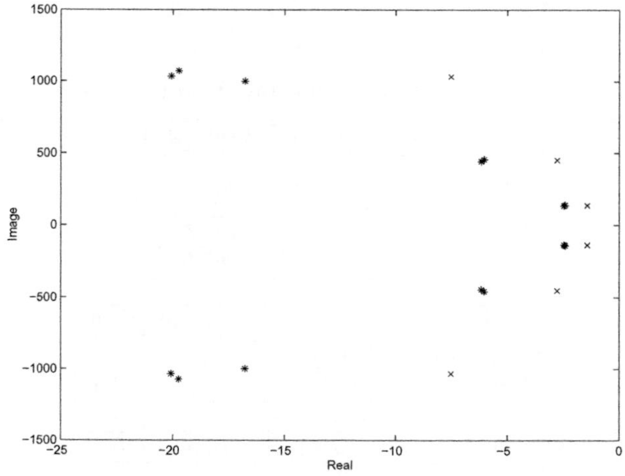

Figure 7.12. Simulated open- ($\times$) and closed-loop ($*$) pole locations

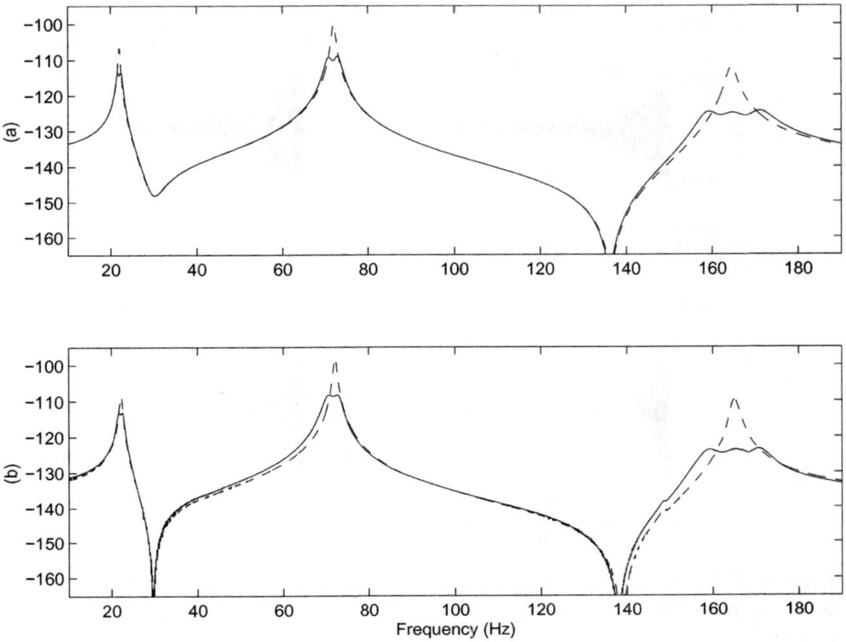

Figure 7.13. The simulated (a) and experimental (b) open-loop (- -) and closed-loop (—) frequency responses from V_{in_1} (volts) to the displacement measured at $x = 0.17\ m$ (meters) in dB.

To examine the time domain performance of the damped system, a 200 Hz low-pass filtered step was applied to V_{in_1}. The simulated and experimental displacement responses measured at $x = 0.17\ m$ are plotted in Figure 7.14. Note that the response is dominated by the first mode of vibration. This is a result of the lower damping achieved for this mode, and the comparatively greater low frequency Fourier components contained in a step function.

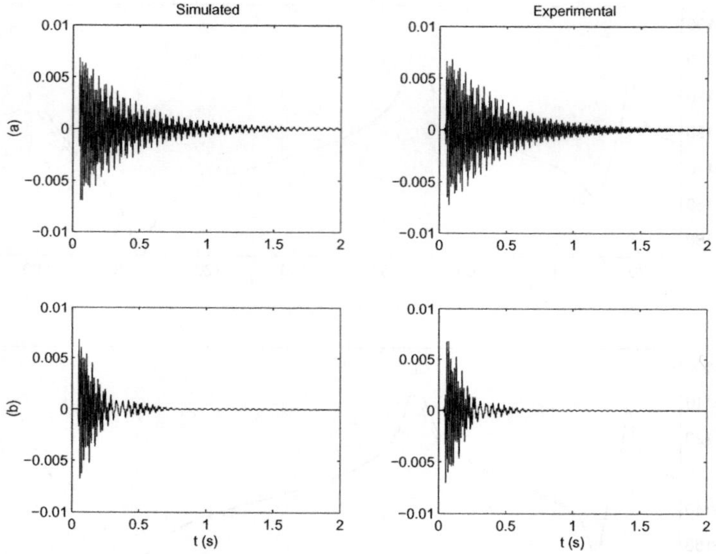

Figure 7.14. Open- (a) and closed-loop (b) displacement response at $x = 0.17\ m$ to a low-pass filtered step input applied to V_{in_1}. The displacement is measured in meters.

8

Adaptive Shunt Damping

8.1 Introduction

Resonant shunt damping circuits, introduced in Chapters 4 and 5, are simple in conception, achieve excellent damping performance and are practically implementable by synthetic admittance.

The greatest disadvantage associated with resonant shunt damping circuits is their high sensitivity to slight variations in the piezoelectric capacitance and structural resonance frequency. In most circumstances, temperature will be the greatest mitigating factor. The dielectric constant, and hence the piezoelectric capacitance, is known to vary significantly with temperature changes [86, 148, 18, 151]. A capacitance increase of 100% is possible over the temperature range of the transducer. As shunt circuit inductance values are inverse functions of capacitance, such variation can easily result in significant performance degradation. In [148], a simple passive circuit was shown to provide better performance than a resonant shunt when considering operation over a moderate temperature range. Although the resonant shunt circuit was able to achieve 14 dB vibration reduction, this drops to only 4 dB after a 40 degree increase in environmental temperature. The passive circuit, however, provides 6 dB attenuation regardless of temperature.

With the aim of eliminating dependence on environmental conditions, this chapter describes a technique for the online adaptation of multi-mode resonant piezoelectric shunt damping circuits.

The synthetic admittance was introduced in Chapter 6 as a simplified technique for the implementation of piezoelectric shunt impedances. As the impedance seen by the transducer can be determined by a digital signal filter, the component values are easily tuned online. A technique exploiting this flexibility was presented in [61], a multi-mode circuit was tuned online to minimize an RMS strain signal estimated from the terminal voltage. Although this method requires no vibration sensor, it is slow to converge and dependent on the disturbance spectrum.

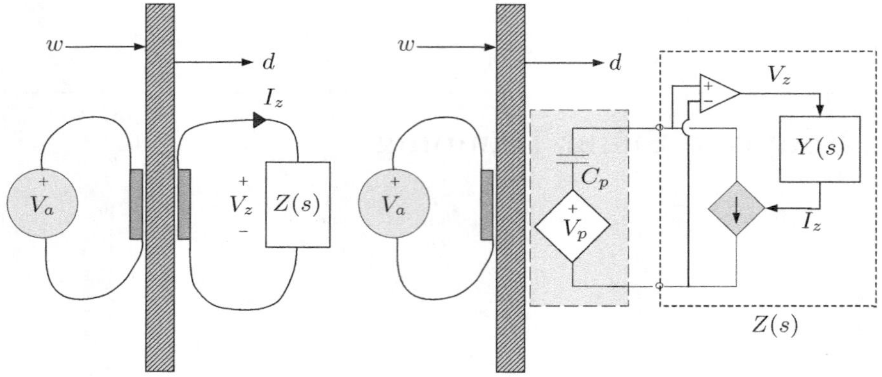

Figure 8.1. A structure disturbed by an applied actuator voltage V_a and external force w. The resulting vibration d is suppressed by the presence of a shunt impedance $Z(s)$. On the right, the shunt circuit is implemented by synthetic admittance as discussed in Section 4.4.

A new adaptive technique based on relative phase shift was presented in [151]. It was shown that optimal tuning occurs when the relative phase difference is minimized between a vibration reference signal and the shunt current. This technique is not based on a time averaged RMS estimate and is thus faster to converge and displays significantly less mis-adjustment at the minima. An extension presented in [150] considers the relative phase adaptation of multi-mode shunt circuits using a synthetic admittance, results from this work are summarized in the following.

8.2 Adaptation Law

In this section, the relative phase adaptation of multi-mode resonant shunts is described [150]. First, the relative phase adaptation of single-mode RL shunts is reviewed, then extended to multi-mode resonant shunts.

8.2.1 Adaptive Single-mode RL Shunt

As discussed in Chapter 5, an electrical shunt impedance can be viewed as parameterizing an equivalent collocated strain feedback controller. If the impedance $Z(s)$ is a series inductor-resistor network (RL shunt), a resonant controller is obtained that significantly reduces the vibration associated with a single structural mode [82]. As described in references [86, 148, 18, 151], the damping performance of resonant shunts is highly sensitive to environmental variation in the structural resonance frequencies and transducer capacitance. Online tuning of the inductance L is required to maintain optimal damping.

To this end, adaptive techniques have been proposed that are based on Root Mean Square (RMS) minimization [89] and relative phase shift [151]. These two approaches applied to an electromagnetic system are compared in [149]. In that instance, relative phase adaptation tunes faster with less misadjustment than the RMS methodology. In the following, a review and extension of relative phase adaptation is presented.

Relative phase adaptation is based on adjusting the relative phase difference between velocity and shunt current to $-\pi/2$. A simple multiplication and filter operation evaluates the relative phase difference. Consider Figure 8.1 with a single-mode resonant shunt, *i.e.* $Z(s) = R + sL$. The shunt current I_z is equal to

$$I_z(s) = V_p(s)\frac{sC_p}{s^2LC_p + sC_pR + 1}. \tag{8.1}$$

As V_p is dynamically proportional to the strain $x(s)$ experienced by the piezoelectric transducer, *i.e.* $V_p(s) = cx(s) = c\nu(s)/s$, where c is a constant and $\nu(s)$ is the velocity, the transfer function $G_{Iv}(s)$ from the velocity $\nu(s)$ to the current $I_z(s)$ can be expressed as

$$G_{I\nu}(s) = \frac{I_z(s)}{\nu(s)} = \frac{cC_p}{s^2LC_p + sC_pR + 1}. \tag{8.2}$$

The phase of $G_{I\nu}(j\omega)$ is

$$\angle\left(G_{I\nu}(j\omega)\right) = \phi_n = -tan^{-1}\left(\frac{\omega C_pR}{1 - LC_p\omega^2}\right). \tag{8.3}$$

According to [82], optimal tuning of the RL shunt is achieved when $\omega_n = 1/\sqrt{LC_p}$, where ω_n is the structural resonance frequency of the nth mode. From Equation 8.3, this tuning condition can be reformulated as $\angle G_{I\nu}(j\omega_n) = -\pi/2$. A function $f_p(L, \omega_n) = sign\left(\angle\left(G_{I\nu}(j\omega_n)\right) + \frac{\pi}{2}\right)$ can be defined that reveals the required tuning direction of the inductance value. The discrete adaptation of L that tunes the nth mode is given by

$$L_{k+1} = L_k + \alpha\cdot\left(f_p(L_k, \omega_n)\right) = L_k + \alpha\cdot sign\left(\angle\left(G_{I\nu}(j\omega_n) + \frac{\pi}{2}\right)\right),$$

where α is the tuning constant. A direct evaluation of the phase angle is not straightforward and complicates the adaptation scheme shown above. A practical alternative for evaluating the tuning direction is shown in the following. If we assume that the velocity $\nu(t)$ and current $I_z(t)$ are tonal, a reasonable assumption as the resonances are very lightly damped, the multiplication of $\nu(t) = sin(\omega_n t)$ with $I_z(t)$ can be written as

$$\nu(t)\cdot I_z(t) = sin(\omega_n t)\cdot A_n sin(\omega_n t + \phi_n),$$

where ϕ_n is the phase shift equal to $\angle\left(G_{I\nu}(j\omega_n)\right)$ and $A_n = |G_{I\nu}(j\omega_n)|$. After some manipulations, the following expression can be obtained

$$\nu(t) \cdot I_z(t) = A_n \left(\frac{1}{2} \left(cos(\phi_n) - cos(2\omega_n t + \phi_n) \right) \right). \tag{8.4}$$

After applying a low-pass filter with cutoff below $2\omega_n$ to the product of $\nu(t)$ and $I_z(t)$, the second term can be neglected and one obtains

$$g_{LP}(t) * [\nu(t) \cdot I_z(t)] = \frac{A_n}{2} \cdot \cos(\phi_n), \tag{8.5}$$

where $g_{LP}(t)$ represents the low-pass filter impulse response and $*$ denotes the time domain convolution operator. It can be seen that

$$sign\left(g_{LP}(t) * [\nu(t) \cdot I_z(t)]\right) = sign(\cos(\phi_n)) = sign\left(\angle\left(G_{I\nu}(j\omega_n)\right) + \frac{\pi}{2}\right),$$

for $-\frac{3\pi}{2} < \phi_n < \frac{\pi}{2}$. This technique constitutes a new means for evaluating the tuning direction. The discrete adaptation law can be rewritten as

$$L_{k+1} = L_k + \alpha \cdot sign\left(g_{LP}(t) * [\nu(t) \cdot I_z(t)]\right). \tag{8.6}$$

By removing the *sign* operator, effectively allowing the tuning rate to vary, the following continuous tuning law can be obtained

$$\frac{dL(t)}{dt} = \beta \left(g_{LP}(t) * [\nu(t) \cdot I_z(t)]\right), \tag{8.7}$$

where β is the tuning rate. Equation 8.7 is the relative phase adaptation law for single-mode RL shunts. A more detailed convergence analysis can be found in [150].

8.2.2 Adaptive Resonant Multi-mode Shunts

Multi-mode shunt damping circuits were developed to allow the control of multiple structural modes with a single piezoelectric transducer. An appraisal of the various techniques can be found in Chapter 4. Referring to Section 4.2.4, current-flowing circuits as pictured in Figure 8.2, are intuitive in design, provide high levels of damping and are low-in-order compared to other designs. These characteristics make them an ideal structure for adaptive multi-mode shunt damping.

The filter network $C_i L_{fi}$ allows current to flow only at the targeted modal frequency ω_i for $i = 1 \ldots n$, and at all other frequencies the branch appears approximately as an open circuit. The damping inductor and resistor $L_{di} R_i$ are designed analogous to a single-mode shunt circuit at ω_i. In Figure 8.2 (b) the circuit is simplified by combining the series inductors L_{fi} and L_{di} into L_i.

To account for variations in structural resonance frequencies and transducer capacitance, the inductive elements must be adapted online. In previous multi-mode techniques, the inductors have been tuned by minimizing a signal related to the RMS strain [61]; the branch inductor values were updated using

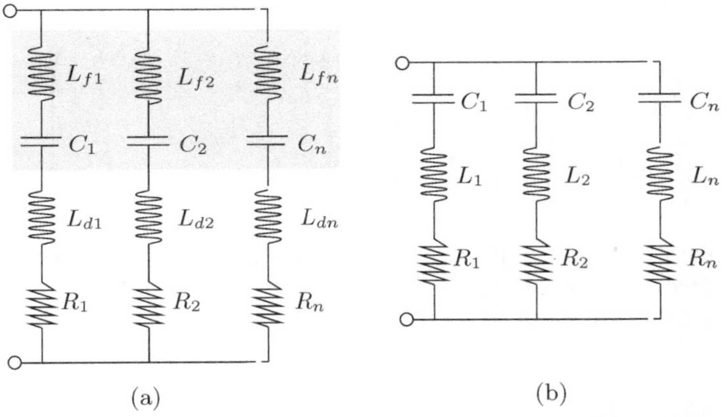

Figure 8.2. (a) Current-flowing circuit, (b) its simplification. The shaded area represents the band-bass filter.

a gradient search algorithm. As the performance function is rather flat around the optimum, this strategy is slow to converge and prone to misadjustment after the minimum has been reached. Relative phase adaptation can be applied to multi-mode shunt circuits as follows [150, 148].

Due to the filter network $C_i L_{fi}$, each circuit branch can be regarded independently. The filter network has zero impedance at $\omega = 1/\sqrt{C_i L_{fi}}$ and high impedance at all other frequencies. Therefore, the transfer function from structural velocity to the current flowing downwards in the ith branch can be written around the ith modal resonance frequency ($s \approx j\omega_i$) as

$$G_i(s) = \frac{I(s)}{\nu(s)} = \frac{c}{s} \cdot \frac{s/L_i}{s^2 + s\frac{R_i}{L_i} + \frac{C_i+C_p}{C_p C_i L_i}}, \tag{8.8}$$

where the electrical resonance frequency is

$$\omega_{i_{el}}^2 = \frac{C_i + C_p}{C_i C_p L_i} = \frac{1}{L_i C_{eq_i}}. \tag{8.9}$$

The phase of $G_i(j\omega)$ becomes

$$\angle(G_i(j\omega)) = -tan^{-1}\left(\frac{\omega R_i/(L_i)}{(C_i + C_p)/(C_i C_p L_i) - \omega^2}\right), \tag{8.10}$$

and is $-\pi/2$ for $\omega_i = 1/\sqrt{C_{eq_i} L_i}$ with $C_{eq_i} = (C_i + C_p)/(C_i C_p)$. Thus relative phase adaptation can be employed to tune each branch to the corresponding mechanical mode. As the impedance of each branch naturally bandpass filters the current, no additional filters for the current signal are necessary. Velocity bandpass filters are neither required as long as the mechanical resonance frequencies are not harmonic. The adaptation of the inductor vector $\bar{L}(s)$ is

$$\frac{\partial \bar{L}(s)}{\partial t} = \begin{pmatrix} \dot{L}_1(t) \\ \dot{L}_2(t) \\ \vdots \\ \dot{L}_n(t) \end{pmatrix} = \begin{pmatrix} \alpha_1 \cdot g_{LP}^1(t) * [\nu(t) \cdot I_1(t)] \\ \alpha_2 \cdot g_{LP}^2(t) * [\nu(t) \cdot I_2(t)] \\ \vdots \\ \alpha_n \cdot g_{LP}^n(t) * [\nu(t) \cdot I_n(t)] \end{pmatrix}, \tag{8.11}$$

where α_i is the ith tuning rate, and $g_{LP}^i(t)$ is the ith low-pass filter with a cut-off frequency below $2\omega_i$.

An analysis of convergence can be found in [150] or [148].

8.3 Experiments

In these experiments, the method of relative phase adaptation is employed to tune and maintain the optimal performance of a two- and four-mode resonant shunt damping circuit.

8.3.1 Implementation

The high-voltage current-source/voltage amplifier discussed in Section 6.6 was employed to implement the admittance. All filtering and processing tasks were performed by a dSPACE 1005 digital signal processing system.

Details on the disassembly of a current-flowing circuit into its constituent branch-admittances can be found in Section 6.8.4. Implementation of the adaptation law is simplified by the availability of the individual branch currents in software.

A Polytec PI PSV300 Laser Scanning Vibrometer was employed to measure the beam tip velocity. This signal is used for performance analysis and by the adaptation law. Alternate vibration reference sensors could include: accelerometers, strain gauges, additional piezoelectric transducers, in fact any sensor providing an output proportional to strain, displacement, velocity, or acceleration. As the adaptation law is based on velocity, the phase reference angle must be adjusted in the case of acceleration or strain measurement.

8.3.2 Test Structure

The experimental structure is a uniform aluminum beam with a rectangular cross section. The dimensions and physical properties are displayed in Figure 8.3 and Table 8.1. Two identical piezoelectric patches are laminated symmetrically onto the front and back faces of the beam. One patch acts as an artificial strain disturbance while the adaptive impedance is connected to the other. The mechanical and piezoelectric properties of the two patches are summarized in Table 8.1.

Table 8.1. Properties of the cantilever beam, the piezoelectric patch and the additional weight

Cantilever Beam		
Name	Symbol	Value
Length	L	450 mm
Width	W	50 mm
Thickness	h	3 mm
Young's modulus	E	$65 \cdot 10^9$ N/m^2
Density	ρ	10.6 kg/m^2
Piezoelectric Patch		
Name	Symbol	Value
Actuator Location	x_1	130 mm
Shunt Location	x_1	130 mm
Length	L_p	75 mm
Width	W_p	25 mm
Thickness	h_p	2.5 mm
Capacitance	C_p	104 nF
Young's Modulus	E_p	$62 \cdot 10^9$ N/m^2
Strain Constant	d_{31}	$-320 \cdot 10^{-12}$ m/V
Electromechanical coupling factor	k_{31}	0.44
Stress Constant / Voltage Coefficient	g_{31}	$-9.5 \cdot 10^{-3}$ m/N

Table 8.2. Properties of the multi-mode shunt circuit

$f_1 = 11.62$ Hz	
Property	Value
L_{d1} [kH]	[5 ↔ 8]
R_{d1} [Ω]	2
L_{f1} [H]	9576
C_{f1} [nF]	20
α_1	1000

$f_2 = 69.5$ Hz	
Property	Value
L_{d2} [H]	[25 ↔ 45]
R_{d2} [Ω]	1350
L_{f2} [H]	237.6
C_{f2} [nF]	20
α_2	1.5

$f_3 = 201.2$ Hz	
Property	Value
L_{d3} [H]	[7 ↔ 8]
R_{d3} [Ω]	900
L_{f3} [H]	29.69
C_{f3} [nF]	20
α_3	1

$f_4 = 401.5$ Hz	
Property	Value
L_{d4} [H]	[0.8 ↔ 2]
R_{d4} [Ω]	210
L_{f4} [H]	7.696
C_{f4} [nF]	20
α_4	0.5

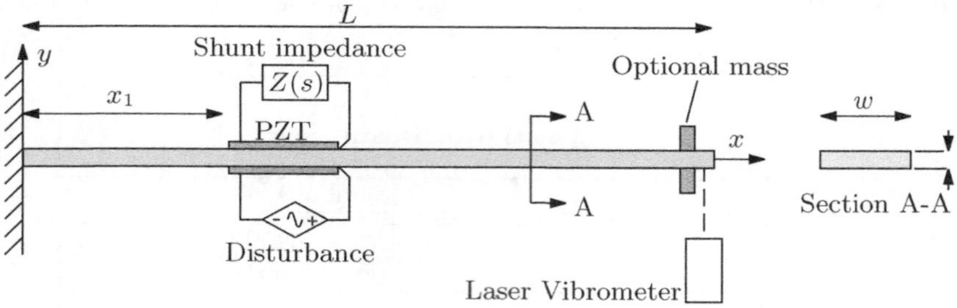

Figure 8.3. Experimental piezoelectric laminated cantilever structure

8.3.3 Two-mode Damping

In a first study, the second and third modes of the cantilever beam are targeted for reduction. The nominal shunt circuit component values are summarized in Table 8.2 (f_2 and f_3). The tuning values α_i are found experimentally to provide fast adaptation while maintaining acceptable levels of misadjustment.

Modal Frequency Variations

In this experiment, the adaptive shunt is subject to a step change in the structural resonance frequencies. The modal frequencies are disturbed by attaching an additional mass to the cantilever beam. The second mode moves from 69.5 Hz to 71.2 Hz, and the third from 201.2 Hz to 192.6 Hz. In Figure 8.5-1, the multi-mode shunt is optimally tuned with the inductance values: $L_{d2} = 44.5$ H and $L_{d3} = 4$ H (see Figure 8.4 (a)). The additional mass is then attached detuning the shunt circuit, peak damping decreases by 11 dB and 14 dB. In Figure 8.5 plots 3 to 6, the adaptation law re-tunes the circuit to the new optimum. The tuning behavior can also be observed in Figure 8.4 (a). After the additional mass is attached, the inductance values tune to $L_{d2} = 34.5$ H and $L_{d3} = 7.5$ H. In the frequency domain, adaptation can be observed by studying the time evolution of the transfer function from applied disturbance to velocity in Figure 8.4 (b).

Piezoelectric Capacitor Changes

In this experiment, the effect of a variation in piezoelectric capacitance C_p is investigated. In Figure 8.6-2, the shunt circuit is de-tuned by artificially decreasing C_p. In plots 3 to 5, a counteractive adaptation of L_{d2} and L_{d3} can be observed.

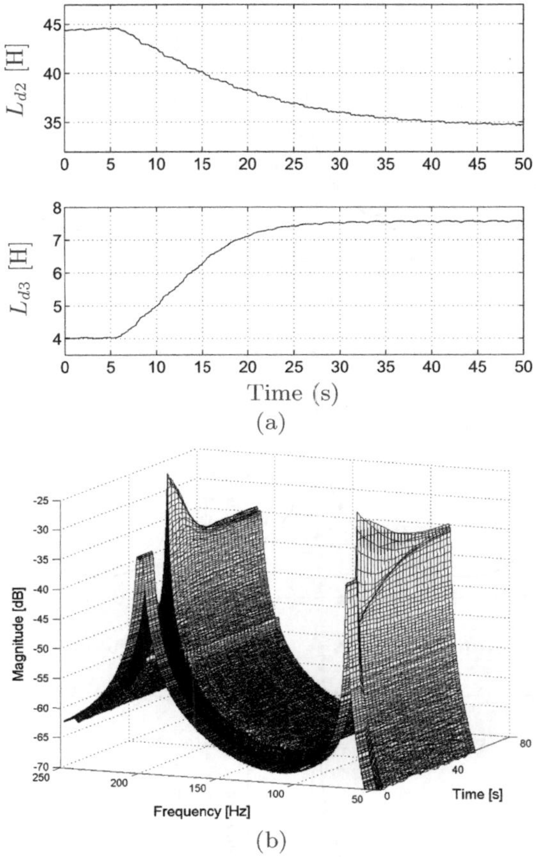

Figure 8.4. (a) Time evolution of the inductance values L_{d2} and L_{d3} after a step change in the resonance frequencies. (b) Time evolution of the transfer function from applied disturbance voltage to tip velocity ($G_{\nu w}(s)$).

8.3.4 Four-mode Damping

Both relative phase adaptation and current-flowing shunt circuits are readily extended to damp a greater number of modes. This section briefly demonstrates four-mode adaptive shunt damping.

The nominal component values of a four-mode current-flowing shunt circuit are listed in Table 8.2. Aside from the circuit order, all other experimental parameters remain unchanged.

In Figure 8.7 the behavior of each inductance value is shown subject to step changes in the structural resonance frequencies. At time 60s the position of an additional point mass is changed, then returned at time 300s.

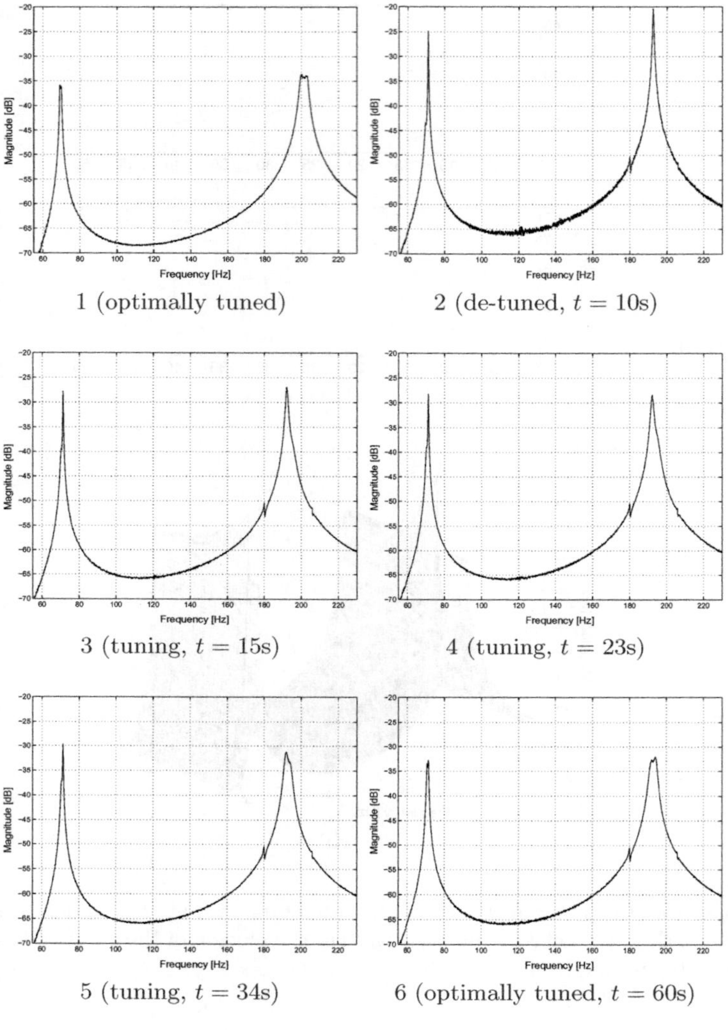

1 (optimally tuned) 2 (de-tuned, $t = 10$s)

3 (tuning, $t = 15$s) 4 (tuning, $t = 23$s)

5 (tuning, $t = 34$s) 6 (optimally tuned, $t = 60$s)

Figure 8.5. Magnitude of the transfer functions ($G_{\nu w}(s)$) during adaptation. In plot 2, the shunt circuit is de-tuned by adding an additional tip mass. Adaptation occurs in plots 3 to 6.

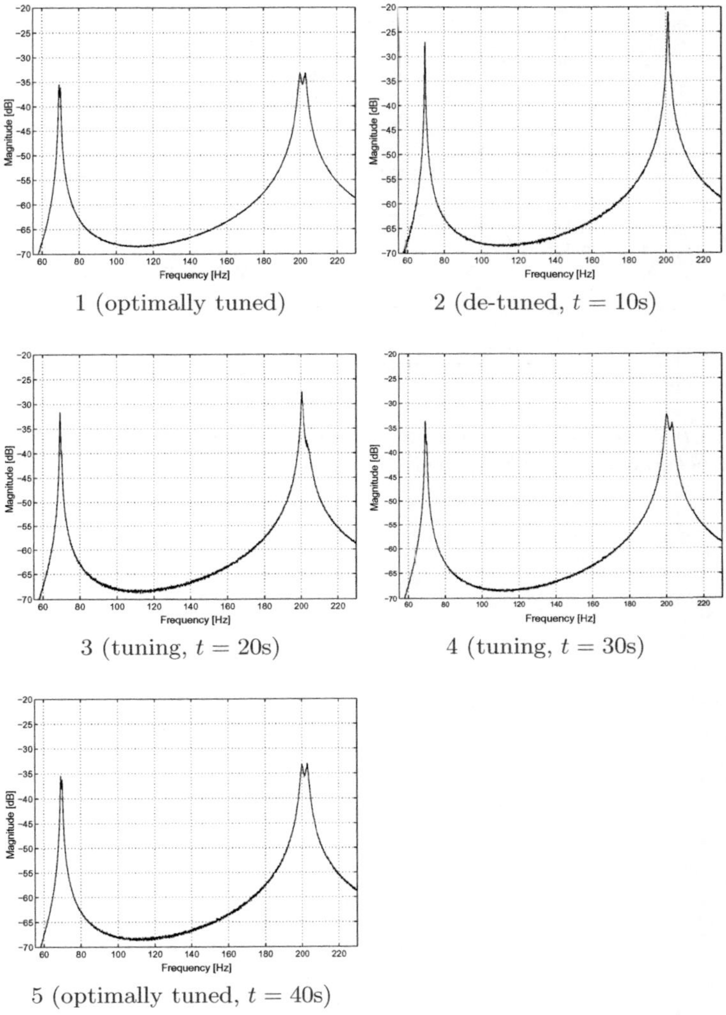

Figure 8.6. Magnitude of the transfer functions ($G_{\nu w}(s)$) during adaptation. In plot 2, the shunt circuit is de-tuned by artificially decreasing C_p. Adaptation occurs in plots 3 to 5.

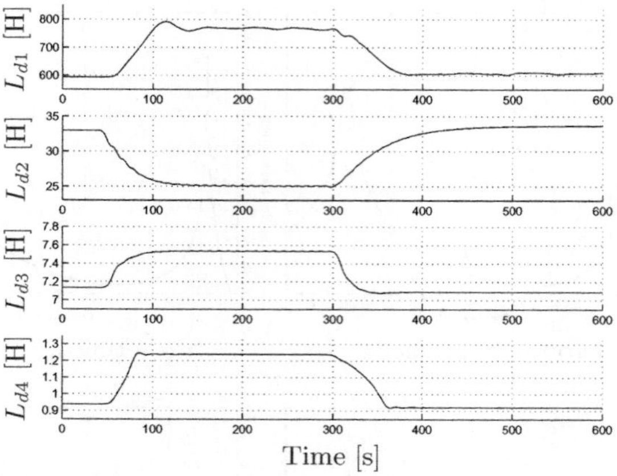

Figure 8.7. Four-mode shunt circuit inductance values subject to step changes in the structural resonance frequencies.

9

Negative Capacitor Shunt Impedances

9.1 Introduction

Resonant shunt damping circuits, as discussed in Chapters 4 and 5, are often favored for their good damping performance and desirable stability characteristics. The foremost disadvantage of resonant shunts is their high sensitivity to slight variations in the structural resonance frequencies. Maximum amplitude reduction is achieved only if the shunt absorber is precisely tuned to the frequency of the targeted mechanical mode. Adaptive shunt damping was discussed as a solution to this problem in Chapter 8. Detrimentally, adaptive shunt damping requires an auxiliary piezoelectric transducer and increased system complexity.

With a view to reducing implementation complexity and performance sensitivity to variations in structural resonance frequencies, this chapter introduces an alternative approach to piezoelectric shunt damping. Negative capacitor shunt circuits, as introduced in Section 4.3, constitute an active technique for structural vibration control. Although negative capacitors cannot be constructed from passive components and do not guarantee unconditional closed-loop stability, they are simple in conception and are known to provide good performance with little dependence on structural resonance frequencies. Their greatest disadvantage is the sensitivity to variations in the transducer capacitance, a 10% change in transducer capacitance can significantly degrade performance or result in instability. This sensitivity may complicate their use in applications that involve large temperature variations. In spite of the associated problems, negative capacitors have been used to suppress under-water structural and acoustic vibration [205].

By treating the internal piezoelectric voltage source as a supply, and the shunt impedance as a load, the traditional concept of maximum power transfer was applied in [202]. The optimal impedance is equal in magnitude to the source impedance, but opposite in phase; hence the negative capacitor. Although this approach arrives at a desirable solution, the derivation is not technically correct as the principle of maximum power transfer does not apply to circuits with internally dependent sources. Another derivation taking into account the dependent piezoelectric voltage was presented in [15] it is this approach that will be followed in the next section.

9.2 Negative Capacitor Shunt Controllers

In Section 5.2 the dynamics of a shunted piezoelectric laminate structure were derived and likened to a collocated strain-voltage feedback control system. Where $G_{vv}(s)$ is the open-loop transfer function from an applied actuator voltage to the resulting collocated strain voltage, the closed-loop or shunted response is

$$\hat{G}_{vv}(s) = \frac{G_{vv}(s)}{1 + G_{vv}(s)K(s)}, \tag{9.1}$$

where $\hat{G}_{vv}(s)$ is the shunted transfer function and $K(s)$ is the equivalent feedback controller shown below,

$$K(s) = \frac{C_p s Z(s)}{1 + C_p s Z(s)}, \tag{9.2}$$

where C_p is the piezoelectric capacitance and $Z(s)$ is the impedance of the shunt network. By substituting (9.2) into (9.1), the shunted dynamics can be expressed as

$$\hat{G}_{vv}(s) = \frac{G_{vv}(s)\left[C_p s Z(s) + 1\right]}{C_p s Z(s)\left[1 + G_{vv}(s)\right] + 1}. \tag{9.3}$$

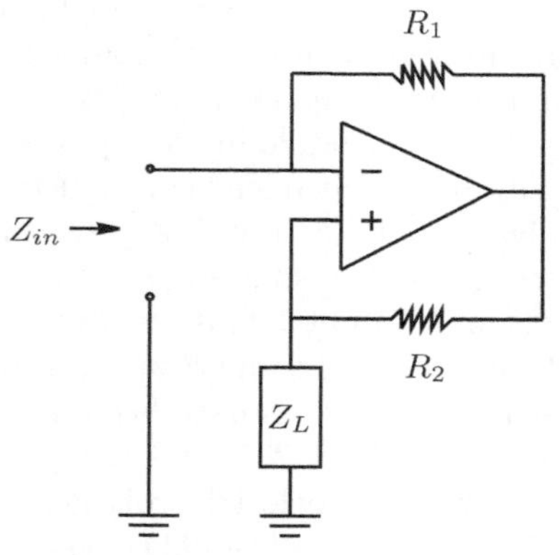

Figure 9.1. A negative impedance converter - the impedance seen from the terminals is proportional to $-Z_L$

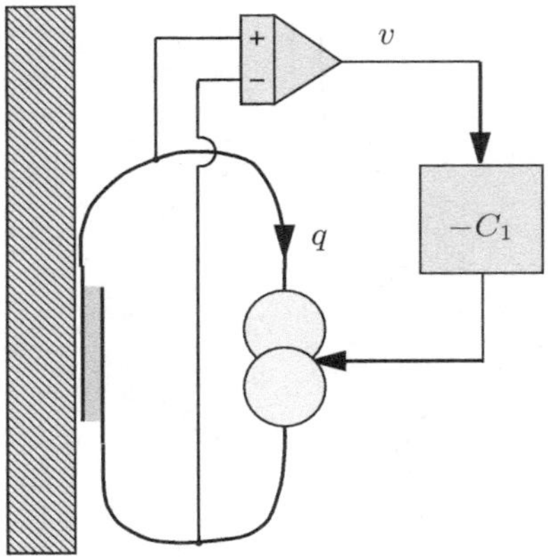

Figure 9.2. Synthetic implementation of a negative capacitor

An important observation is that the numerator of the closed-loop transfer function $\hat{G}_{vv}(s)$ is parameterized by the impedance $Z(s)$. Thus, the numerator, and hence closed-loop dynamics, can be equated to zero by selecting $Z(s) = -\frac{1}{C_p s}$. This choice of $Z(s)$ is equivalent to a proportional strain-feedback controller of infinite gain, and consequently, not realizable. The control effort can be moderated by choosing a negative capacitor slightly greater than C_p, *i.e.*

$$Z(s) = \frac{-1}{C_1 s} \qquad C_1 > C_p. \tag{9.4}$$

Aside from its simplicity, the greatest benefit of a negative capacitor shunt circuit is the independence from structural resonance frequencies. This can also be a downfall as a constant control effort is applied over the entire bandwidth regardless of whether it is required or not. Although the performance of this class of shunts is virtually independent of the structural resonance frequencies, it is strongly influenced by the magnitude of piezoelectric capacitance. A decrease in the piezoelectric capacitance will result in decreased damping performance. Conversely, an increase in C_p results in increased control effort and, eventually, instability. Generally, a negative capacitance 2-5% greater than the piezoelectric capacitance will provide good damping performance with adequate robustness to small changes in environmental temperature.

The tendency to instability when $C_1 < C_p$ can be understood intuitively by studying the nature of the equivalent feedback controller. With $Z(s) = \frac{-1}{C_1 s}$, the equivalent feedback controller is

$$K(s) = \frac{C_p}{C_p - C_1}, \tag{9.5}$$

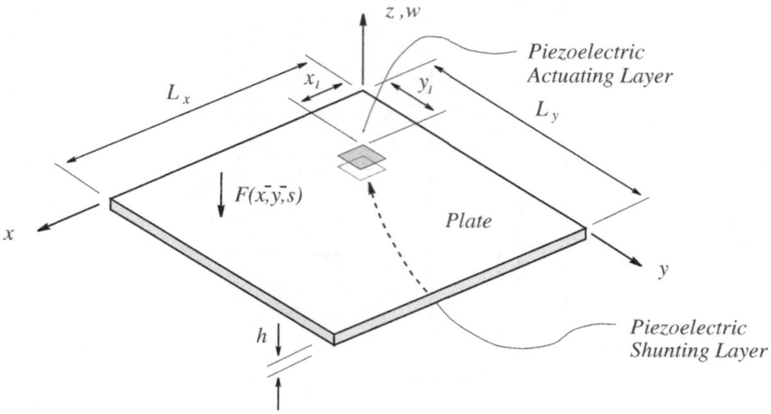

Figure 9.3. A thin simply-supported plate with actuator and shunting piezoelectric layers

when $C_1 > C_p$, $K(s)$ has a phase of 180 degrees and a gain that increases as C_1 approaches C_p. If C_1 becomes slightly less than C_p the equivalent controller experiences a 180 degree phase change in combination with a very large gain, resulting in closed-loop instability.

9.3 Implementation

As negative capacitors are active devices, they cannot be implemented by passive components. In the following, a negative impedance converter and synthetic admittance are compared for the implementation of negative capacitor shunt circuits.

9.3.1 Negative Impedance Converter

As illustrated in Figure 9.1, a negative impedance converter consists of a single op-amp, two resistors R_1 and R_2, and the impedance that is to be negated Z_L. It is straightforward to verify that the input impedance, $Z_{in}(s)$, is equal to

$$Z_{in}(s) = -\frac{R_1}{R_2} Z_L. \tag{9.6}$$

By choosing $Z_L = \frac{1}{C_1 s}$, *i.e.* a capacitance of C_1, the resistance ratio can be varied to fine-tune the negative capacitance value. Practical difficulties such as bias-current and offset-voltage induced errors are usually solved by placing a large parallel resistance across Z_L. A resistor in series with the input has

Table 9.1. Plate dimensions

Name	Symbol	Unit
Length	L_x	0.8 m
Length	L_y	0.6 m
Thickness	h	0.004 m
Young's modulus	E	$65 \times 10^9\ N/m^2$
Poisson's ratio	v	0.3
Mass / unit area	ρ	10.6 kg/m^2

also been used to overcome problems due to stray negative resistance and for use as a tuning parameter [15]. In this case, an expression for the resulting impedance and equivalent controller can be found in reference [15].

9.3.2 Synthetic Admittance

The synthetic admittance was introduced in Section 4.4.2 for the implementation of piezoelectric shunt damping circuits. A variant described in Section 6.6 utilizes a charge source in place of a current source, in this case the admittance transfer function to be implemented is $Y(s)/s$, *i.e.* the time integral of the desired terminal admittance. In order to implement a negative capacitor by this technique $Y(s)/s$ is simply a constant gain equal to $-C_1$. A diagrammatic representation is shown in Figure 9.2.

9.4 Experimental Application

In this section, a negative capacitor shunt is applied to augment the damping of a plate and beam structure. A negative impedance converter is used to implement the plate shunt impedance while the second experiment utilizes a synthetic admittance.

9.4.1 Control of a Plate Structure

Pictured in Figure 5.10, the experimental plate is of uniform thickness and experimentally pinned on all edges. A pair of piezoelectric ceramic patches are attached symmetrically to either side of the plate surface. While one patch is used as an actuator the other is used as a shunting layer. Dimensions of the plate and physical properties of the piezoelectric layers are summarized in Figure 9.3 and Tables 9.1 and 9.2.

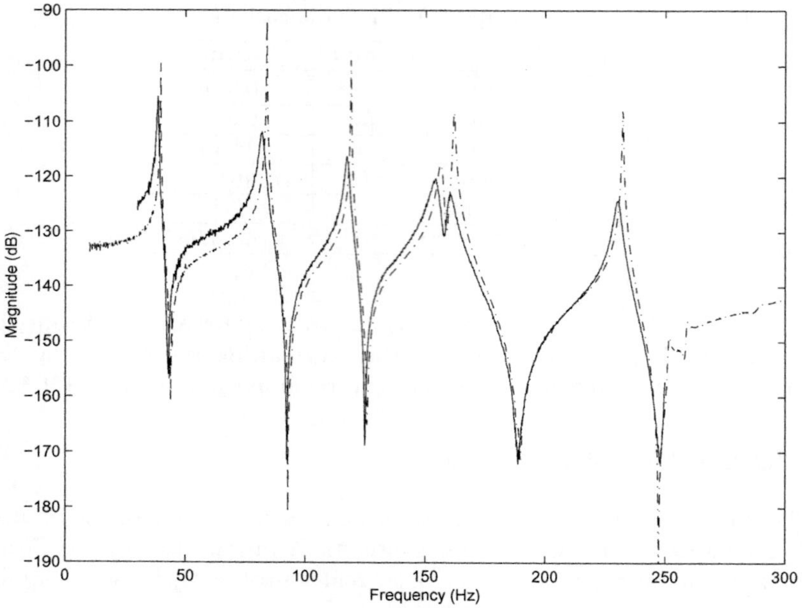

Figure 9.4. The open-loop (— · —) and shunted (—) velocity response to an applied disturbance voltage. The velocity was measured at $(x, y) = (0.2, 0.2)$.

The negative impedance converter used to implement the plate shunt impedance was constructed from a Texas Instruments OPA445 high voltage op-amp. Further circuit details are contained in reference [15].

In Figure 9.4 the shunted frequency response with a 485 nF negative capacitor shows a significant amount of additional damping. Modal amplitude reductions of up to 20 dB are summarized in Table 9.3.

Table 9.2. Plate transducer parameters

Name	Symbol	Unit
Location x-direction	x_1	$0.1536\ m$
Location y-direction	y_1	$0.1418\ m$
Length	$L_{px}\ L_{py}$	$0.0724\ m$
Thickness	h_p	$0.00191\ m$
Capacitance	C_p	$471 \times 10^{-9}\ F$
Young's modulus	E_p	$62 \times 10^9\ N/m^2$
Poisson's ratio	v_p	0.3
Strain Constant	d_{31}	$-320 \times 10^{-12}\ m/V$
Electromechanical coupling factor	k_{31}	0.44
Stress constant / voltage coefficient	g_{31}	$-9.5 \times 10^{-3}\ V\ m/N$

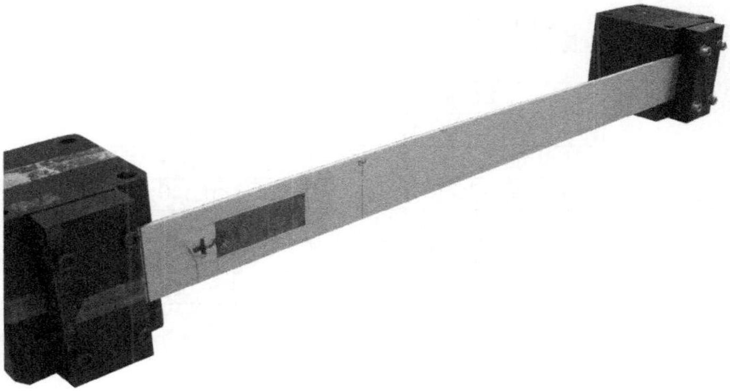

Figure 9.5. The experimental beam structure

9.4.2 Control of a Beam Structure

The experimental beam, pictured in Figure 9.5, is a uniform aluminum bar with rectangular cross section and experimentally pinned boundary conditions at both ends. A pair of piezoelectric ceramic patches are attached symmetrically to either side of the beam surface 5 cm from one end. One patch is used as an actuator and the other as a shunting layer. Experimental beam and piezoelectric parameters are summarized in Tables 9.4 and 9.5.

A grounded-load charge amplifier as discussed in Section 12.5.1 was used to implement a negative capacitor of approximately $-110\ nF$. The open and shunted frequency response from an applied actuator voltage to the velocity

Table 9.3. Amplitude reduction

Mode	Experimental (dB)
1	5.8
2	20.1
3	18.2
4	3.8
5	16.7
6	17.2

Table 9.4. Beam Dimenstions

Beam Parameters	
Length	0.6 m
Width	0.05 m
Thickness	0.003 m
Youngs Modulus	$65 \times 10^9\ N/m^2$
Density,	$2650\ kg/m^3$

Table 9.5. Beam transducer parameters

PIC151 Parameters	
Length	$0.070\ m$
Charge Constant, d_{31}	$-210 \times 10^{-12}\ m/V$
Voltage Constant, g_{31}	$-11.5 \times 10^{-3}\ Vm/N$
Coupling Coefficient, k_{31}	0.340
Capacitance, C_p	$0.105\ \mu F$
Width	$0.025\ m$
Thickness	$0.25 \times 10^{-3}\ m$
Youngs Modulus	$63 \times 10^9\ N/m^2$

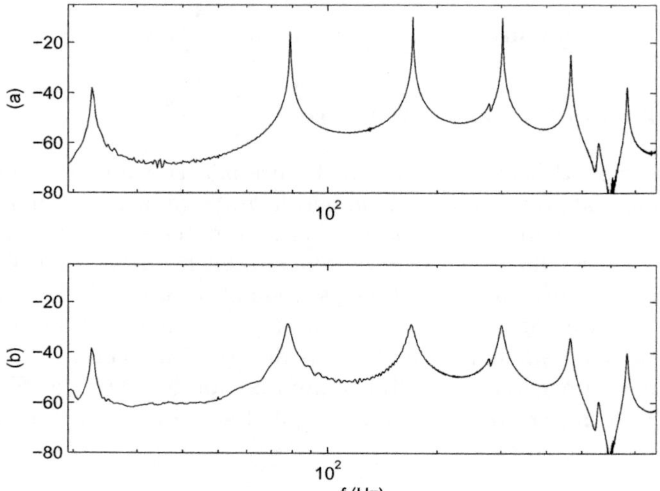

Figure 9.6. The (a) open and (b) shunted frequency responses from an applied actuator voltage to the measured velocity 17 cm from the end without the transducers

17 cm from the side without the transducers is shown in Figure 9.6. The modal amplitudes of the first 6 modes are reduced by 0.9, 13, 20, 19, 10 and 3dB. The torsional mode at 550 Hz is not considered. The poor performance associated with the first mode is due to the placement of the transducers. The strain corresponding to the first mode is extremely low, hence the poor coupling at this frequency. A step response test illustrating the improvement in settling time is plotted in Figure 9.7.

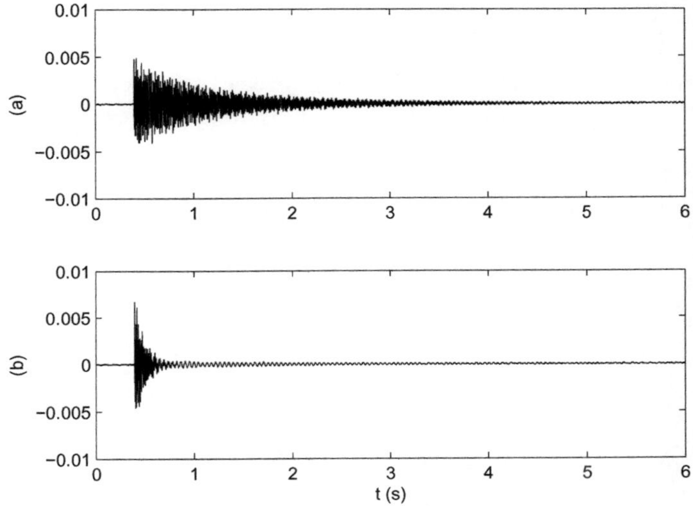

Figure 9.7. The (a) open and (b) shunted velocity responses to a step change in the actuator voltage. The velocity was measured 17 *cm* from the end without the transducers.

10

Optimal Shunt Synthesis

10.1 Introduction

So far we have established that the problem of piezoelectric shunt damping can be viewed as a feedback control problem and based on this observation we have proposed a number of piezoelectric shunts, both SISO and multivariable, that are quite effective in adding damping to the system. Resonant shunts as proposed in previous chapters have performance limitations which is due to their passive structures.

Having established the feedback structure of the shunt damped systems, it is no longer necessary to restrict ourselves solely to passive shunts. Rather, it is possible to take advantage of the identified feedback structure and design very efficient shunts using standard feedback control design tools such as the LQG, H_∞, H_2, etc. [5, 110, 80, 206].

An important first step in realizing this is to choose appropriate variables for measurement and control signals. Charge, current or voltage could be used for this purpose. For example, the voltage across a piezoelectric transducer can be measured and the current flowing out of the same transducer can be controlled (Figure 4.12), or the situation can be reversed (Figure 4.13). In any case, it is possible to devise the necessary instrumentation, as explained in Section 6.

Modern and robust control design tools depend on state-space models of the system that is to be controlled. In the first part of this chapter the state-space structure of MIMO shunt damping is derived depending on the choice of sensing and actuation signals. Furthermore, we will explain how an H_∞ s-impedance can be designed and implemented on a piezoelectric laminated cantilever beam. Similar ideas have been employed to design and implement a number of other shunts. Details are reported in references [65] and [66].

10.2 Abstracted Model of Shunted Systems

The main thrust of this chapter is to illustrate that the piezoelectric shunt damping problem, once cast as a feedback control problem, can be solved using standard feedback control tools. Given that the majority of modern and robust control methods are based on state-space models, this section is devoted to developing state-space models of shunt damping systems.

We consider a flexible structure and a number of piezoelectric transducers that are bonded to it. Furthermore, we assume that the structure is subject to a disturbance vector, W. Essentially, we are assuming a setup as illustrated in Figure 7.1. Using a similar terminology to (7.1) and representing the vector of voltages at the piezoelectric transducers as V and the vector of piezoelectric voltages as V_p, and taking into account the linear nature of the shunted system we may write

$$\begin{aligned} \dot{x}(t) &= Ax(t) + B_1W(t) + B_2V(t) \\ V_p(t) &= Cx(t) + D_1W(t) + D_2V(t). \end{aligned} \tag{10.1}$$

This equation describes the voltages generated inside piezoelectric transducers due to the applied disturbances and the impedance that is seen at each piezoelectric patch. Naturally, the effect of disturbance signals is destructive as they generate unwanted vibrations throughout the structure. Once the piezos are shunted by a multi-port impedance $Z(s)$, the impedance should then be designed with a view to minimizing the effect of disturbances on the structure.

As argued earlier, the impedance $Z(s)$ establishes a specific relationship between the current flowing through and the voltage across each terminal. This means that the impedance $Z(s)$ may be replaced by a current-controlled voltage source. This suggests the feedback structure depicted in Figure 10.1 (a). That is, the impedance, $Z(s)$ can be viewed as the feedback controller acting on the system Γ.

There is a difficulty associated with viewing the impedance as a controller since the underlying system will be improper. It still may be possible to use feedback control design tools developed for descriptor systems [180, 181, 103, 96] to tackle this issue. Nonetheless, it turns out that this hurdle can be avoided altogether if the electrical charge is used as the measurement signal, as shown in Figure 10.1 (b). If so done, the generalized system $\tilde{\Gamma}(s)$ can be written as

$$\left[\begin{array}{c} \dot{x}(t) \\ \hline V_p(t) \\ Q(t) \end{array}\right] = \left[\begin{array}{c|cc} A & B_1 & B_2 \\ \hline C & D_1 & D_2 \\ -\Lambda^{-1}C & -\Lambda^{-1}D_1 & -\Lambda^{-1}(I+D_2) \end{array}\right] \left[\begin{array}{c} x(t) \\ \hline W(t) \\ V(t) \end{array}\right]. \tag{10.2}$$

This equation follows from (10.1) and

$$Q(t) = -\Lambda^{-1}(V(t) + V_p(t)) \tag{10.3}$$

which is the KVL (7.3) re-written to express Q in terms of V and V_p.

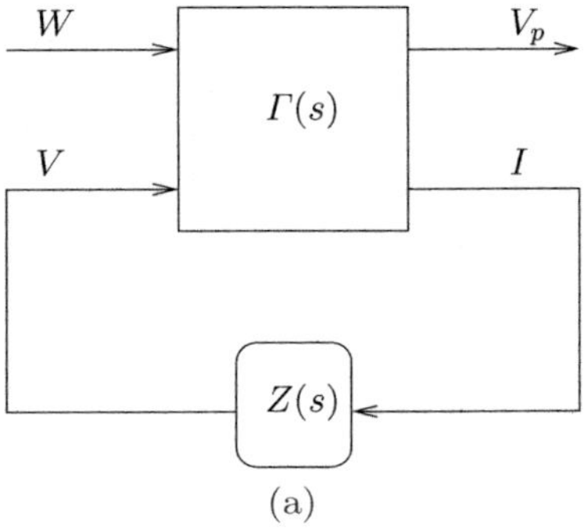

Figure 10.1. The control problem: (a) with impedance as the controller; (b) with s-impedance as the controller

Having reduced the problem to this level, a range of feedback control tools can be employed to design an s-impedance to achieve the necessary damping. There is, of course, no reason to design an impedance, or a s-impedance for that matter, which is strictly passive. It is indeed possible to design an active impedance that offers higher levels of performance compared to passive shunts, such as those studied in Chapter 4.

In the remainder of this chapter, we will illustrate that an s-impedance can be designed to add significant damping to the structure. We will also show that standard control design tools can be employed for this purpose, and we will illustrate how an s-impedance can be implemented digitally. But, before proceeding further we wish to clarify that there are alternatives to s-impedance design. These alternative frameworks are illustrated in Figure 10.2 (a) and (b).

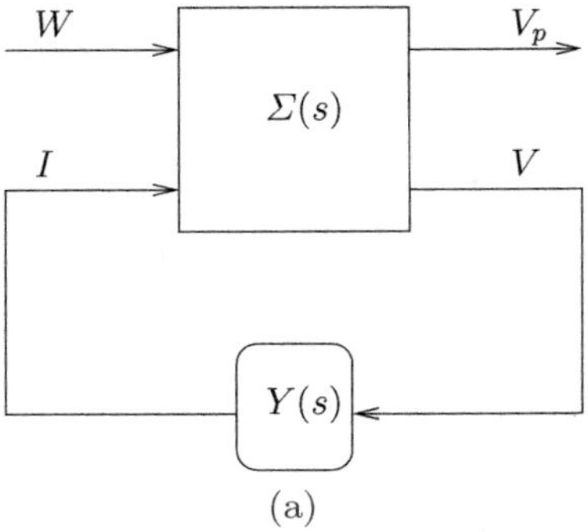

Figure 10.2. The control problem: (a) with admittance as the controller; (b) with s^{-1}-admittance as the controller

If, as shown in Figure 10.2 (a), the vector of voltages across the piezoelectric transducers is taken as the measurement signal, and the vector of currents flowing out of the piezos and into the multi-port impedance is used as the control signal, we can show that the augmented plant, $\Sigma(s)$, is characterized as follows:

$$\left[\begin{array}{c}\dot{x}(t)\\ \dot{Q}(t)\\ \hline V_p(t)\\ V(t)\end{array}\right] = \left[\begin{array}{cc|cc} A - B_2(I+D_2)^{-1}C & -B_2(I+D_2)^{-1}\Lambda & B_1 - B_2(I+D_2)^{-1}D_1 & 0\\ 0 & 0 & 0 & I\\ \hline (I+D_2)^{-1}C & -D_2(I+D_2)^{-1}\Lambda & (I+D_2)^{-1}D_1 & 0\\ -(I+D_2)^{-1}C & -(I+D_2)^{-1}\Lambda & -(I+D_2)^{-1}D_1 & 0\end{array}\right] \times \left[\begin{array}{c}x(t)\\ Q(t)\\ \hline W(t)\\ \dot{Q}(t)\end{array}\right].$$

Validity of the above equation can be verified starting from (10.1) and by incorporating (10.3) and augmenting the states by $\dot{Q}(t)$. It can be observed that in this case, the dynamics of the plant is heavily dominated by the integrator included in (10.4).

It is also straightforward to obtain system dynamics, if the current is replaced with charge in the above, as illustrated in Figure 10.2 (b). In this case, the augmented system $\tilde{\Sigma}(s)$ can be described by

$$\left[\begin{array}{c}\dot{x}(t)\\ \hline V_p(t)\\ V(t)\end{array}\right] = \left[\begin{array}{c|cc} A - B_2(I+D_2)^{-1}C & B_1 - B_2(I+D_2)^{-1}D_1 & -B_2(I+D_2)^{-1}\Lambda\\ \hline (I+D_2)^{-1}C & (I+D_2)^{-1}D_1 & -D_2(I+D_2)^{-1}\Lambda\\ -(I+D_2)^{-1}C & -(I+D_2)^{-1}D_1 & -(I+D_2)^{-1}\Lambda\end{array}\right] \times \left[\begin{array}{c}x(t)\\ \hline W(t)\\ Q(t)\end{array}\right].$$

10.3 Experimental Apparatus

To illustrate the possibility and effectiveness of active impedance synthesis, an experimental apparatus was built and used for experiments in the Laboratory

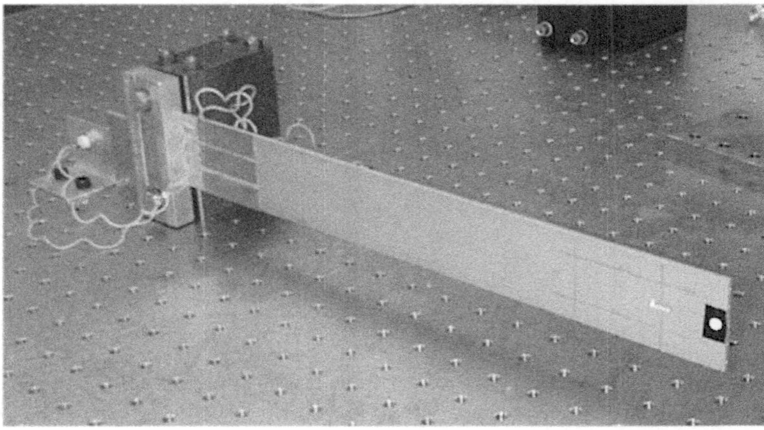

Figure 10.3. The cantilever beam

for Dynamics and Control of Smart Structures at the University of Newcastle. The experimental apparatus, shown in Figure 10.4 and pictured in Figure 10.3, consists of a uniform aluminum cantilever beam. Three piezoelectric transducers are laminated onto the front face and connected electrically in series to the voltage source V. A single collocated disturbance transducer, identical to each of the shunt transducers, is also mounted on the back face and driven with the disturbance voltage W. Physical parameters of the beam and piezoelectric transducers can be found in Tables 10.1 and 10.2.

As illustrated in Figure 10.4, the charge deposited on the piezoelectric transducer is measured and the voltage applied to the same transducer is controlled. This amounts to the shunt synthesis scenario sketched in Figure 10.1 (b) and formulated by (10.2). The problem, therefore, boils down to designing an s-impedance transfer function. A variety of control design techniques can now be used to design a high performance s-impedance shunt. Before we delve into this topic, we need to clarify how the instrumentation for implementing a given s-impedance circuit can be put together.

Table 10.1. Beam Parameters

Length, L	376 mm
Thickness, h	3 mm
Width, W	50 mm
Density, ρ	2.770×10^3 kg/m^3
Young's Mod., E	7.00×10^{10} N/m^2

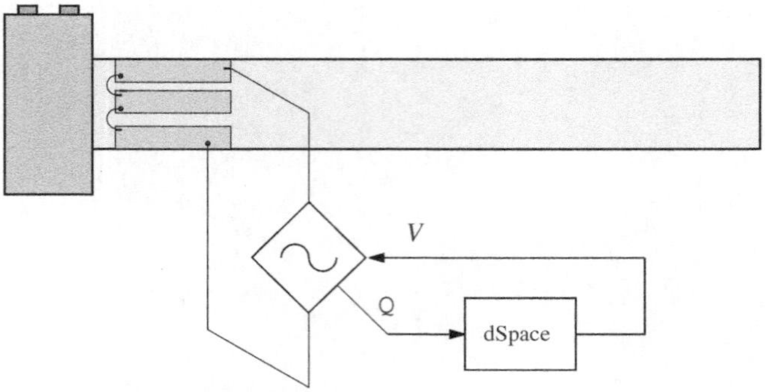

Figure 10.4. A front elevation of the cantilever beam. A single co-located disturbance transducer excited by the voltage V_a, is also mounted on the back face.

10.4 Instrumentation Electronics

The purpose of this section is to explain how the electronic instrumentation to implement a given s-impedance circuit can be devised. The key functionality that is needed is the ability to first measure the charge accumulated on the electrodes of a piezoelectric transducer and then to command the voltage applied to the same piezoelectric transducer. The charge measurement signal (a voltage signal) will be provided to digital signal processing system, in this case the dSPACE rapid prototyping system, and the DSP system will command the output voltage, in a way that the requisite s-impedance is presented to the piezoelectric transducer.

Schematics of a high-voltage power amplifier with charge instrumentation is illustrated in Figure 10.5. As shown in Figure 10.5, a high-gain op-amp is used to maintain a reference voltage V_{ref} across the load $Z_L(s)$. An arbitrary voltage gain can be implemented by controlling attenuation in the feedback path. The voltage V_s, measured across the sensing capacitor C_s, is propor-

Table 10.2. Properties of the Physik Instruments Transducers (PIC151 Ceramic)

Length, L_{pz}	50 mm
Thickness, h_{pz}	0.25 mm
Width, W_{pz}	15 mm
Charge Constant, d_{31}	-210×10^{-12} m/V
Voltage Constant, g_{31}	-11.5×10^{-3} Vm/N
Coupling Coefficient, k_{31}	0.34
Capacitance, C_p	43 nF
Young's Mod., E_{pz}	63×10^{9} N/m^2

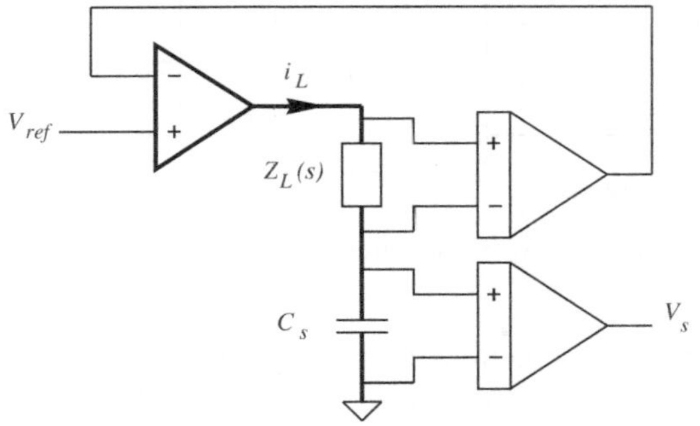

Figure 10.5. A voltage amplifier with charge measurement

tional to the load charge Q. The charge gain in Volts per Coulomb is equal to $\frac{1}{C_s}$, *i.e.*

$$\frac{V_s}{Q} = \frac{1}{C_s}\ (v/c). \tag{10.4}$$

For implementation of s-impedance controllers, the charge q is defined flowing out of the load, in this case the charge instrumentation gain is negative. An alternative to the circuit shown in Figure 10.5 is to interchange the load and sensing impedances. In this case, the feedback voltage is taken directly across the grounded load.

10.5 Active S-impedance Shunt Design

The purpose of this section is to explain how a high-performance s-impedance shunt can be designed to add damping to the cantilever beam apparatus described in the previous section. The first step involves obtaining a model of the augmented plant.

10.5.1 Modeling and Parameter Identification

In Section 10.2 we described how a model of the controlled system could be derived. The developed model lends itself to multivariable s-impedance design problems. The problem at hand, however, is a SISO s-impedance design. Furthermore, given the specific structure of the problem, the augmented plant model as described by (10.2) needs to be slightly modified.

Given that there is only one piezoelectric transducer shunted to the s-impedance, Λ^{-1} in (10.2) can be replaced with C_p, the capacitance of the

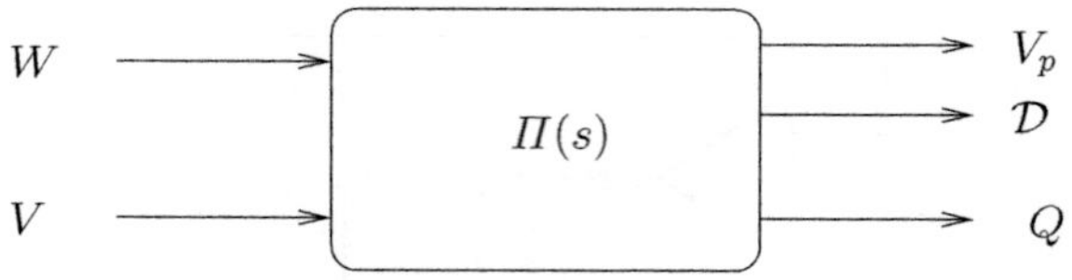

Figure 10.6. Augmented plant to be identified

shunted patch[1]. The exogenous output of the augmented plant, as described by (10.2) and illustrated in Figure 10.1, is the piezoelectric voltage V_p. Although V_p could serve as a good indication of structural vibration, - V_p is proportional to mechanical strain - it cannot be measured directly, nor is it the physical variable that is to be controlled. Here, our intention is to minimize tip displacement of the beam. Subsequently, the augmented plant will need to be modeled accordingly. Having said this, V_p needs to be identified since the expression for the charge, Q, depends on the the parameters that make up V_p. Therefore, we need to identify a 2-input 3-output model as sketched in Figure 10.6 and described by

$$\left[\begin{array}{c} \dot{x}(t) \\ \hline \left\{\begin{array}{c} V_p(t) \\ \mathcal{D}(t) \end{array}\right\} \\ Q(t) \end{array}\right] = \left[\begin{array}{c|cc} A & B_1 & B_2 \\ \hline C & D_1 & D_2 \\ C_d & D_{d_1} & D_{d_2} \\ -C_p C & -C_p D_1 & -C_p(1+D_2) \end{array}\right] \left[\begin{array}{c} x(t) \\ \hline W(t) \\ V(t) \end{array}\right], \tag{10.5}$$

where $\mathcal{D}$ represents the tip displacement.

The next step involves identifying important parameters of (10.5). This is done by taking advantage of the well-defined structure of the system. That is,

$$A = \begin{bmatrix} 0 & 1 & & & \\ -\omega_1^2 & -2\varsigma_i\omega_1^2 & & & \\ & & \ddots & & \\ & & & 0 & 1 \\ & & & -\omega_N^2 & -2\varsigma_N\omega_N^2 \end{bmatrix} \tag{10.6}$$

$$\tilde{B} = \begin{bmatrix} B_1 & B_2 \end{bmatrix} = \begin{bmatrix} 0 & 0 \\ F_1 & H_1 \\ \vdots & \vdots \\ 0 & 0 \\ F_N & H_N \end{bmatrix} \tag{10.7}$$

and

[1] Note that C_p represents the equivalent capacitance of the three identical patches put in series, *i.e.* one third of the capacitance of each patch.

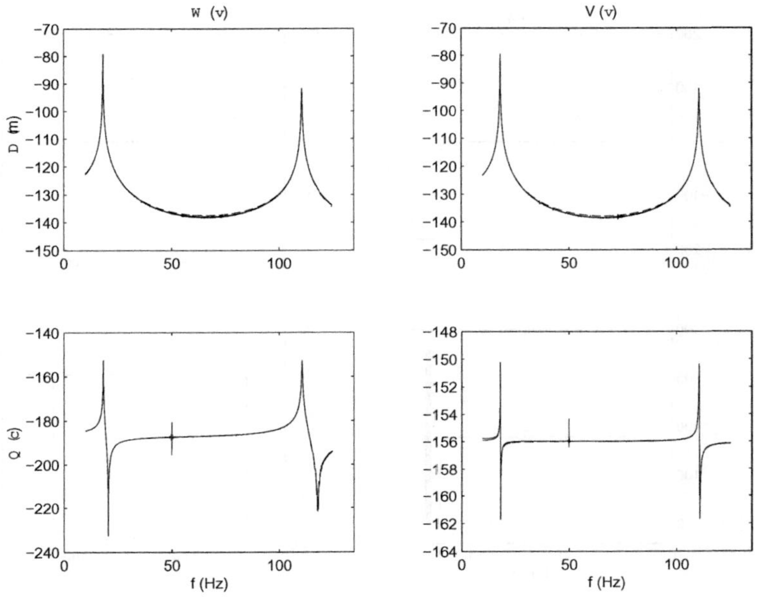

Figure 10.7. The simulated (- -) and experimental (—) magnitude frequency response (in decibels) of the shunt voltage controlled piezoelectric beam

$$\tilde{C} = \begin{bmatrix} C \\ C_d \\ C_p C \end{bmatrix} = \begin{bmatrix} E_1 & 0 \cdots & E_N & 0 \\ 1 & 0 \cdots & 1 & 0 \\ C_p E_1 & 0 \cdots & C_p E_N & 0 \end{bmatrix} \tag{10.8}$$

$$\tilde{D} = \begin{bmatrix} D_1 & D_2 \\ D_{d_1} & D_{d_2} \\ -C_p D_1 & -C_p(1 + D_2) \end{bmatrix}. \tag{10.9}$$

To determine the model parameters we employ a simple optimization scheme. From an initial guess, parameters ω_i and ζ_i are found through a simplex optimization based on the measured disturbance to displacement transfer function $\frac{\mathcal{D}(s)}{W(s)}$, *i.e.*

$$\begin{bmatrix} \omega_k \ \zeta_k \end{bmatrix} = \arg \ \min \ \left\| \tilde{\Pi}_{\mathcal{D}W}(s) - \Pi_{\mathcal{D}W}(s) \right\|_2, \tag{10.10}$$

where $\tilde{\Pi}_{\mathcal{D}W}(s)$ is the measured transfer function from an applied disturbance $W(s)$ to the displacement $\mathcal{D}(s)$. With these parameters in hand, those remaining are determined from a final optimization, by solving

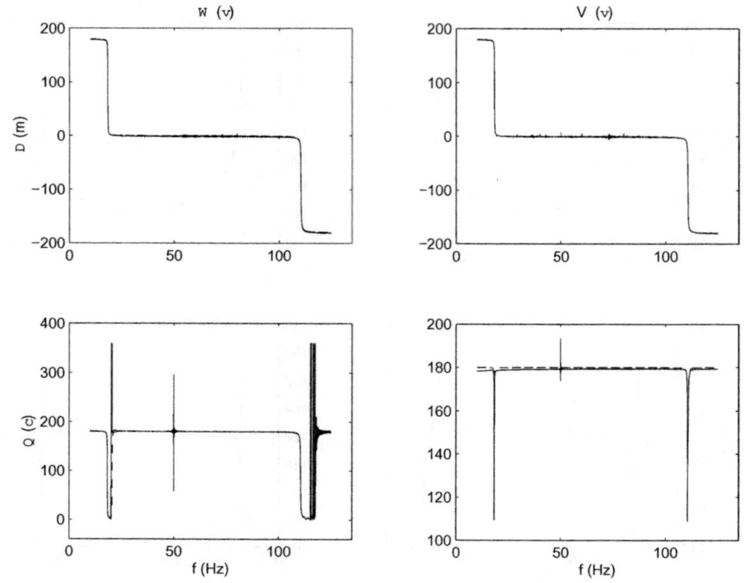

Figure 10.8. The simulated (- -) and experimental (—) phase frequency response (in degrees) of the shunt voltage controlled piezoelectric beam

$$\arg\ \min\ \left\| \tilde{\Pi}(s) - \Pi(s) \right\|_{2,\ \mathcal{W}} . \tag{10.11}$$

As gains from channel to channel vary greatly, a multivariable frequency weight $\mathcal{W}$ is required to normalize the cost of each error transfer function.

The magnitude and phase response of the measured system and resulting model are shown in Figures 10.7 and 10.8. The model is an accurate representation of the measured system that contains the first two vibration modes of the composite structure. Note the close pole-zero spacing in the transfer function from an applied shunt voltage V to the charge Q. Referring to (10.9), this behavior is due to the transducer capacitance which results in a large direct feed-through term.

10.5.2 H_∞ S-impedance Design

Having access to an accurate multivariable plant model enables us to design an s-impedance using standard control design tools. Here, we endeavor to design an H_∞ s-impedance shunt.

In order to limit the level of control effort, the voltage applied to the piezoelectric transducer, the exogenous output needs to be augmented by the

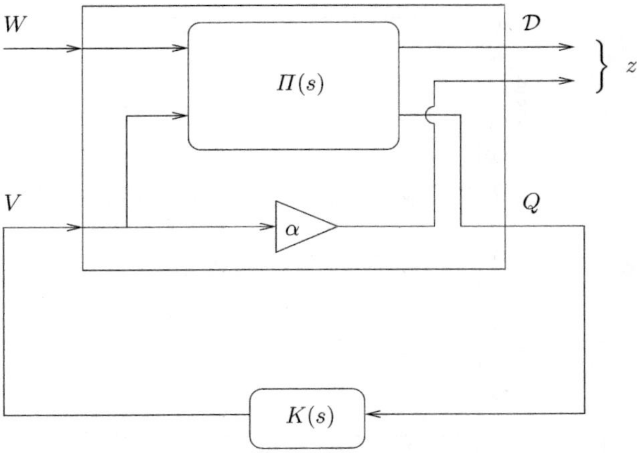

Figure 10.9. Augmented feedback control problem for H_∞ s-impedance design

control signal. This amounts to the control problem sketched in Figure 10.9, where the augmented plant is described by

$$\left[\begin{array}{c}\dot{x}(t)\\ \hline z(t)\\ Q(t)\end{array}\right]=\left[\begin{array}{c|cc}A & B_1 & B_2\\ \hline \begin{Bmatrix}C_d\\0\end{Bmatrix} & \begin{Bmatrix}D_{d_1}\\ \alpha\end{Bmatrix} & \begin{Bmatrix}D_{d_2}\\0\end{Bmatrix}\\ C_d & D_{d_1} & D_{d_2}\\ -C_pC & -C_pD_1 & -C_p(1+D_2)\end{array}\right]\left[\begin{array}{c}x(t)\\ \hline W(t)\\ V(t)\end{array}\right], \tag{10.12}$$

where

$$z(t)=\begin{bmatrix}\mathcal{D}(t)\\ \alpha\end{bmatrix} \tag{10.13}$$

is the exogenous output.

The H_∞ s-impedance synthesis then involves finding a controller $K(s)$ that minimizes

$$J_\infty=\left\|\hat{\Pi}_{zW}(s)\right\|_\infty, \tag{10.14}$$

where $\hat{\Pi}$ is the augmented plant in Figure 10.9.

The control signal weighting α enables the control designer to make a compromise between the level of performance that is required and control effort. In this experiment α was chosen to be 3.2×10^{-7}.

The bode plot of the s-impedance associated with the resulting H_∞ controller is plotted in Figure 10.10.

Examining the open- and closed-loop pole locations shown in Figure 10.11 (a), the controller is clearly enhancing the system damping. Corresponding mitigation of the transfer function from an applied disturbance to the measured displacement can be seen in both the frequency domain, Figure 10.11 (b), and time domain, Figure 10.13. The magnitude of the first and second

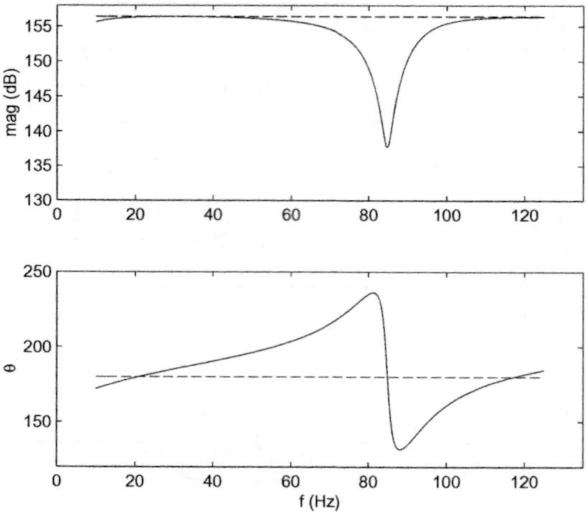

Figure 10.10. Complex s-impedance of the H_∞ (—), and ideal negative capacitor (- -) shunt controller

structural modes are reduced by 30.3 dB and 24.0 dB respectively. Damping ratios are increased from 0.00246 and 0.0011 to 0.0288 and 0.00766.

An unexpected feature of the H_∞ s-impedance is its smooth frequency response; there are no localized peaks at the resonance frequencies. In contrast, active strain-, velocity-, or acceleration-feedback controllers characteristically apply a highly localized gain at the frequencies of structural resonance. In the advent of model variation, such localized behavior can result in considerable performance degradation.

In order to examine the closed-loop robustness with respect to varying resonance frequencies of the underlying structure, the nominal system is perturbed by adding a mass 60 mm from the beam tip. To make a concrete comparison, a two-mode passive current-flowing shunt as described in Section 4.2.4 was also designed and implemented on the beam. Aside from the disturbance to the underlying partial differential equation, the first and second resonance frequencies are decreased by 13.5 % and 2.2 % respectively.

The consequences on both passive and active shunt circuits are shown in Figure 10.12. While the H_∞ controller loses 3.3 and 0.8 dB from its nominal closed-loop attenuation of the first and second modes, the passive shunt loses 13.4 dB and 4.8 dB respectively. Corresponding time domain results are shown in Figure 10.13.

In a final test to validate the H_∞ s-impedance, an acoustic loud speaker was used to spatially excite the structure. The measured frequency response, shown in Figure 10.15, verifies that the achieved performance is disturbance-channel independent. Time-domain experimental results corresponding to this disturbance are illustrated in 10.14.

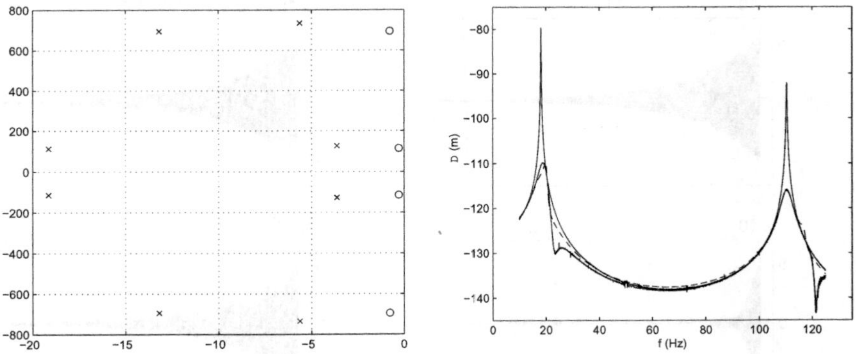

(a) (b)

Figure 10.11. (a) The open- (○), and closed-loop (×) pole locations of the H_∞ shunt controlled system; (b) The experimental (—), and simulated (- -), H_∞ shunt controlled frequency responses (in decibels) from an applied disturbance voltage W (V) to the resulting tip displacement $\mathcal{D}$ (m). The open-loop response is also shown (—).

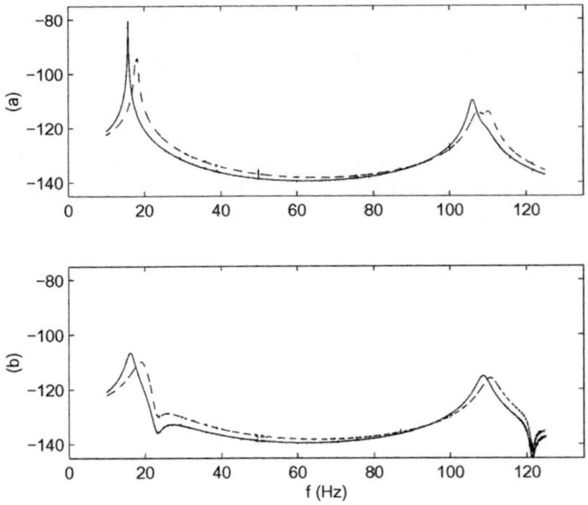

Figure 10.12. The free (- -), and with-mass (—), passive (a) and H_∞ shunt controlled (b) experimental frequency responses (in decibels) from an applied disturbance voltage W (v) to the resulting tip displacement $\mathcal{D}$ (m)

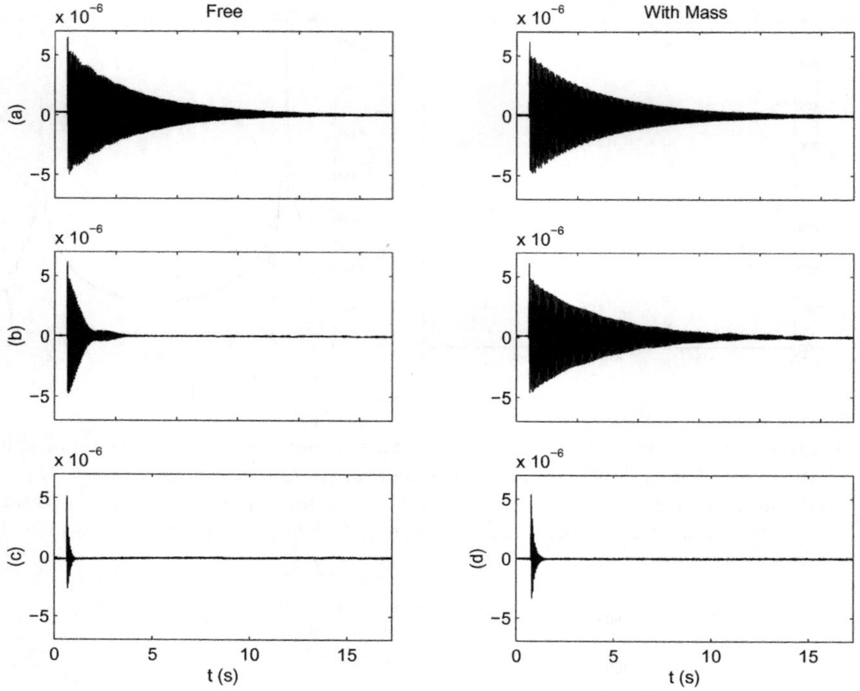

Figure 10.13. The free (left column) and with-mass (right column) tip displacement response $\mathcal{D}$ (m) to a step disturbance in W. Experimental open-loop (a), passive shunt controlled (b), and H_∞ shunt controlled (c) systems.

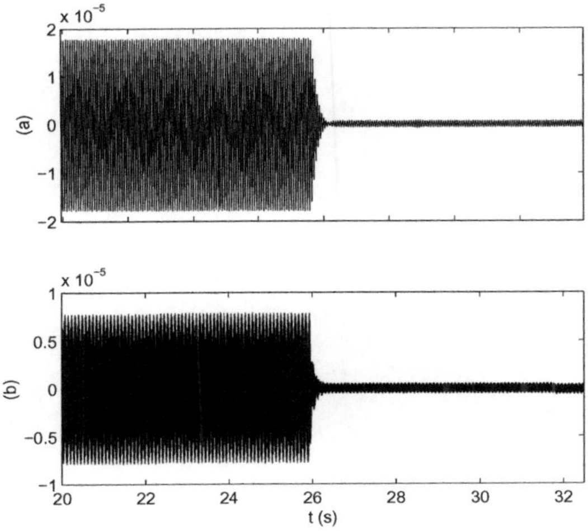

Figure 10.14. The open-loop and H_∞ shunt controlled tip displacement response $\mathcal{D}$ (m) to an acoustic sinusoidal disturbance at the 1st (a) and 2nd (b) structural resonance frequencies. Control is applied at approximately time 24.8 s.

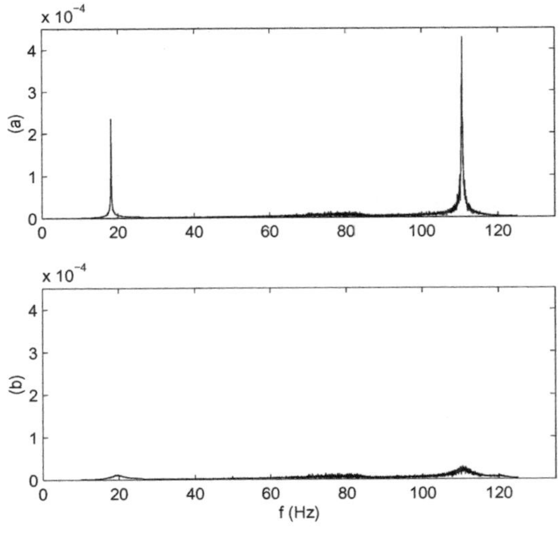

Figure 10.15. The open-loop (a), and H_∞ shunt controlled (b) linear magnitude response from an applied acoustic disturbance to the resulting tip displacement $\mathcal{D}$ (m)

11

Dealing with Hysteresis

11.1 Introduction

So far in this book we have assumed that piezoelectric transducers are linear devices. However, it is well known that piezoelectric transducers suffer from a form of nonlinearity known as hysteresis [127, 74]. Hysteresis is less of an issue at low drives. However, when a piezoelectric device is operated at higher drives it becomes more profound. Hysteresis can contribute to loss of robustness, performance degradation or even instabilities in feedback controlled piezoelectric devices [122, 123].

The existence of hysteresis in piezoelectric materials is generally attributed to residual misalignment of crystal grains in the poled ceramic [74, 94]. A large number of techniques have been developed that are aimed at reducing the hysteresis associated with voltage-driven piezoelectric actuators (see [71], and references therein). In particular, methods such as inversion-based Preisach modeling [127] and phase control [47] are two examples of the proposed techniques.

It has been argued that hysteresis is an electrical property of piezoelectric materials which mainly exists between the applied electric field and the resulting electrical charge [78]. Indeed, it has been demonstrated that by controlling electrical charge, or current rather than the applied voltage, the hysteresis effect can be substantially reduced [147]. As another example, in reference [75] a five-fold reduction in hysteresis was achieved by regulating charge rather than voltage.

Even though this approach has been known for some time, it has not been widely used due to the perceived difficulty of driving highly capacitive loads such as piezoelectric actuators. The main problem being the existence of offset voltages in the charge or current source circuit, which will eventually charge up the capacitive load. This will then distort the control signal being applied to the piezoelectric load. A quote from a recent paper [47] typifies the sentiment toward this technique: "While hysteresis in a piezoelectric actuator is reduced if the charge is regulated instead of the voltage [147], the implementation complexity of this technique prevents a wide acceptance [102]". This issue has been pointed out by a number of authors [122, 124]. Recent research, reported in [62] and detailed in Sections 6.4 and 6.5, proposes a new structure for charge and current sources capable of regulating the DC profile of the actuator. Due to this development, it is now possible to use electrical charge as the driving control signal for piezoelectric actuators in structural control applications.

In this chapter we study the dynamics of charge-driven piezoelectric actuators and we compare that with voltage-driven piezoelectric actuators. Furthermore, we will explain how the structure of resonant controllers can be altered to suit charge-driven piezos.

11.2 Hysteresis

Hysteresis is the major form of nonlinearity present in piezoelectric transducers. The original meaning of the word refers to "lagging behind" or "coming after", however, it must not be confused with "phase lag" which is not a nonlinearity and is present in many linear system. These two behaviors can be distinguished by plotting the input-output signals coming in and out of a system against each other.

For a piezoelectric actuator, hysteresis can be observed by applying an input voltage and measuring elongation of the device. The hysteretic system will show sharp reversal peaks at its extrema values similar to what is illustrated in Figure 11.1 (a). This means that the derivative of the input and output signals always have the same sign, whereas the tips of a similar plot obtained from a linear system will be more rounded and will display the overall shape of an ellipse such as illustrated in Figure 11.1 (b). It is important to note that

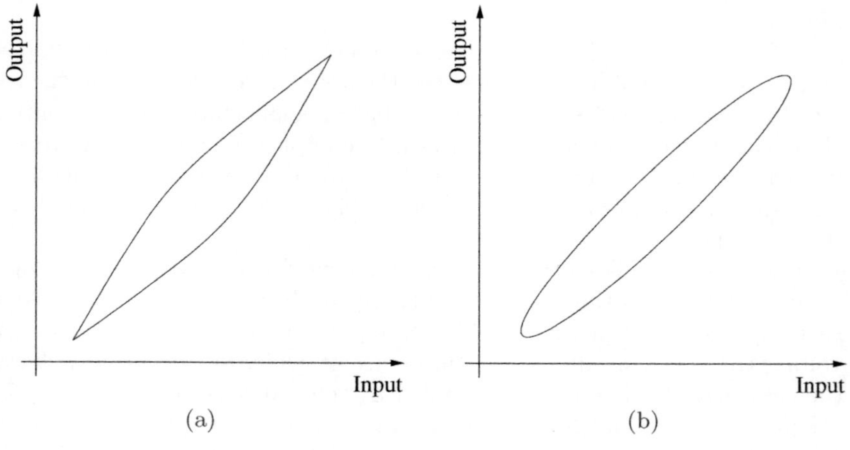

Figure 11.1. Input-Ouput plots displaying a) Hysteresis b) Phase Lag

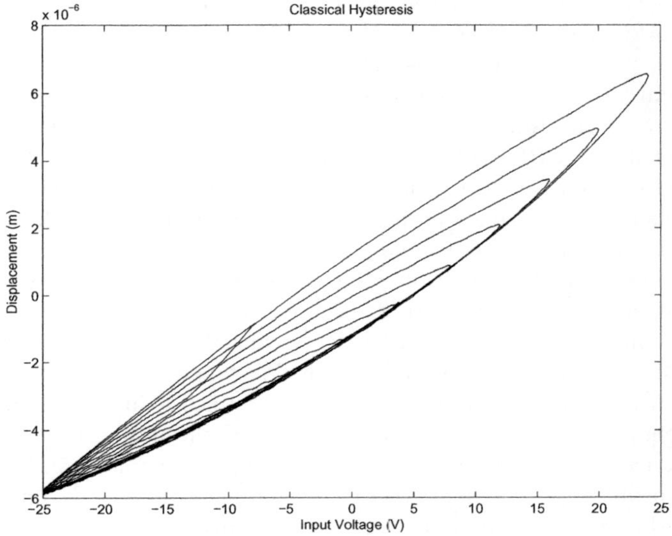

(a)

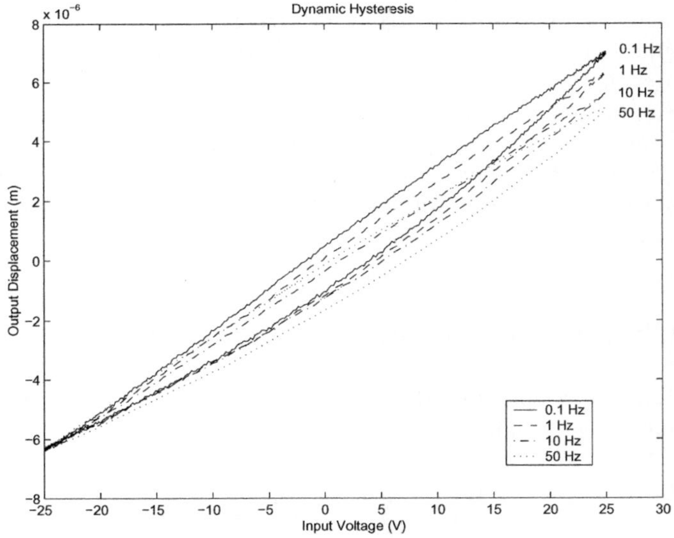

(b)

Figure 11.2. (a) Classical hysteresis (b) Dynamic hysteresis

these two behaviors are not mutually exclusive. In fact they often occur together, especially when piezoelectric transducers are bonded or attached to a mechanical structure. Nevertheless a clear distinction must be made between the two. The hysteretic distortion does not affect the whole input cycle, but will manifest itself between signal reversals.

The level of hysteretic distortion will also vary depending on either the maximum value of the input voltage being applied (Figure 11.2 (a)) or the frequency of the input signal (Figure 11.2 (b)) or both. These plots were obtained using a PI P-844.60 [1] piezoelectric stack actuator. Graph (a) was obtained by applying a modulated sinusoidal input voltage signal to the stack and measuring its corresponding elongation. Graph (b) was obtained by using a sinusoidal input voltage signal at different frequencies and measuring the corresponding stack elongation. The latter case is referred to as "dynamic" or "rate dependent" hysteresis [173, 175, 92, 146].

On a macroscopic level hysteresis is caused by internal energy losses (or power dissipation) in piezoelectric materials when expanding or contracting.

The hysteresis associated with piezoceramics exhibits nonlocal memory. This means that the current output state of the actuator not only depend on the current voltage input but also on its past history [75]. This can cause serious problems in feedback control systems if it is not accounted for, since several different output states can be obtained from the same input value depending on the "memory" (or past history) stored in the material.

As the voltage is increased the capacitance of the ceramic changes, and the energy loss increases, resulting in a change of the major axis angle. The area bounded by the hysteresis loop is directly proportional to the energy loss for each cycle [127].

The main idea promoted in this chapter is that by controlling charge rather than voltage, the effect of hysteresis nonlinearity can be reduced to the level that it can be effectively ignored. To control charge, however, one needs to first identify the dynamics of the underlying system which is covered in the next section.

11.3 Charge Control versus Voltage Control

In Section 3.5 we introduced two resonant controller structures for voltage-driven piezoelectric laminates. These are

$$K_v^\alpha = \sum_{i=1}^{\tilde{N}} \frac{\alpha_i \alpha_i' s^2}{s^2 + 2d_i\tilde{\omega}_i s + \tilde{\omega}_i^2}, \tag{11.1}$$

and

$$K_v^\beta = \sum_{i=1}^{\tilde{N}} \frac{\beta_i \beta_i' s(s + 2d_i\tilde{\omega}_i)}{s^2 + 2d_i\tilde{\omega}_i s + \tilde{\omega}_i^2}, \tag{11.2}$$

where both α_i and β_i are $m \times 1$ vectors.

[1] Physik Instrument (PI) GmbH and Co.

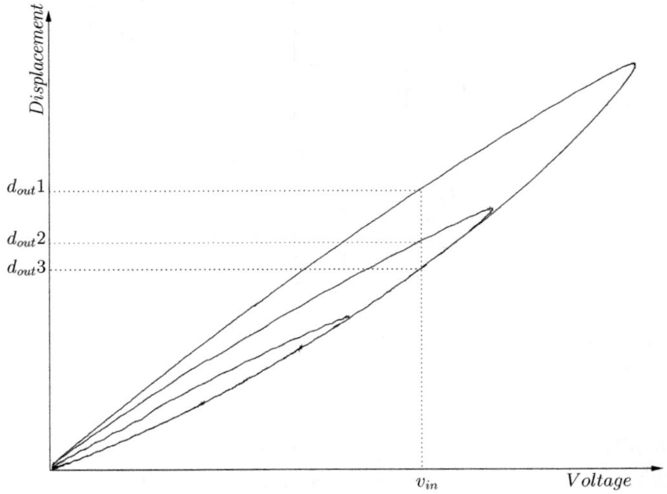

(a) Several output states for the same input

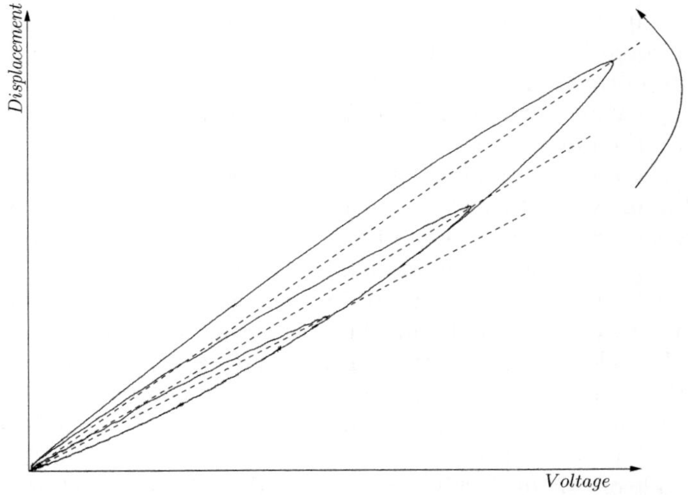

(b) Major axis angle change with increasing input magnitudes

Figure 11.3. Nonlocal memory examples

We also proved that the closed-loop system consisting of any of the above controllers and a system of the form described by (3.1) is guaranteed to be stable.

The resonant controllers (11.1) and (11.2) will raise two significant problems when voltage is used to drive the piezoelectric actuators.

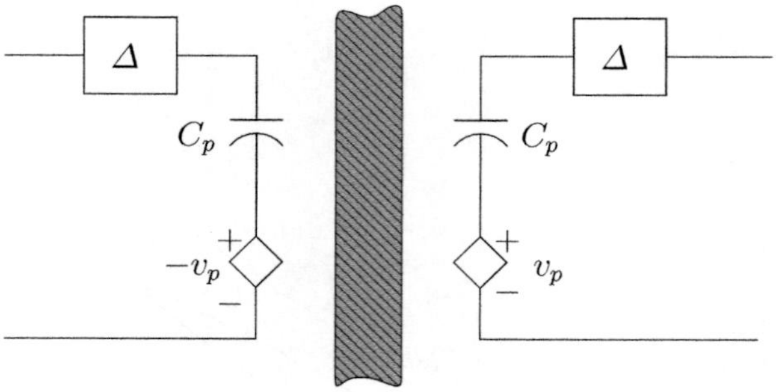

Figure 11.4. Electrical equivalent of a collocated piezoelectric actuator/sensor pair incorporating the effect of hysteresis

Typically the frequency response of a flexible structure with collocated piezoelectric transducers, as described by (3.1), rolls off very slowly at high frequencies[2]. A controller that rolls off is therefore preferable. The first problem therefore arises from the fact that voltage driven resonant controllers do not roll off quickly enough at higher frequencies. Although, the stability of the closed-loop system under (11.1) and (11.2) is guaranteed, because of the asymptotic behavior of (11.1) and (11.2), the performance of the closed-loop system may suffer from high-frequency noise and the inevitable phase lag introduced by the sensor.

The second complication is due to the hysteretic nature of piezoelectric actuators when driven by voltage amplifiers. As illustrated in Figure 11.4, each piezoelectric transducer can be modeled as the series connection of a voltage source, v_p which is proportional to the total strain in the piezo, a capacitor C_p and a nonlinear element Δ [78]. The nonlinearity is of the hysteresis type and is much more profound when the actuator is operated at high voltages[3].

This clearly suggests that when possible electrical charge rather than voltage should be used to drive piezoelectric actuators. The reason why charge or current sources have not been extensively used to drive piezoelectric transducers can be attributed to the perceived difficulty in implementing devices capable of driving highly capacitive loads such as piezoelectric actuators. Existence of offsets in conjunction with the uncontrolled nature of the output voltage generally results in the capacitor being charged up. Once the output voltage reaches the power supply rail, the output becomes saturated, and the amplifier fails to perform properly. In Section 6.5 we demonstrated that

[2] This situation can be improved by reducing the dimensions of piezoelectric transducers [35]. This would come at the price of reduced controller authority over the structure and cannot be recommended for all applications.

[3] In fact, when the piezoelectric actuator is being operated at low voltages, hysteresis becomes almost negligible. Restricting the level of control voltage, however, is not advisable since the actuator is not being used to its full potential.

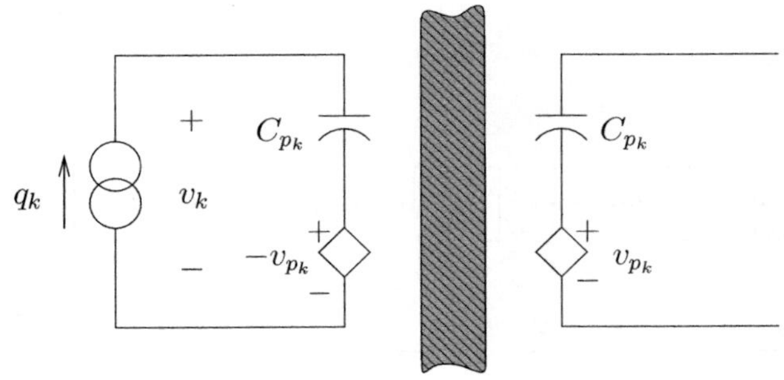

Figure 11.5. Collocated charge control of one piezoelectric actuator/sensor pair

this difficulty can be overcome by careful modification of the structure of a standard charge controller[4].

In order to analyze the effect of using charge rather than voltage in vibration control applications let us define the following parameters:

$$V = \begin{bmatrix} v_1 \\ v_2 \\ \vdots \\ v_m \end{bmatrix}; \quad V_p = \begin{bmatrix} v_{p1} \\ v_{p2} \\ \vdots \\ v_{pm} \end{bmatrix}; \quad W = \begin{bmatrix} w_1 \\ w_2 \\ \vdots \\ w_\ell \end{bmatrix};$$

$$\Lambda = \begin{bmatrix} \frac{1}{C_{p1}} & & & \\ & \frac{1}{C_{p1}} & & \\ & & \ddots & \\ & & & \frac{1}{C_{pm}} \end{bmatrix}; \quad Q = \begin{bmatrix} q_1 \\ q_2 \\ \vdots \\ q_m \end{bmatrix},$$

where V is the vector of voltages applied to the piezoelectric actuators, V_p is the vector of voltages measured at the piezoelectric sensors, W is the vector of disturbances acting on the beam, C_{p_i} is the capacitance associated with each collocated piezoelectric patch and Q represents the vector of electrical charges.

Some of the assumptions made here are: (i) m piezoelectric actuator/sensor pairs are bonded to the structure; (ii) each pair consists of two identical transducers, however, not all transducers are necessarily identical; (iii) ℓ disturbances are acting on the structure; (iv) the controller is fully multivariable and (v) C_{p_i} represents capacitance of the ith transducer.

Writing the KVL around the kth loop, as illustrated in Figure 11.5, we obtain:

$$v_k = -v_{p_k} + \frac{1}{C_{p_k}} q_k.$$

[4] Another approach will be discussed in Chapter 12.

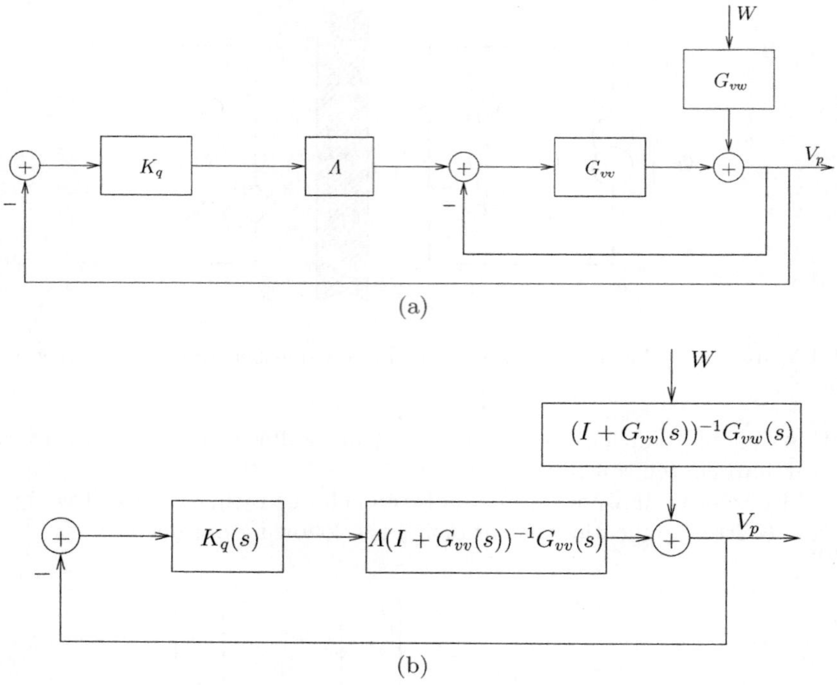

Figure 11.6. Feedback structure associated with charge-driven piezoelectric actuator/sensor pairs (a); and its equivalent (b)

Using the above notation, this implies

$$V = -V_p + \Lambda Q. \tag{11.3}$$

Since

$$Q = -K_q V_p \tag{11.4}$$

and

$$V_p = G_{vw} W + G_{vv} V, \tag{11.5}$$

which follows from the linearity of the system, the feedback structure illustrated in Figure 11.6 (a) can be obtained. Here, G_{vv} is the multivariable transfer function matrix of the collocated system (3.1). G_{vw} is also a multivariable $m \times \ell$ transfer function matrix, with a structure specified in (3.2).

It can be verified that the multivariable transfer function matrix relating V_p to W is given by

$$V_p(s) = \left[I + G_{vv}(s)\left(I + \Lambda K_q(s)\right)\right]^{-1} G_{vw}(s) W(s). \tag{11.6}$$

A point that is needed to be clarified here is what happens when $K_q = 0$ in (11.6). When the piezoelectric actuators are driven by voltage amplifiers, and K_v is set to zero, the closed-loop transfer function matrix of the system reduces to G_{vw}. However, if K_q in (11.6) is set to zero, the closed-loop transfer function matrix reduces to

$$[I + G_{vv}(s)]^{-1} G_{vw}(s). \tag{11.7}$$

When $K_v = 0$, the piezoelectric actuators are effectively short-circuited. However, $K_q = 0$ means that the actuators are left open-circuited. Although the difference between the response in these two cases may come as a surprise, it does make a difference if piezoelectric transducers are open- or short-circuited. To appreciate this, we point out that (11.6) can be re-written as

$$\begin{aligned} V_p(s) = \big[I + \Lambda(I + G_{vv}(s))^{-1} G_{vv}(s) K_q(s)\big]^{-1} \\ \times\ (I + G_{vv}(s))^{-1} G_{vw}(s) W(s), \end{aligned} \tag{11.8}$$

which suggests the feedback structure in Figure 11.6 (b).

To this end we remind the reader of a similar discussion, in the context of shunted piezoelectric transducers that are left open circuited, on page 97.

11.4 Resonant Controllers for Charge-driven Piezoelectric Actuators

Identification of the feedback structure associated with a charge-driven piezoelectric laminate leads to an important observation: If the piezoelectric actuators were to be driven by voltage amplifiers, rather than charge amplifiers, and if the underlying system were linear, then the two closed-loop systems would be identical as long as

$$K_v(s) = I + \Lambda K_q(s). \tag{11.9}$$

This observation becomes especially useful in terms of developing resonant control structures for charge-driven systems. Given that for an arbitrary $K_v(s)$ the equivalent charge controller is

$$K_q(s) = \Lambda^{-1}(K_v(s) - I), \tag{11.10}$$

if $K_v(s)$ is chosen to be either $K_v^{\alpha}(s)$ or $K_v^{\beta}(s)$, as defined in (3.12) and (3.13), with the additional constraints

$$\sum_{i=1}^{\tilde{N}} \alpha_i \alpha_i' = I \tag{11.11}$$

and

$$\sum_{i=1}^{\tilde{N}} \beta_i \beta_i' = I \tag{11.12}$$

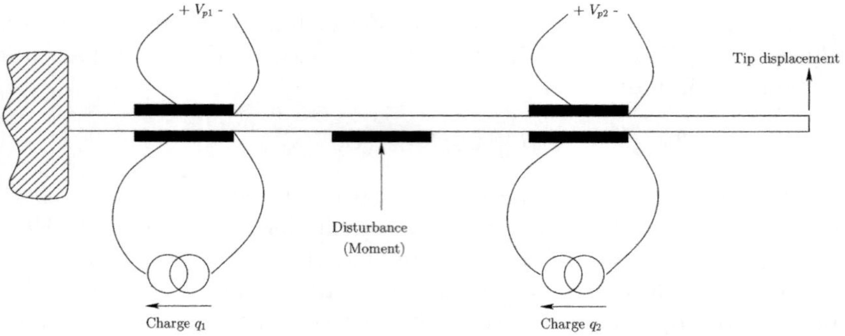

Figure 11.7. Beam Arrangement

we obtain the following resonant controllers for equivalent charge-driven systems:

$$K_q^{\alpha}(s) = -\Lambda^{-1} \sum_{i=1}^{\tilde{N}} \frac{\alpha_i \alpha_i'(s + 2d_i\tilde{\omega}_i)}{s^2 + 2d_i\tilde{\omega}_i s + \tilde{\omega}_i^2} \tag{11.13}$$

and

$$K_q^{\beta}(s) = -\Lambda^{-1} \sum_{i=1}^{\tilde{N}} \frac{\beta_i \beta_i' \tilde{\omega}_i^2}{s^2 + 2d_i\tilde{\omega}_i s + \tilde{\omega}_i^2}. \tag{11.14}$$

It can be observed, from (11.13) and (11.14), that resonant controllers for charge-driven piezoelectric actuators will be strictly proper as long as Conditions (11.11) and (11.12) are enforced. And in the SISO case, Controllers (11.13) and (11.14) will roll off quickly at 20 dB and 40 dB respectively. This is a favorable property that is not present for the voltage driven case as discussed in Section 3.5.

It should be noticed that Conditions (11.11) and (11.12) could limit the closed-loop performance of resonant controllers since they impose a hard constraint on the structure of the controller. Despite this, good performance can still be obtained using these controllers as illustrated, experimentally, in Section 11.5.

11.5 Experimental Implementation of a Multivariable Resonant Controller

To evaluate the performance of proposed resonant controllers, an experimental testbed consisting of a cantilever beam and two collocated piezoelectric pairs was constructed. One pair was located close to the clamped end and the other closer to the free end of the beam. For each collocated pair, one piezoelectric patch was used as an actuator and was driven by a charge amplifier, while

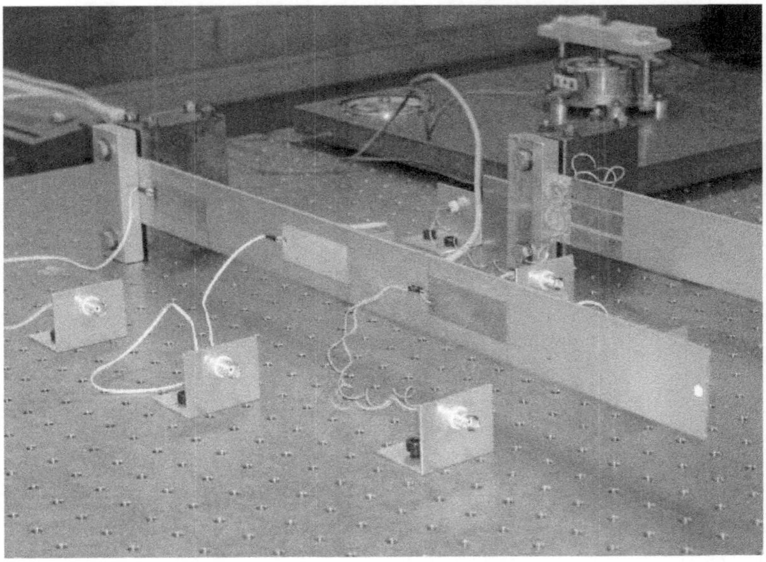

Figure 11.8. Picture of the Cantilever Beam

the voltage induced in the other patch was used as the measurement. Another piezoelectric actuator was bonded to the beam, somewhere between the two actuating patches. This transducer was driven by a voltage source to apply a disturbance to the beam. The transducer collocated with this actuator was short circuited so that it would not add any loading on the structure. A schematic of the experimental testbed is demonstrated in Figure 11.7 and a picture of the actual beam is shown in Figure 11.8. The collocated control transducers are 0.25mm in thickness while the disturbance transducers are 0.5mm.

The purpose of the experiment was to design and implement a two-input-two-output resonant controller to regulate the tip displacement of the beam in face of vibrations arising from a disturbance voltage applied to the third actuator. For the purpose of designing such a controller an accurate model of the cantilever beam was needed. The modeling process is detailed in the remainder of this section.

11.5.1 The Hysteresis Effect

To demonstrate the presence of hysteresis one of the piezoelectric actuators was first driven by a voltage source and then subsequently by a charge source. The voltage induced in the collocated piezoelectric transducer was measured and recorded. The actuating signal in each case was a linearly decaying single-tone sinusoid of 24 Hz. At this frequency the corresponding transfer function

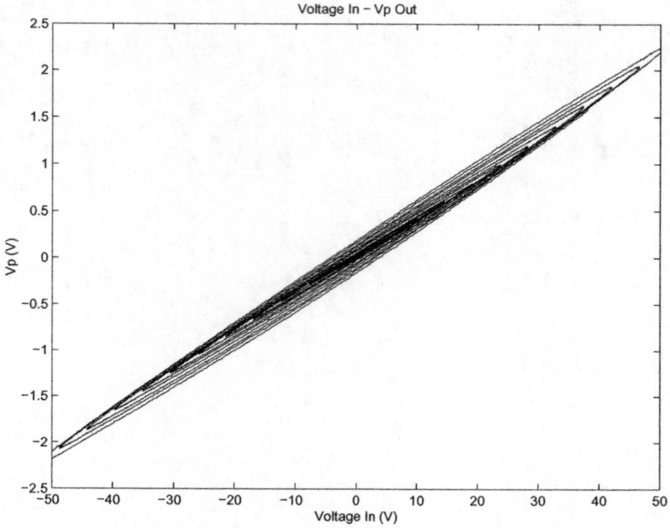

(a)

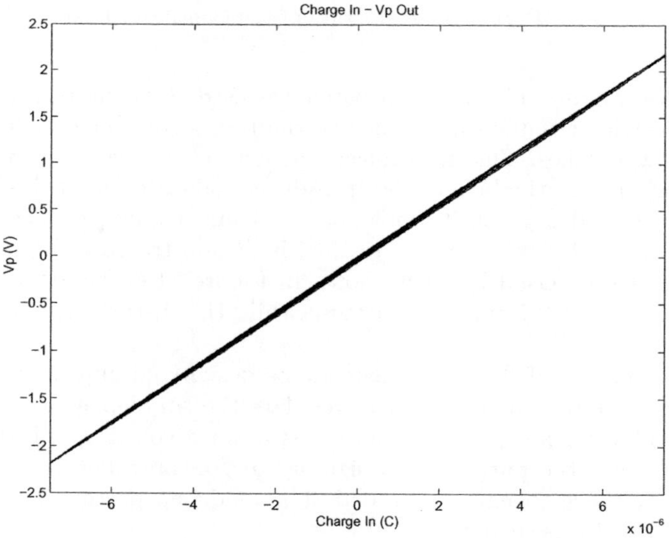

(b)

Figure 11.9. Plot of the output voltage versus (a) the input voltage and (b) the input charge

has zero phase. Therefore, any deviation from a straight line on the input-output plot is purely due to hysteresis.

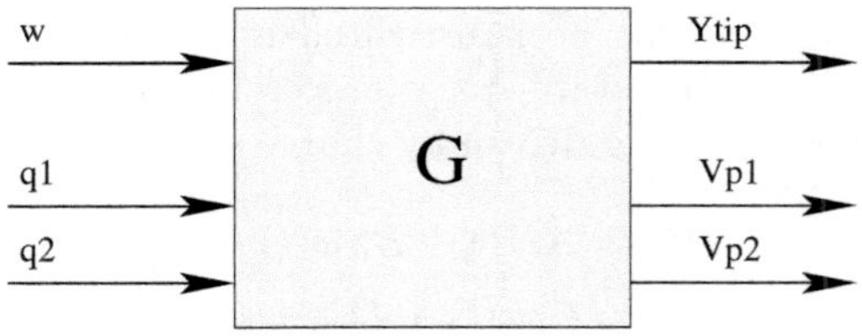

Figure 11.10. Augmented MIMO Plant

In this experiment, the amplitude of the charge signal was adjusted to ensure that the measured voltage at the collocated piezoelectric transducer was at a comparable level to that measured when the actuator was driven by a voltage source. The results are illustrated in Figure 11.9 (a) and (b). The presence of hysteresis when the actuator is driven by a voltage source is evident from Figure 11.9 (a). However, when a charge source is used, hardly any hysteresis can be observed. This agrees with similar results discussed in [122] and [124]. In particular, it has been reported that piezoelectric actuators exhibit up to 80% less hysteresis when driven by charge amplifiers [147, 39]. In other words, in a piezoelectric transducer hysteresis mainly exists between the voltage and mechanical strain, rather than the charge (or current) and strain [78].

It should be pointed out that hysteresis is not a major source of difficulty with piezoelectric sensors. The use of a buffer circuit with very high input impedance will significantly reduce the effect of hysteresis in the sensor. This should be evident from Figure 11.4. A similar strategy for the actuator does not exist.

11.5.2 State-space Model of the Composite System

Assuming for a moment that all actuators are being driven by voltage amplifiers, the multivariable plant can be represented by the following state space equations:

$$\dot{x}(t) = Ax(t) + B_w w(t) + B_v V(t) \tag{11.15}$$
$$Y_{tip} = C_y x(t) + D_{yw} w(t) + D_{yv} V(t) \tag{11.16}$$
$$V_p(t) = C_v x(t) + D_{vw} w(t) + D_{vv} V(t), \tag{11.17}$$

where $x \in \mathbf{R}^{2N}$ represents the state vector of the system, N is the number of modes included in the model, and w and $V = \begin{bmatrix} v_1 & v_2 \end{bmatrix}$ represent the disturbance and control input voltages respectively.

Substituting (11.3) into (11.17) we obtain:

$$V(t) = -(I + D_{vv})^{-1} C_v x(t) - (I + D_{vv})^{-1} D_{vw} w(t) + (I + D_{vv})^{-1} \Lambda Q(t). \tag{11.18}$$

Equation (11.18) can now be substituted into Equations (11.15)-(11.17) to obtain the multivariable state-space representation for the plant when the actuating patches are being driven by charge sources. The resulting system is:

$$\dot{x}(t) = \tilde{A}x(t) + \tilde{B}_w w(t) + \tilde{B}_q Q(t) \tag{11.19}$$
$$Y_{tip}(t) = \tilde{C}_y x(t) + \tilde{D}_{yw} w(t) + \tilde{D}_{yq} Q(t) \tag{11.20}$$
$$V_p(t) = \tilde{C}_v x(t) + \tilde{D}_{vw} w(t) + \tilde{D}_{vq} Q(t), \tag{11.21}$$

where

$$\begin{aligned}
\tilde{A} &= A - B_v(I + D_{vv})^{-1}C_v \\
\tilde{B}_w &= B_w - B_v(I + D_{vv})^{-1}D_{vw} \\
\tilde{B}_q &= B_v(I + D_{vv})^{-1}\Lambda \\
\tilde{C}_y &= C_y - D_{yv}(I + D_{vv})^{-1}C_v \\
\tilde{D}_{yw} &= D_{yw} - D_{yv}(I + D_{vv})^{-1}D_{vw} \\
\tilde{D}_{yq} &= D_{yv}(I + D_{vv})^{-1}\Lambda \\
\tilde{C}_v &= C_v - D_{vv}(I + D_{vv})^{-1}C_v \\
\tilde{D}_{vw} &= D_{vw} - D_{vv}(I + D_{vv})^{-1}D_{vw} \\
\tilde{D}_{vq} &= D_{vv}(I + D_{vv})^{-1}\Lambda.
\end{aligned}$$

The reader should notice that the dynamics of the plant are different depending on whether charge or voltage is used to drive the actuating patches. This is reflected in the different state matrices associated with the systems represented by (11.15) and (11.19). Furthermore, as evidenced from Equations (11.15) and (11.19), the charge driven state-space equations are expressed in terms of A, B, C and D matrices associated with the voltage-driven case. Therefore, to successfully construct the charge-driven equations, it is necessary to accurately identify these parameters.

These values can be derived from the structural properties and dimensions of the beam and its boundary conditions as demonstrated in [83]. Here, we use a system identification approach, which is detailed in Section 11.5.4.

11.5.3 Structure of the State-space Model

The physical modeling for the voltage driven scenario is well understood [142, 83] and uses derivations from the Euler-Bernoulli beam equation to determine the transfer function from input voltages (V) to output voltages (V_p) (or output tip displacement Y_{tip}). From these derivations it has been shown that the A, B and C matrices (for the voltage driven case) are of the form:

$$A = \begin{bmatrix} 0 & 1 & & 0 & 0 \\ -\omega_1^2 & -2\zeta_1\omega_1 & & 0 & 0 \\ & & \ddots & & \\ 0 & 0 & & 0 & 1 \\ 0 & 0 & & -\omega_N^2 & -2\zeta_N\omega_N \end{bmatrix} \tag{11.22}$$

$$B = [B_w \quad B_{v_1} \quad B_{v_2}] = \begin{bmatrix} 0 & 0 & 0 \\ g_1^w & g_1^{v_1} & g_1^{v_2} \\ \vdots & \vdots & \vdots \\ 0 & 0 & 0 \\ g_N^w & g_N^{v_1} & g_N^{v_2} \end{bmatrix} \tag{11.23}$$

$$C = \begin{bmatrix} C_y \\ C_{v_1} \\ Cv_2 \end{bmatrix} = \begin{bmatrix} f_1^y & 0 \ldots & f_N^y & 0 \\ f_1^{v_1} & 0 \ldots & f_N^{v_1} & 0 \\ f_1^{v_2} & 0 \ldots & f_N^{v_2} & 0 \end{bmatrix}, \tag{11.24}$$

where ω_i and ζ_i correspond to the natural frequency and damping ratio of the ith vibrational mode of the structure, g_i^w, $g_i^{v_1}$ and $g_i^{v_2}$ are values that correspond to the input disturbance (w) and control input voltages (v_1 and v_2) respectively, and f_i^y, $f_i^{v_1}$ and $f_i^{v_2}$ are values that correspond to the output tip displacement and induced output voltages (v_{p_1} and v_{p_2}) respectively. Note that (11.23) and (11.24) can be equally expressed in terms of the notation used in (3.1) and (3.2).

Since the cantilever beam has two collocated pairs of piezoelectric transducers (Figure 11.7) we can make the following simplification and set $g_i^{v_1} = f_i^{v_1}$ and $g_i^{v_2} = f_i^{v_2}$ [142, 35].

The matrices (A, B and C) are then used in Equations (11.19), (11.20) and (11.21) to obtain the charge driven state-space equations for the plant.

In the experiment only the first three vibrational modes of the beam were considered for damping, *i.e.* $N = 3$. We therefore had to include a feed-through term D to compensate for the misalignment of the in-bandwidth zero locations that arises from the inability to model the higher order modes of the structure [133, 34].

11.5.4 System Identification

To obtain a model of the plant, suitable for control design purposes, a three-input three-output model as illustrated in Figure 11.10 was identified. The first input corresponds to the disturbance voltage (w) applied to the middle patch. The second and third inputs are the charges (q_1 and q_2) applied to the first and second actuators respectively. The first output corresponds to the displacement measured at the tip of the cantilever (Y_{tip}). The second and third outputs are the voltages (V_{p_1} and V_{p_2}) measured at the first and second piezoelectric transducers, respectively.

To model the plant we measured all nine frequency responses for each input-output combination. These frequency responses were obtained by ap-

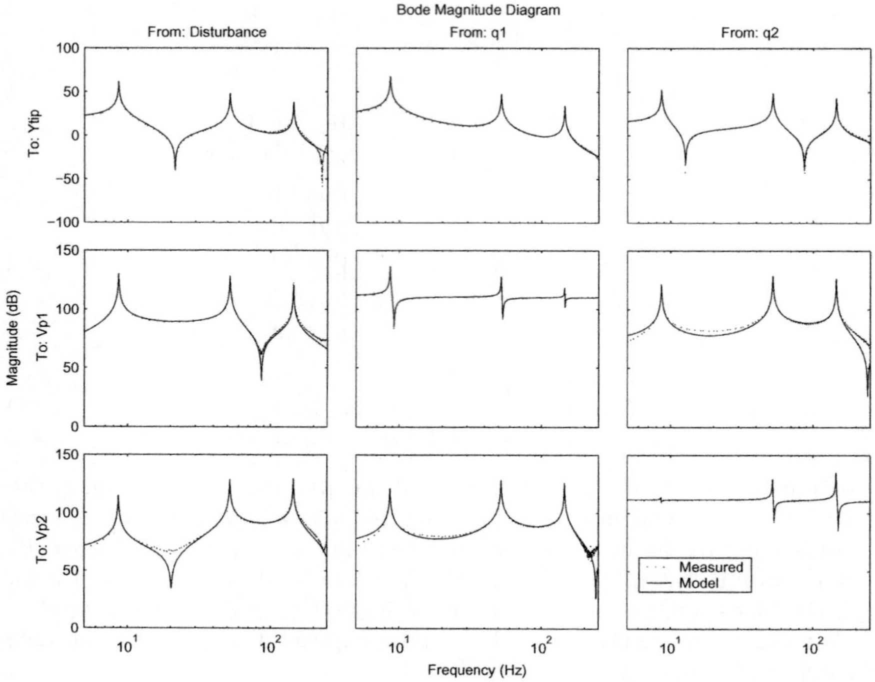

Figure 11.11. Identified Model with Measured data

plying a sinusoidal chirp signal of varying frequency (from 5 to 250 Hz) to the piezoelectric actuators and measuring the corresponding output signals of interest (namely the output voltages V_p from the collocated sensors and the displacement at the tip of the beam Y_{tip}). The input/output data was processed in real time by the Polytec laser scanning vibrometer (PSV-300) software to obtain the desired frequency responses. An optimization problem was set up and solved to obtain the "best fit" state space model by minimizing the normalized least squared error between the simulated and measured data.

Prior to running the optimization, the values of ω and ζ for each mode were fixed (*i.e.* the A matrix is fixed). This significantly reduces the numerical complexity of the optimization problem that is to be solved. In fact, the natural frequencies ω_i and damping ratios ζ_i can be obtained, and fine-tuned with relative ease by measuring the location and height of the peaks from the measured frequency responses. Another approach is to perform a sub-space system identification on one of the channels, which would yield a good initial guess for the values of ω and ζ for each mode. Once the A matrix is fixed, only numerical values of the B, C and D matrices need to be determined using the optimization.

The results of the magnitude bode plots are illustrated in Figure 11.11 and demonstrate that the identified model closely matches the experimentally measured data. Furthermore some interesting observations can be drawn from these plots. The Bode plots corresponding to the first and second collocated actuator/sensor pairs are found in the [2,2] and [3,3] positions respectively. A closer look at these plots reveals that the first mode is clearly the dominant mode in the first collocated pair ([2,2]). Therefore, one would expect the controller to have significant authority over the first vibratory mode of the structure through this actuator/sensor pair. However, the controller would have little authority over the second and third modes as evident from the low profile of these two modes in the ([2,2]) plot. This situation is reversed for the the second collocated pair ([3,3]). The reason for this can be explained by examining the first three vibrational mode shapes of a cantilever beam in Figure 11.12 [142]. For an actuator to have maximum authority over a given mode, it must be placed at the location where the curvature[5] is greatest for that particular vibrational mode [35].

It can also be inferred that it would be nearly impossible to control a large number of vibration modes with only one actuator/sensor pair. To ef-

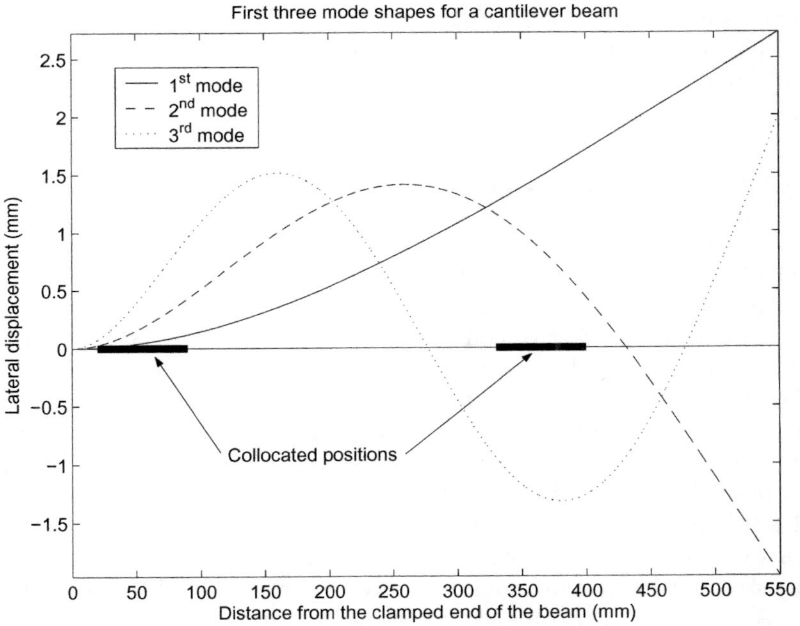

Figure 11.12. First three mode shapes for a cantilever beam

---

[5] For a one-dimensional structure such as a beam strain is proportional to the curvature [160]

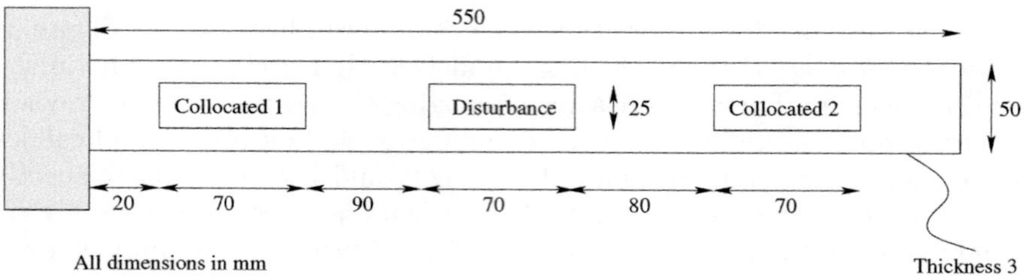

Figure 11.13. Beam setup and dimensions

fectively control a number of modes, one may need to use several collocated actuator/sensor pairs along with a multivariable controller.

If we now return to the plots [2,2] and [3,3], the reader will notice that the collocated arrangement for these two plots can easily be deduced from their peak-trough alternating pattern. Further evidence to such a collocated arrangement is demonstrated because the plots [2,3], [3,2] are nearly identical.

The largest mismatch is found in the [3,1] bode magnitude plot, where the model fails to properly match the measured data around the first anti-node. The reason for this is not quite clear but may be due to the low signal to noise ratio as measured by the sensors.

The identified model was then used to design a number of resonant controllers, which were subsequently implemented on the system. This is detailed in the following sub-section.

11.5.5 Controller Design

Two resonant controllers were proposed in Section 11.4. Of the two controllers, (11.14) rolls off faster at higher frequencies and is therefore chosen as the candidate controller for the above structure.

To obtain an effective controller, appropriate values for d_i, $\tilde{\omega}_i$ and β_i for each mode need to be selected. This can be achieved in a number of ways. For example, by minimizing the H_2 or H_∞ norm of the closed-loop transfer function from disturbance voltage to tip displacement ($T_{Y_{tip}w}$). However, as argued earlier, more than simply minimizing a specific measure, the ultimate purpose of the controller is to add extra damping to the system by shifting the closed-loop poles of the plant deeper into the left half of the complex plane. This will ensure that no matter where a disturbance is entering the system, the structural vibration is minimized within the controlled bandwidth.

A number of performance measures were used to determine appropriate parameters for the controller. Only two of these methods are reported here. The first controller was obtained by minimizing the H_2 norm of $T_{Y_{tip}w}$ and the second by placing the poles of the closed-loop system as far to the left of the $j\omega$ axis as possible.

The optimizations were carried out using the simplex search algorithm in MATLAB®. Before performing the optimization, the values of β were fixed such that Condition (11.12) was satisfied.

The values of β were chosen manually taking into account the authority of the controller over each specific mode, through either actuator. In practice this means the need to select relatively large absolute numbers for the first mode associated with the first actuator, and for the second and third modes associated with the second actuator. The numbers must be chosen such that Condition (11.12) is satisfied. The values used in the experiments are

$$\beta = \begin{bmatrix} 0.9722 & -0.2308 & 0.0399 \\ 0.2046 & 0.7541 & -0.6241 \end{bmatrix}. \tag{11.25}$$

The first and second row correspond to the first and second collocated arrangements respectively. The column entries correspond to their respective modes as well.

Since the β values are fixed, only the $\tilde{\omega}$ and d values needed to be determined using the optimization routine.

Minimizing the H_2 norm of the closed-loop system yields a resonant controller that adds damping to the structure, in a roundabout way. The pole placement technique, however, is a more straightforward approach since the physical location of the closed-loop poles is intrinsically included in the cost function. In this approach the absolute distance between a pre-specified pole location and the actual closed-loop poles of the system is minimized.

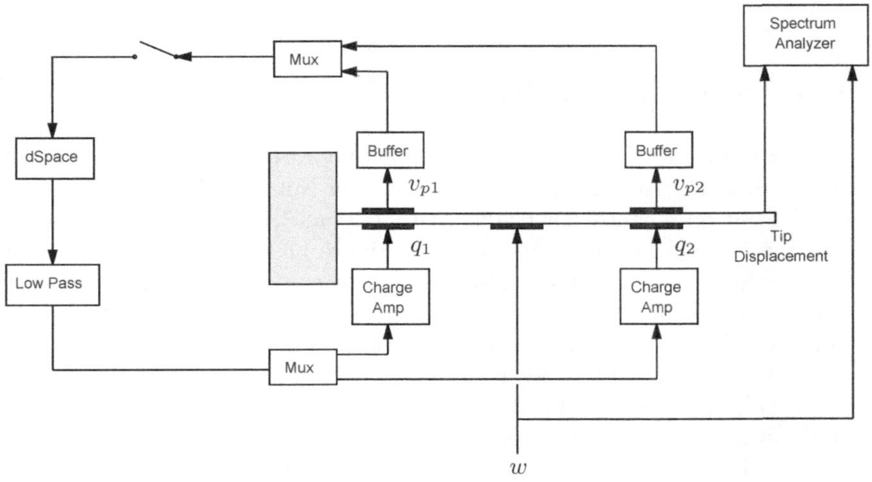

Figure 11.14. Experimental Schematic

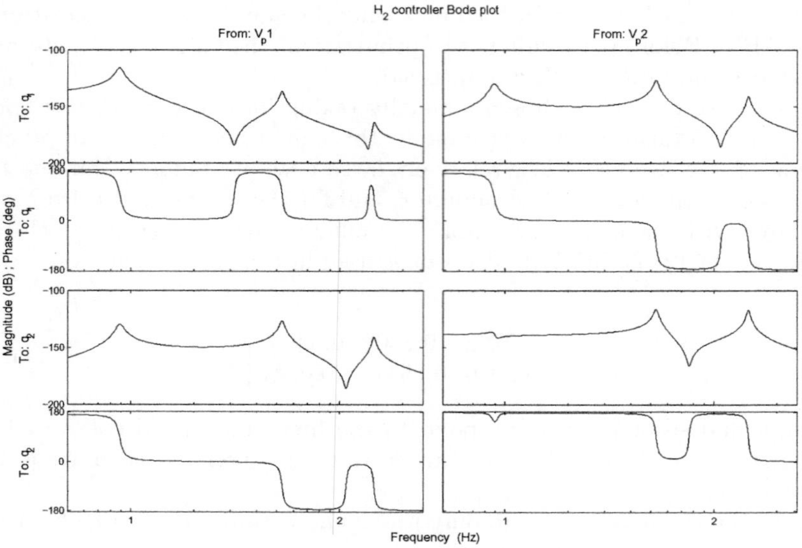

Figure 11.15. Bode plot of the resonant H_2 controller

A Bode plot of the multivariable H_2 controller is shown in Figure 11.15. As can be noted, the diagonal entries correspond to the collocated transfer functions. Also due to (11.12) the cross diagonal transfer functions are identical.

11.6 Experimental Results

The experiments were performed in the Laboratory for Dynamics and Control of Smart Structures at the University of Newcastle, Australia, and were carried out on a cantilever beam with identical collocated PIC 151 piezoelectric patches, with dimensions shown in Figure 11.13. The disturbance voltage was applied to a secondary patch located at the center of the beam and the two collocated actuator-sensor pairs were used for feedback control purposes only. A Polytec laser scanning vibrometer (PSV-300) was used to measure the velocity at the tip of the beam.

In this experiment, the frequency responses were obtained by applying a sinusoidal voltage signal of varying frequency to the "disturbance" piezoelectric patch and measuring the corresponding output tip displacement of the beam (Y_{tip}), with and without the controller being switched on. The Polytec software was again used to determine the corresponding frequency responses.

Table 11.1. Damping ratios

	Damping ratios (10^{-3})		
Modes	1st	2nd	3rd
Open loop	3.29	3.26	2.35
H_2 norm	18.9	12.5	9.15
Pole placement	39.8	26.2	19.2

A Schematic of the experimental testbed is shown in Figure 11.14. To carry out the experiments, the controller was downloaded from Simulink® onto a dSPACE DS-1103 DSP board. Low-pass anti-aliasing and reconstruction filters were added to the system. The measured voltage for each piezoelectric sensor was passed through a high-impedance buffer to avoid the measurement device loading of the piezoelectric sensor, and hence to minimize hysteresis. This ensured the sensor signals were not distorted at low frequencies.

The measured and simulated open- and closed-loop bode magnitude plots for the H_2 optimized resonant controller are given in Figure 11.16 (a) and (b) respectively. The same plots for the pole placement optimization method are shown in Figure 11.17. On average these controllers reduced the resonant peak of the input disturbance (w) to output displacement (Y_{tip}) transfer function by 13 dB for each mode.

Time domain displacement step responses for the H_2 and pole placement optimized resonant controllers are displayed in Figure 11.18 and 11.19 respectively. These plots were obtained by applying a low-pass filtered (250 Hz) step signal to the disturbance patch and measuring the velocity at the tip of the beam. The velocity was then integrated off line to obtain the corresponding displacement. The open-loop step response is also included in these plots as a comparison. Both control schemes performed well and reduced the settling time of the beam by a factor of 5.

11.7 Some Observations

The two optimization schemes produced acceptable results. Indeed the closed-loop response for these controllers demonstrate significant damping. The damping ratios (ζ_i) associated with each mode of the closed-loop system were determined for each controller. The results are given in Table 11.1. Notice that in each case only the damping ratio associated with the pole closest to the $j\omega$ axis is shown.

It can be observed that the pole placement optimization method provides the most damping for each mode. This can be verified by looking at the closed-loop pole locations for each optimization scheme (Figure 11.20) since the closed-loop poles associated with the pole placement method are located furthest to the left from the imaginary axis.

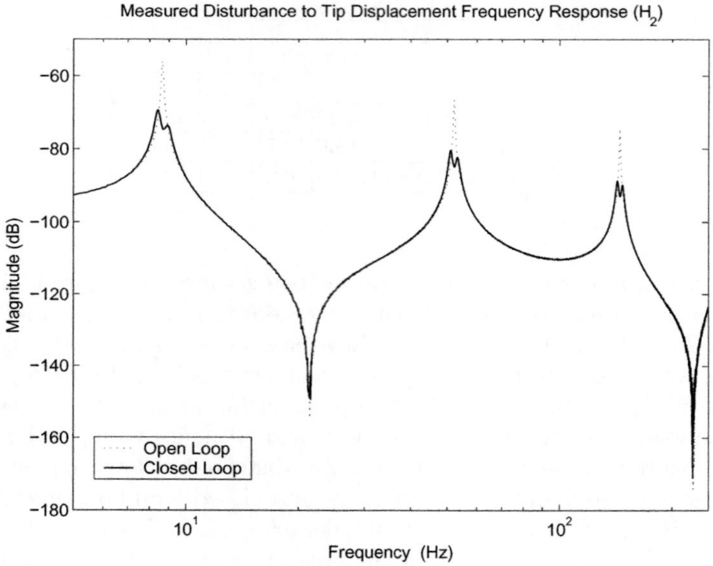

(a) Measured Data

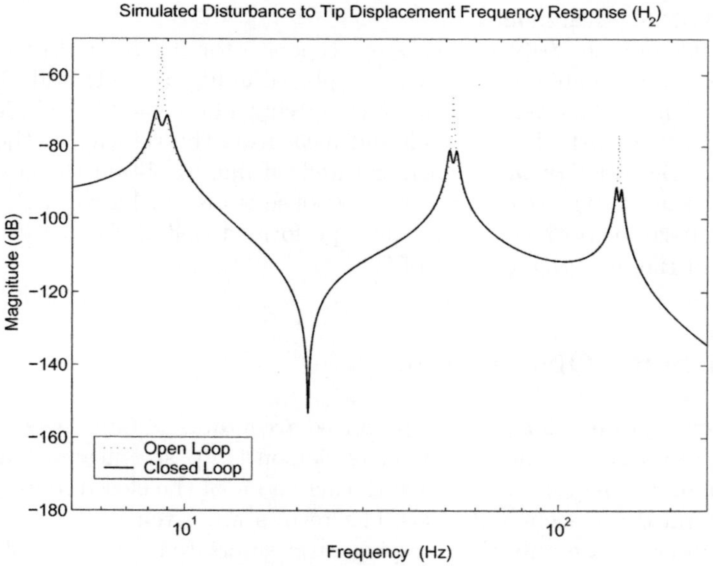

(b) Simulation Results

Figure 11.16. Open- and closed-loop frequency response for input disturbance w to output tip displacement Y_{tip} using the H_2 norm optimization

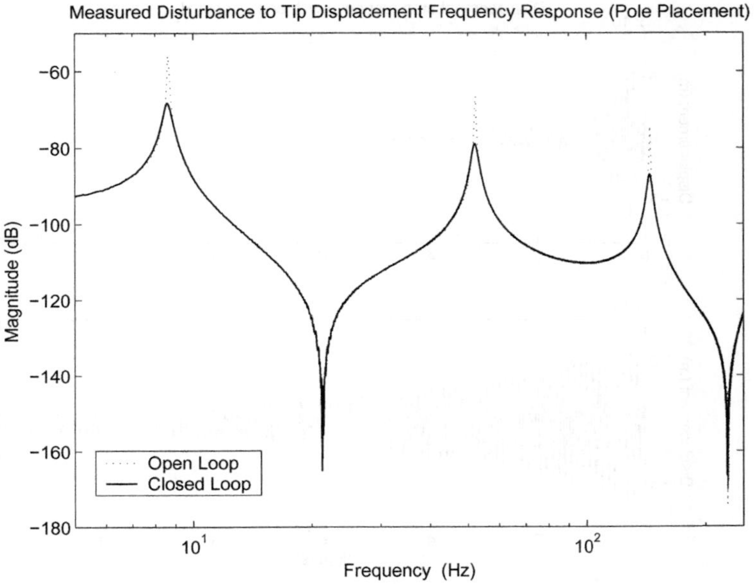

(a) Measured Data

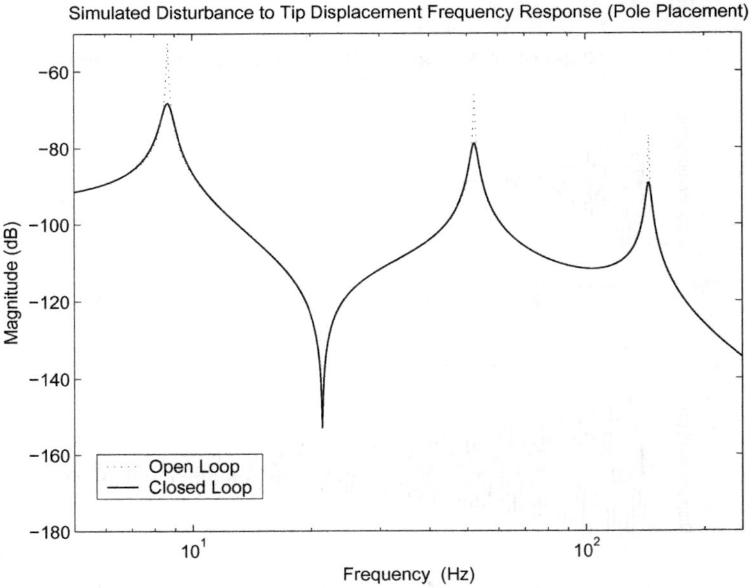

(b) Simulation Results

Figure 11.17. Open- and closed-loop frequency response for input disturbance w to output tip displacement Y_{tip} using the pole placement optimization

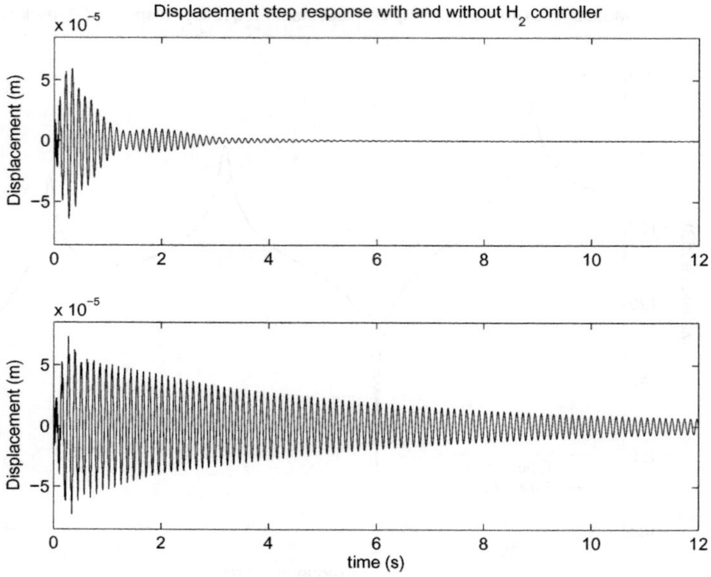

Figure 11.18. Displacement Step Response at the tip of the beam with and without the H_2 controller

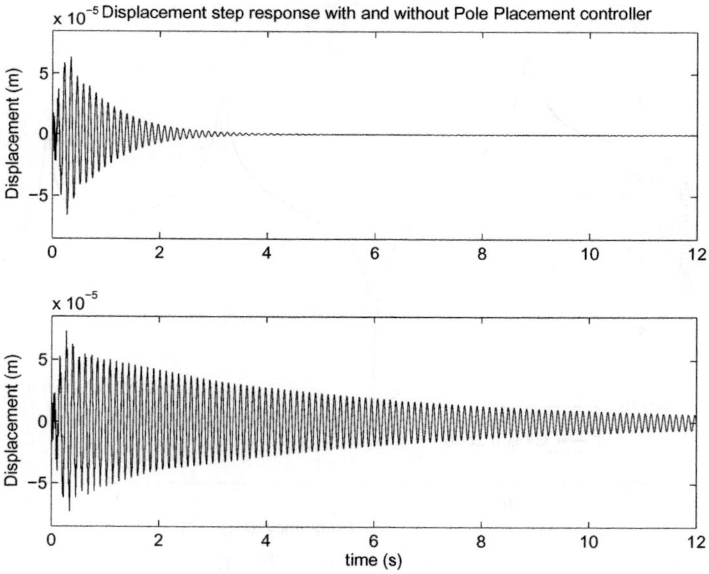

Figure 11.19. Displacement Step Response at the tip of the beam with and without the pole placement controller

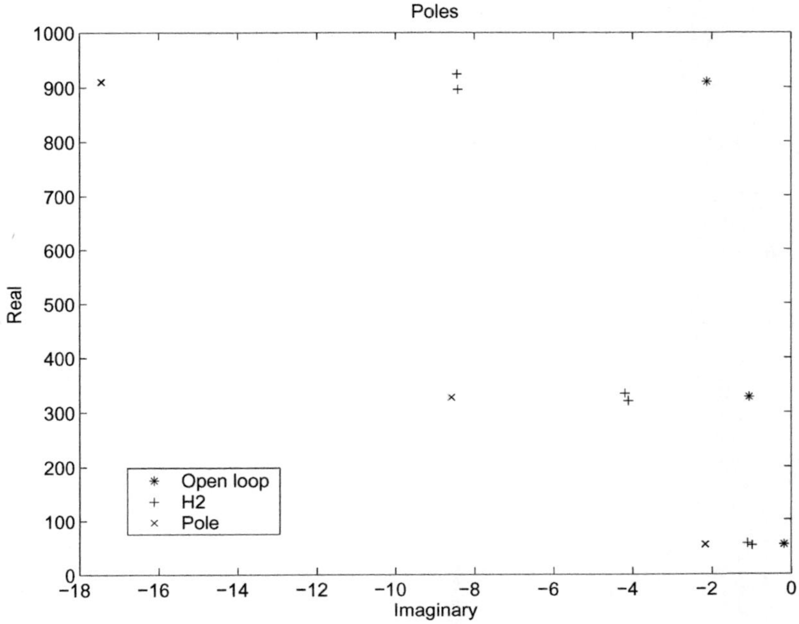

Figure 11.20. Comparison of the closed-loop pole locations for the different resonant controllers (2nd quadrant shown only.)

12

Nanopositioning

12.1 Introduction

Piezoelectric tube scanners were first reported in [25] for use in scanning tunneling microscopes [131]. They were found to provide a higher positioning resolution and greater bandwidth than traditional tripod positioners whilst being simple to manufacture and easier to integrate into a microscope. Piezoelectric tube scanners are now used extensively in scanning probe microscopes and many other applications requiring precision positioning, *e.g.* nanomachining [44, 72] etc.

The performance of a piezoelectric tube scanner is dictated by the accuracy and speed in which it can position a probe about a sample or *vice versa.* The three major causes of positioning inaccuracy are: excitation of resonant mechanical modes by external noise and the scanning trajectory; and actuator non-linearity. In this chapter, scan- and noise-induced vibration is reduced by employing a piezoelectric shunt circuit to augment the damping of the influential structural mode. This technique is combined with charge drive to provide both vibration reduction and significant linearity improvements.

12.2 Scanning Probe Microscopes

Scanning Probe Microscopes (SPMs) utilize quantum electrical and mechanical forces to analyze nano-scale surface properties and topography. Scanning Tunneling Microscopy (STM) was first demonstrated by Binnig and Rohrer in 1982 [23]. The quantum mechanical phenomena of free-electron tunnelling was exploited to map the topography of conductive sample surfaces. As shown in Figure 12.1, STM essentially involves the movement of a sample about an ultra-sharp tip. A common mode of operation is to scan the sample under the tip, reproducing topography by mapping the tip-sample tunnelling current. Since the original inventors demonstrated the mapping of Silicon(111) in

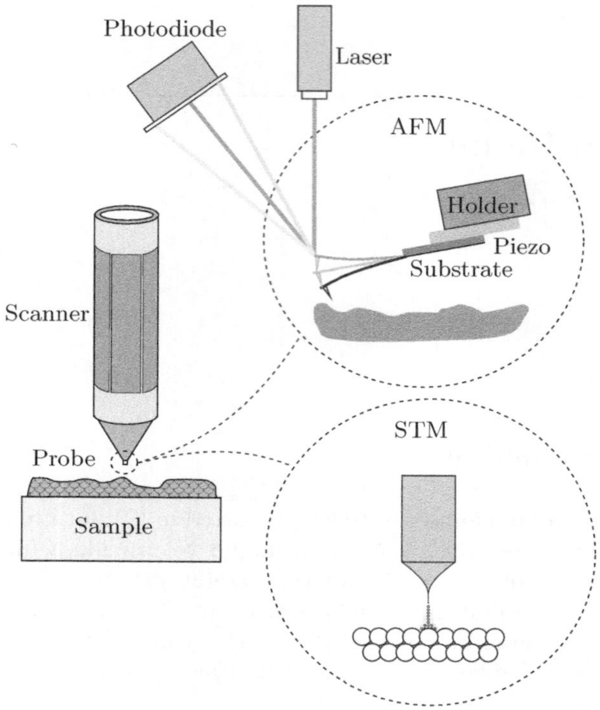

Figure 12.1. A scanning probe microscope comprises a probe scanned over the sample surface to map the particular property measured by the chosen probe. Atomic Force Microscopy (AFM) utilizes a vibrating lever to detect changes in sample topography while Scanning Tunnelling Microscopy (STM) employs a sharpened tip and electron tunelling current to measure electron density or atomic topography.

1983 [24], the field of scanning tunnelling microscopy has experienced a geometric expansion with dramatic consequences to the engineering and physical sciences.

To allow imaging of non-conductive samples, Binnig and co-workers demonstrated the Atomic Force Microscope (AFM) in 1986 [22]. Rather than solely topographic mapping, the vibration measured in a specifically coated AFM cantilever can be interpreted to map a large variety of physical characteristics including: thermal gradients, magnetic forces, photon emission, electrostriction etc.

A laser beam reflected off the cantilever back surface and focused on to a four-quadrant photodiode is the accepted technique for measuring the static and dynamic probe deflection.

Contact mode is presently the most common technique for atomic force microscopy. In contact mode, the cantilever is mounted parallel to the surface

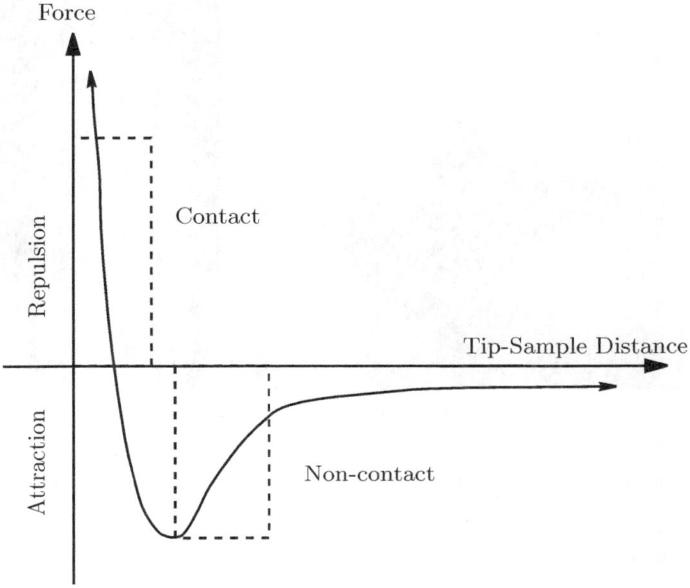

Figure 12.2. A typical force profile and operating modes of an AFM

with a sharp tip on the bottom face. As the cantilever is brought closer to the sample, interactive forces result in a deflection of the cantilever. When scanned over the sample, the probe deflection is representative of the sample topography. Typically the contacting force is held constant by regulating the height of the cantilever during each scan as this reduces sample and tip wear and provides a greater dynamic range in the measurement. In constant force mode, topographic information is contained in the vertical positioning of the cantilever.

Soon after the introduction of the Atomic Force Microscope, a plethora of derivatives and improvements began to emerge. Two readily adopted techniques were the Scanning Electrostatic Microscope and Scanning Magnetic Force Microscope. Both of these techniques involve a preliminary contact mode scan to determine the sample's mechanical topography. This information is then used to hover the probe at a constant height over the surface during a successive scan. Using a magnetically coated, or electrically charged tip, the magnitude of long-range magnetic and electric forces can be mapped independently from varying sample heights and short range atomic forces.

The technique of hovering the probe at a constant distance from the sample surface is referred to as non-contact mode. Rather than a direct measurement of force, the non-contact mode of operation provides a measurement of the

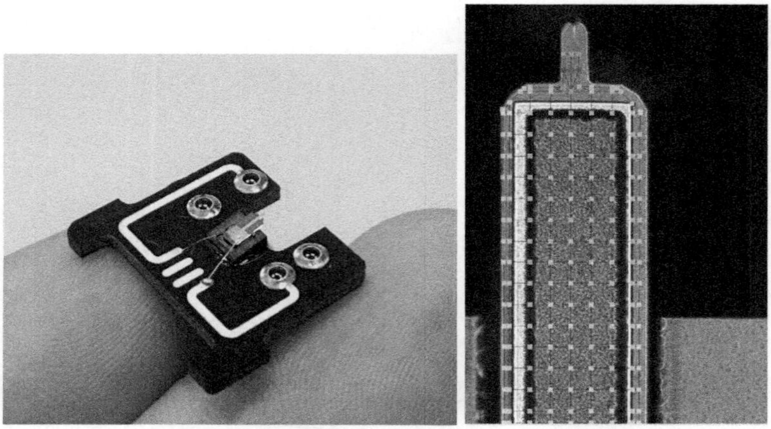

(a) Actual Size (b) 10x Magnification

Figure 12.3. Photographs of a silicon micro-cantilever with ZnO coating

force derivative, or stiffness. A mechanical oscillator, when exposed to a displacement sensitive force, experiences a change in resonance frequency. By forcefully vibrating the cantilever close to a nominal mechanical resonance, small changes in the resonance frequency, and hence force, can be measured with high sensitivity. A phase sensitive detector, the so-called lock-in amplifier, is used to measure small changes in the signal amplitude at the driving frequency. A significant increase in the signal-to-noise ratio is realized due to the removal of low-frequency 1/f noise from the measurement. In addition, the sensitivity of such a scheme can be increased further by artificially increasing the quality factor of the mechanical resonance.

The final mode of interest became popular after the sensitivity benefits of dynamic cantilever operation were discovered and exploited. Intermittent-contact mode, also named Dynamic Force Mode (DFM) and Tapping mode as copyrighted by Digital Instruments, involves a forced vibration of the cantilever in close proximity to the sample. A typical relationship between exerted force and the probe-sample distance is shown in Figure 12.2. Intermittent contact mode operates in the attractive region. As the cantilever is lowered to the sample, the attractive force gradient lowers the effective spring constant and decreases the resonance Non-Contact Repulsion Tip-Sample Distance Attraction Force frequency and vibration amplitude. When the vibrating cantilever begins to impact the sample, *i.e.* experiences a repulsive force gradient, the resonance frequency begins to increase. By maintaining this turning point in resonance frequency and scanning the sample, images of constant force gradient can be obtained. A NanoSensors dynamic mode cantilever is pictured in Figure 12.3.

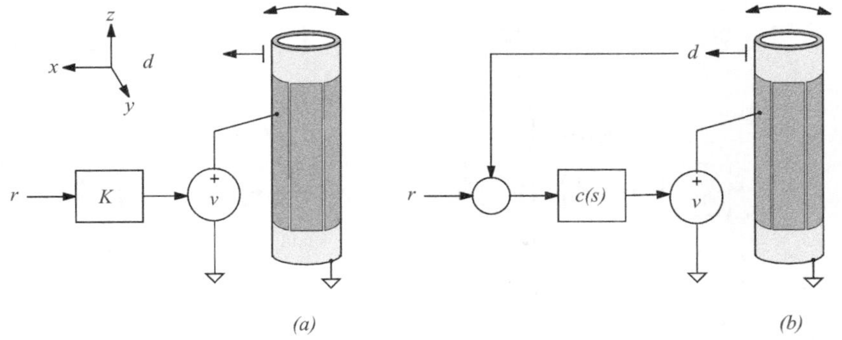

Figure 12.4. Voltage-driven tube scanner (a) Open-loop with signal pre-compensation (b) Closed-loop with displacement feedback

12.3 Piezoelectric Tube Scanners

As shown in Figure 12.4 (a), a piezoelectric scanner comprises a tube of radially poled piezoelectric material, four external electrodes and a grounded internal electrode. Other configurations may include: a circumferential electrode for independent vertical extension or diameter contraction, and/or sectored internal electrodes. Small-deflection expressions for the lateral tip translation, derived from the IEEE Piezoelectricity Standard [154], can be found in [31]. Measured in the same axis (x,or y) as the applied voltage, the tip translation d is approximately

$$d_i = \frac{\sqrt{2} d_{31} L^2}{\pi D h} v_i \qquad i = x, y, \tag{12.1}$$

where d_i is the (x or y axis) deflection, d_{31} is the piezoelectric strain constant, L is the length of the tube, D is the outside diameter, h is the tube thickness, and v_i is the (x or y axis) electrode voltage. Tip deflection can be doubled by applying an equal and opposite voltage to electrodes in the same axis. Vertical translation due to a voltage applied equally to all four quadrants is given approximately by

$$\Delta L = \frac{2 d_{31} L}{h} v. \tag{12.2}$$

Although the statics and dynamics of piezoelectric tubes are inherently nonlinear and three-dimensional, tube geometries with large length/diameter ratios can be highly simplified. For such geometries, and in cases of small deflections, the tube top can be assumed flat with no vertical excursion or tilting due to lateral deflections. Although there has been some recent effort to consider the coupling from lateral to vertical dimensions, tubes are generally designed to minimize such effects. Other design considerations are the

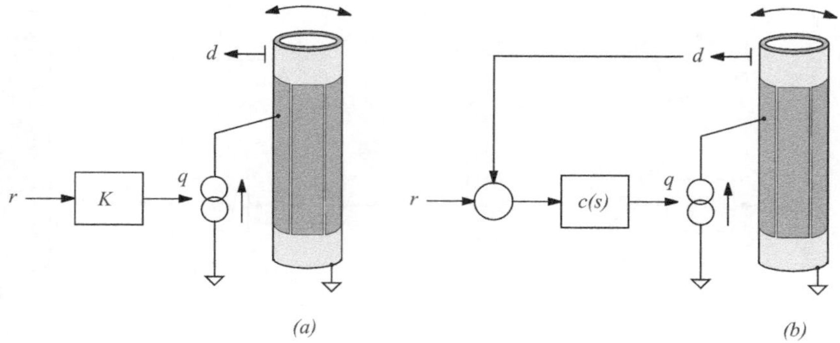

Figure 12.5. Charge-driven tube scanner (a) Open-loop with signal pre-compensation (b) Closed-loop with displacement feedback

deflection sensitivity and maximum deflection; both of these characteristics are optimized by large length to diameter ratios.

A consequence of designing tubes with large length/diameter ratios is low mechanical resonance frequencies. This has been a fundamental problem since the inception of piezoelectric tube scanners. A lightly damped low-frequency mechanical resonance severely limits the maximum achievable scan frequency. A triangular scan rate of around $1/100^{th}$, the first mechanical resonance frequency is usually assumed the upper limit in precision scanning applications.

Nonlinearity is another ongoing difficulty associated with piezoelectric tube scanners (and piezoelectric actuators in general). When employed in an actuating role, piezoelectric transducers display a significant hysteresis in the transfer function from an applied voltage to strain or displacement [2]. A full discussion on the nature of hysteresis can be found in Chapter 11. Due to hysteresis, ideal scanning signals can result in severely distorted tip displacements, and hence poor image quality or nanofabrication defects.

Techniques aimed at addressing both mechanical dynamics and hysteresis can be grouped generally into two broad categories, feedforward and feedback. Feedforward techniques, as shown in Figure 12.4, do not include a sensor but require accurate knowledge of the undesirable dynamics. Feedback systems, although more robust to modeling error, are limited by the noise performance and bandwidth of the sensor. In many cases it is also difficult and/or prohibitively expensive to integrate displacement sensors into the scanning apparatus.

Feedforward and signal compensation approaches have been extensively studied as their implementation requires no additional hardware or sensors. It should be considered, however, that additional hardware such as displacement sensors and DSP processors are required to identify the behavior of each tube prior to implementation. A technique for designing optimal linear

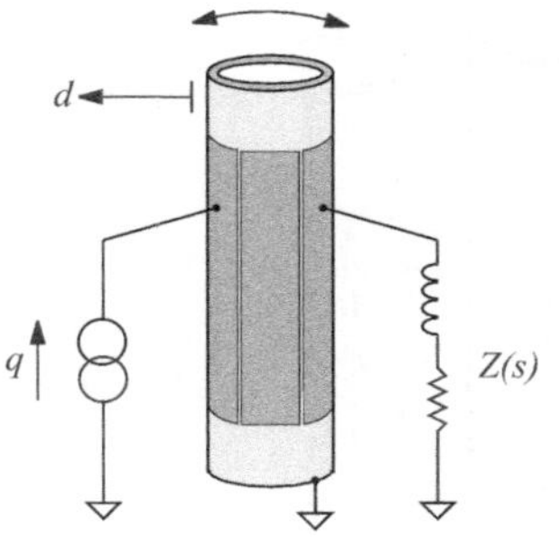

Figure 12.6. Charge-driven tube scanner with piezoelectric shunt damping circuit

feedforward compensators was presented in [46], then later extended to incorporate a PD feedback controller [44]. In these works, the authors identify the main limitation to performance as being modeling error. Another feedforward technique, known as iterative or learning control is aimed at reducing unmodeled hysteresis. In this method the need for a model is essentially annulled with the use of a sensor and online iteration to ascertain the optimal input compensation [112]. The foremost problems with iterative techniques are the time taken to iterate the compensator and difficulties associated with non-monotonic trajectories. Other feedforward approaches have included: Optimal H_∞ compensation [168]; Compensation for creep, Preisach hysteresis, and resonance [45]; Improved iterative Preisach inversion [88]; and various optimal linear feedforward compensation techniques [169, 155].

Feedback techniques can provide excellent low-frequency tracking performance, but depend heavily on the sensor noise performance and bandwidth. As a consequence, such techniques are most applicable to scan ranges in the hundreds of nanometers or greater. Good tracking of a 5 Hz triangle wave, while maintaining robustness to nonlinearity was presented in [167]. With the integration of displacement sensors into the next generation of commercial microscopes, feedback systems will become more feasible.

Considering the breadth of research aimed at improving scan performance, it is surprising to find that commercial microscope manufacturers have been reluctant to adopt the techniques discussed. The majority of commercial scanning systems operate in much the same fashion as they did in the early 90s. Regardless of the potential benefits, the requirement for data acquisition, sophisticated modeling experiments and additional sensors have severely limited the application of feedforward and feedback scan compensation. With this in mind, this chapter introduces two simple non-model-based techniques for the reduction of hysteresis and vibration.

In Section 6.5 a charge amplifier free from DC and low-frequency voltage drift was presented. As discussed in Section 11.1, the linearity of piezoelectric positioning systems can be significantly improved by replacing voltage

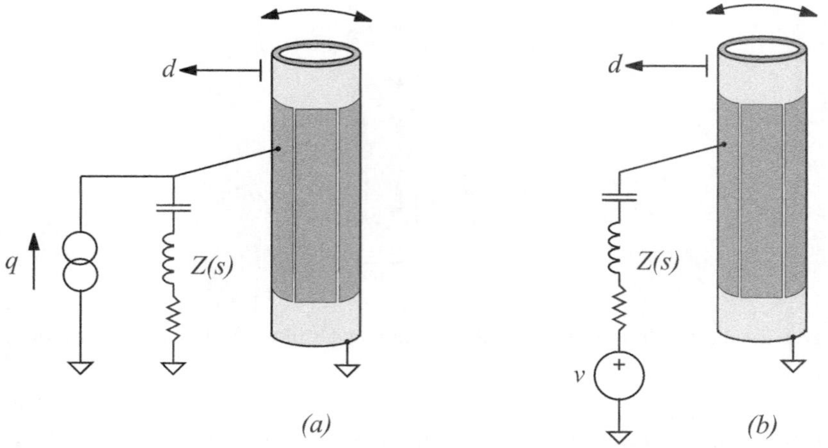

Figure 12.7. (a) Charge-driven tube scanner (b) Voltage driven equivalent

amplifiers with charge amplifiers as shown in Figure 12.5. To simplify the adaptation of previous techniques, an analysis of the relationship between charge and voltage actuation is provided. In most cases, a constant gain will equate the displacement response of a tube driven by either charge or voltage.

The second topic of this chapter is a new technique for the reduction of scan induced and exogenous vibration. Drawn from the field of Smart Structures, the connection of an electrical impedance to the terminals of one x and y electrode is proposed. Referred to as piezoelectric shunt damping in Chapters 4 and 5, this technique is capable of significantly reducing the magnitude of one or more structural modes. Figure 12.6 illustrates an inductor and resistor connected to the terminals of a charge-driven piezoelectric tube. In this configuration, the inductor and resistor are tuned to damp the first x axis cantilever mode. Undesired resonance excitation due to scanning and external disturbance is attenuated.

Piezoelectric shunt damping requires no feedback sensor and is thus immune to the usual problems of low-bandwidth and measurement noise associated with optical and capacitive sensors. Furthermore, as illustrated in Figure 12.7, we demonstrate that the shunt impedance $Z(s)$, can be applied to the same electrode as the driving charge or voltage source. This allows the redundant electrode to be used for increasing the scan range or as a piezoelectric strain sensor. Like the charge amplifier, such a technique can be implemented independently or in conjunction with a previous technique to improve performance.

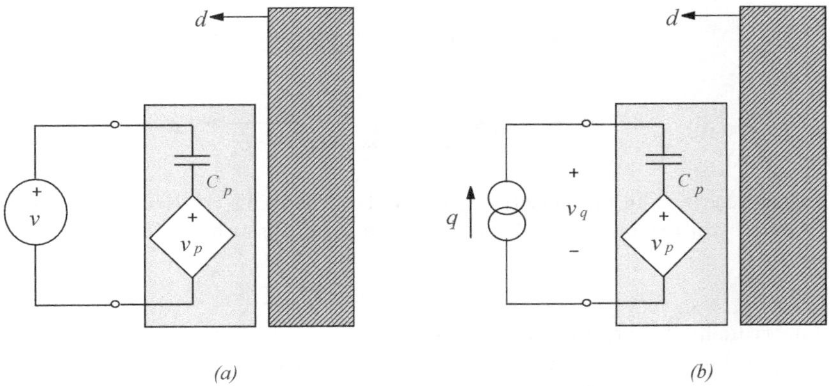

Figure 12.8. (a) A voltage- and (b) charge-driven piezoelectric tube

In the following section, a charge-based analysis is used to model the dynamics of a shunted piezoelectric tube. Implementation issues are then discussed, followed by experimental results.

12.4 Shunt Circuit Modeling

Modeling of piezoelectric transducers with attached resonant shunt circuits has traditionally been performed using voltage driven models (*e.g.* Chapter 5). Here, only charge-driven models are utilized. The following sub-section introduces the models required to simulate the effect of an attached shunt circuit. Traditional voltage driven models are initially discussed then related to their charge-driven equivalents as used throughout.

12.4.1 Open-loop

We first consider the open-loop translational dynamics of a piezoelectric tube. The electrically equivalent model of a voltage and charge driven piezoelectric tube is shown in Figure 12.8. Each electrode acts as a piezoelectric transducer, represented by a strain dependent voltage source v_p and series capacitor C_p. The polarization vector is assumed to be oriented radially outward, in this case, a positive voltage or charge results in a positive deflection. We are interested in the transfer functions from an applied voltage v to the resulting piezoelectric voltage v_p and tip translation d, that is

$$G_{vv}(s) = \frac{v_p(s)}{v(s)} \quad G_{dv}(s) = \frac{d(s)}{v(s)} . \tag{12.3}$$

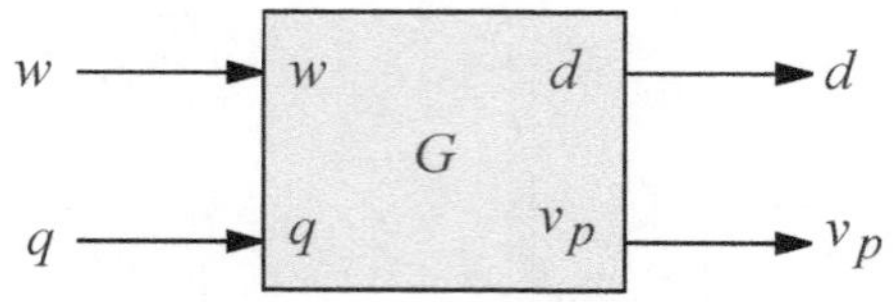

Figure 12.9. The piezoelectric tube model describing the deflection d and strain voltage v_p in response to an applied charge q and disturbance w

The transfer functions $G_{vv}(s)$ and $G_{dv}(s)$ can be derived analytically or determined experimentally. Due to the difficulties involved with modeling complicated geometries from first principles, empirical models obtained through system identification are preferable.

In the case of charge actuation, Figure 12.8 (b), equivalent transfer functions can be derived. Kirchoff's Voltage Law for the loop is,

$$\frac{-q}{C_p} - v_p + v_q = 0. \tag{12.4}$$

Substituting $v_p = G_{vv} v_q$ and simplifying yields

$$G_{vq}(s) = \frac{v_p(s)}{q(s)} = \frac{1}{C_p} \frac{G_{vv}(s)}{1 - G_{vv}(s)}. \tag{12.5}$$

The displacement transfer function can be derived in a similar fashion,

$$G_{dq}(s) = \frac{d(s)}{q(s)} = \frac{1}{C_p} \frac{G_{dv}(s)}{1 - G_{vv}(s)} \tag{12.6}$$

Off resonance, where $G_{vv}(s) \ll 1$

$$G_{vq}(s) \approx \frac{G_{vv}(s)}{C_p} \quad G_{dq}(s) \approx \frac{G_{dv}(s)}{C_p}. \tag{12.7}$$

Thus the relationship between charge and voltage actuation is revealed. Due to the benefits in reducing hysteresis, only charge actuation will be considered in the proceeding sections.

In addition to a charge input, the possibility for a disturbance input w is also desirable. The signal w can be used to study the regulation or rejection of environmental noise. In the following sections the tube system will be referred to as G, a multi-input multi-output system describing the deflection d and piezoelectric voltage v_p in response to a driving charge q and disturbance w. The transfer functions $G_{vq}(s)$ and $G_{dq}(s)$ are contained in G. Such a realization is advantageous as the system G will later be identified directly from experimental data using system identification. The tube model G is shown in Figure 12.9.

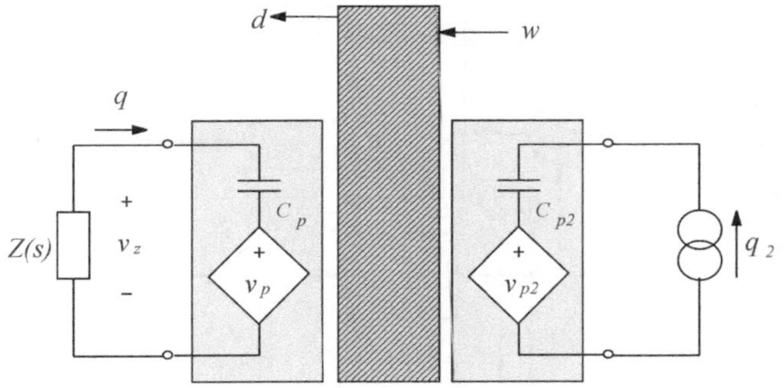

Figure 12.10. The electrical equivalent of a charge-driven piezoelectric tube with attached shunt circuit

12.4.2 Shunt Damping

As discussed in Chapter 4, a series inductor-resistor network, as shown in Figure 12.6, can bring about a significant reduction in the magnitude of a single structural mode. Analogous to a tuned mechanical absorber, additional dynamics introduced by the shunt circuit act to increase the effective structural damping [82].

The equivalent electrical model of a shunted piezoelectric tube (as shown in Figure 12.6) is illustrated in Figure 12.10. To find the transfer function relating displacement d to the driving charge q_2 we start by writing Kirchoff's Voltage Law around the impedance loop and substituting $v_z = -qsZ(s)$,

$$\frac{-q(s)}{C_p} - v_p(s) + -q(s)sZ(s) = 0. \tag{12.8}$$

When the opposing tube electrodes are equal in dimension, the charges q and q_2 have an equal but opposite influence on the tube deflection d and v_p. Furthermore

$$v_p = -v_{p2} \tag{12.9}$$

$$\frac{v_p(s)}{q(s)} = \frac{v_{p2}(s)}{q_2(s)} = \frac{-v_p(s)}{q_2(s)} = G_{vq}(s). \tag{12.10}$$

The principle of superposition can be applied to find an expression for v_p.

$$v_p(s) = G_{vq}(s)q(s) - G_{vq}(s)q_2(s). \tag{12.11}$$

Rearranging (12.11) in terms of q_2 and substituting into (12.8) yields

$$\frac{v_p(s)}{q_2(s)} = \frac{-G_{vq}(s)}{1 + G_{vq}(s)K(s)}, \tag{12.12}$$

where

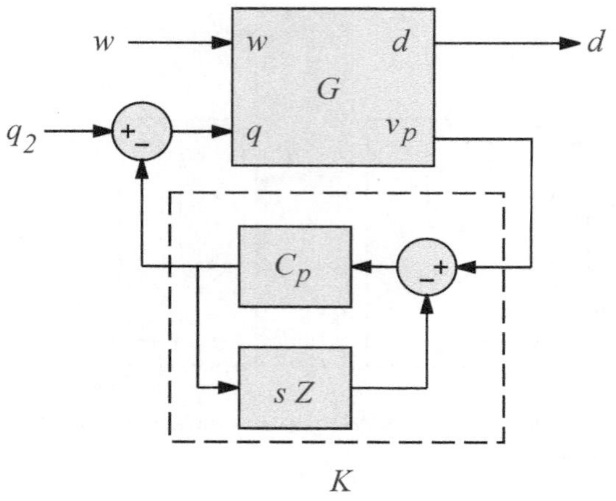

Figure 12.11. The equivalent feedback diagram when an electrical impedance is connected to the terminals of one tube electrode and the other is driven with charge

$$K(s) = \frac{C_p}{1 + C_p s Z(s)}. \tag{12.13}$$

The shunted displacement transfer function can be derived in a similar manner:

$$\frac{d(s)}{q_2(s)} = \frac{G_{dq}(s)}{1 + G_{vq}(s)K(s)}. \tag{12.14}$$

Using the principle of superposition, the influence of an external disturbance w can also be included,

$$d(s) = \frac{1}{1 + G_{vq}(s)K(s)} \left(G_{dq}(s)q_2(s) + G_{dw}(s)w(s)\right), \tag{12.15}$$

where G_{dw} is the transfer function measured from an external force w to the displacement d.

From Equations (12.14) and (12.15) it is concluded that the presence of an electrical shunt impedance can be viewed equivalently as a strain-voltage feedback control system. A diagrammatic representation of Equation (12.15) is shown in Figure 12.11. Further interpretation and analysis can be found in [140].

In some cases (where a second electrode is not available), it may be difficult to obtain a model describing the piezoelectric voltage v_p directly. In such cases, the terminal voltage v_z can also be considered. The equivalent terminal-voltage feedback diagram is shown in Figure 12.12. v_z is related to v_p by

$$v_z = v_p + \frac{1}{C_p} q, \tag{12.16}$$

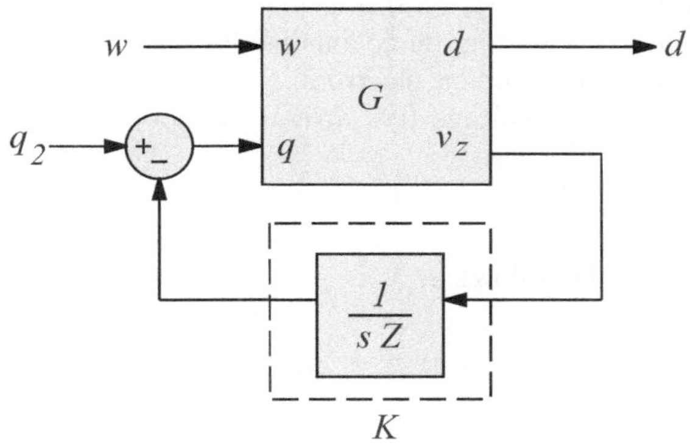

Figure 12.12. An alternative feedback interpretation considering the terminal voltage v_z rather than the piezoelectric voltage v_p

that is,

$$\frac{v_z(s)}{q(s)} = G_{vq}(s) + \frac{1}{C_p}. \tag{12.17}$$

Equations (12.8) to (12.15) can be modified accordingly.

Hybrid Operation

As mentioned in Section 12.3, it is advantageous to connect the shunt impedance and driving charge source to the same electrode. This scenario is depicted in Figure 12.13. In this sub-section, the electrical filtering effect of

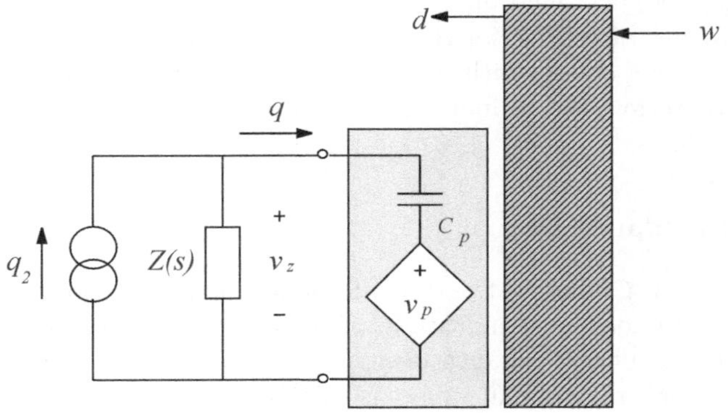

Figure 12.13. A charge-driven tube electrode with attached parallel shunt circuit

$Z(s)$ on q_2 is derived. If such a filtering effect can be inverted, the charge source q_2 can be used for scanning, analogous to the case where a shunt impedance is attached to an independent electrode.

Writing Kirchoff's Voltage Law around the loop,

$$-\frac{q}{C_p} - v_p + v_z = 0, \tag{12.18}$$

and substituting the following,

$$q(s) = -\frac{v_z(s)}{sZ(s)} + q_2(s), \tag{12.19}$$

results in the loop equation

$$-\frac{q(s)}{C_p} - v_p(s) + q(s)sZ(s) + q_2(s)sZ(s) = 0. \tag{12.20}$$

Given that $v_p = G_{vq}q$, we can substitute $q = v_p/G_{vq}$ into (12.20). After simplification, the transfer function from q_2 to v_p can be found:

$$\frac{v_p(s)}{q_2(s)} = \frac{K(s)Z(s)G_{vq}(s)}{1 + G_{vq}(s)K(s)},$$

where K is as given in (12.13). Similarly,

$$\frac{d(s)}{q_2(s)} = \frac{K(s)Z(s)G_{dq}(s)}{1 + G_{vq}(s)K(s)}. \tag{12.21}$$

Unlike the case in Section 12.4.2, the impedance $Z(s)$ distorts the tube transfer function from the driving charge q_2 to the deflection d. Rather than simply adding a strain feedback controller to the mechanical system, the transfer function from q_2 to d now also contains a filter $F(s) = K(s)Z(s)$. An equivalent feedback diagram is shown in Figure 12.14.

An obvious technique for recovering the natural tube dynamics is to pre-filter the driving charge with $F^{-1}(s)$. Fortunately this pre-filtering and inversion is straightforward to implement in practice. This solution is discussed in Section 12.5.

Shunt Impedance Design

As reviewed in Chapters 4 and 5, the Smart Structures and Vibration Control literature contain a multitude of passive, active, linear, and non-linear piezoelectric shunt impedance designs capable of reducing structural vibration. However, only a small subset is suitable for piezoelectric tube damping. The resonant linear shunts meet all of the requisite criteria: they are easy to design, implement and tune; they offer excellent damping performance (especially for single modes of vibration); they are strictly passive and inject

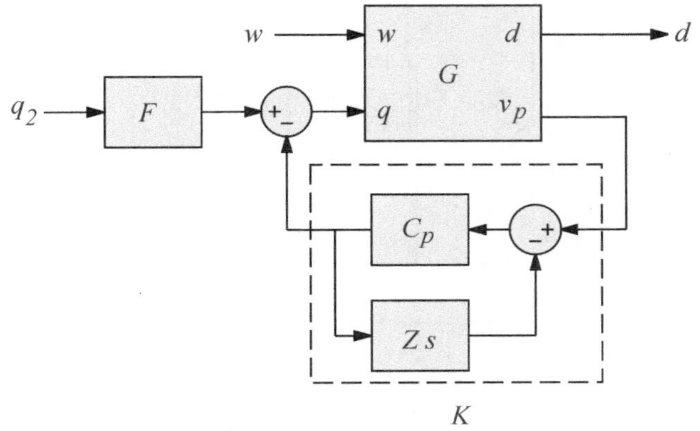

Figure 12.14. The equivalent feedback diagram when the driving charge and shunt impedance are applied to the same electrode (as shown in Figure 12.13)

no harmonics; and finally, their presence influences the mechanical dynamics only over a small frequency range. Resonant linear shunts have been shown to emulate the effect of a tuned-mass mechanical absorber [82].

After examination of various impedance designs, the LCR circuit depicted in Figure 12.7 was found to offer good performance. The presence of a series capacitance is necessitated by the requirement for DC tracking. If the impedance of the network was not infinity at DC, constant tube deflections would require a ramp signal in charge (eventually saturating the amplifier), this is reflected in the scan filter $F(s)$ and its inverse $F(s)^{-1}$.

To damp a single mode of structural vibration, the circuit inductance L, capacitance C, and piezoelectric capacitance C_p are tuned to resonate at the target mechanical frequency ω_1. Although the capacitance value C is essentially arbitrary, values of 1 to 10 times the piezoelectric capacitance have been found suitable. To equate the frequency of electrical resonance to mechanical resonance, the inductor is tuned as follows:

$$L = \frac{C + C_p}{CC_p\omega_1}. \tag{12.22}$$

The resistance value, dependent on the inherent system damping, is most easily found experimentally. For such systems, resistances in the order of 1 $K\Omega$ are typical.

As the required shunt damping circuit contains impractically large values of inductance, the charge amplifier will also be used as part of a synthetic admittance as discussed in Section 4.4. Consider the schematic shown in Figure 12.15, neglecting the filter $F^{-1}(s)$ and input q_2, the charge applied to the piezoelectric tube is equal to

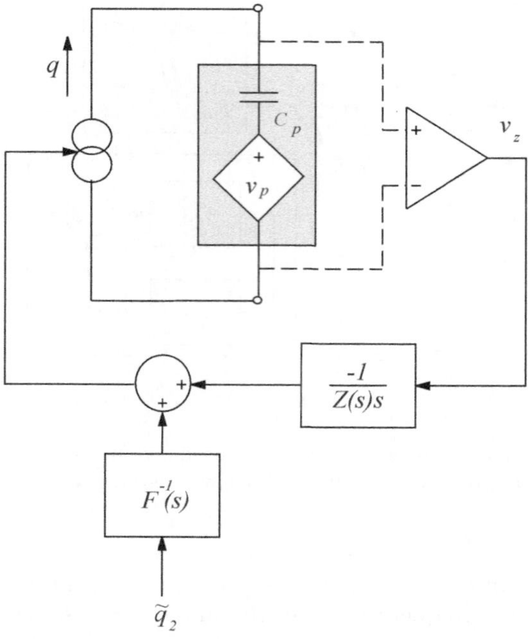

Figure 12.15. Schematic diagram of a charge-driven tube with integrated shunt circuit

$$q(s) = \frac{-1}{sZ(s)} \cdot v_z(s). \tag{12.23}$$

The impedance (or admittance) experienced by the piezoelectric transducer can be calculated by examining the ratio of current to voltage at its terminals. As the current is equal to $-\dot{q}$, and q is defined by (12.23), the impedance presented to the terminals is simply $Z(s)$ (as defined by the filter in Figure 12.15). By implementing the filter $\frac{-1}{sZ(s)}$ any arbitrary impedance can be presented to the terminals of the transducer. Simple techniques for designing analog and digital filters that represent $\frac{1}{Z(s)}$ can be found in Section 6.8. In this work, a dSPACE DSP system is used to implement and tune the filter $\frac{-1}{sZ(s)}$.

In addition to the charge required for shunt impedance synthesis, the additive charge q_2 is used for tube scanning. As mentioned in Section 12.4.2, the additive charge q_2 requires a filter $F^{-1}(s)$ to compensate for the electrical dynamics of the shunt impedance when attached to the same electrode.

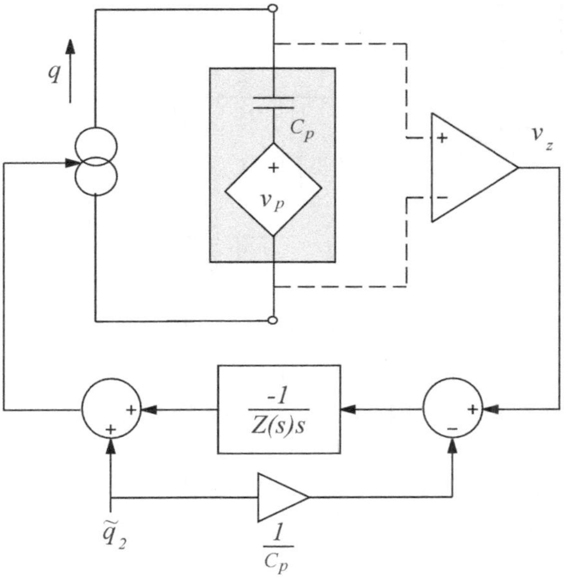

Figure 12.16. Simplified diagram of a charge amplifier with integrated shunt impedance

12.5 Implementation

A substantial simplification of the system shown in Figure 12.15 can be made by studying the structure of the filter $F^{-1}(s)$,

$$F^{-1}(s) = \frac{1}{K(s)Z(s)} = \frac{1 + C_p s Z(s)}{C_p s Z(s)} = \frac{1}{C_p s Z(s)} + 1. \tag{12.24}$$

Considering that the transfer function $\frac{1}{sZ(s)}$ has already been implemented, $F^{-1}(s)$ can be replaced as shown in Figure 12.16.

12.5.1 Grounded-load Charge Amplifier

As piezoelectric tubes contain multiple external electrodes referenced to a single grounded internal electrode, the DC-accurate charge amplifier presented in Section 6.5 is not directly applicable.

A modified amplifier configuration capable of driving grounded loads is shown in Figure 12.17. This circuit incorporates a high common-mode rejection, high common-mode range differential stage consisting of the lower op-amp, voltage bridge and instrumentation amplifier. The amplifier works to

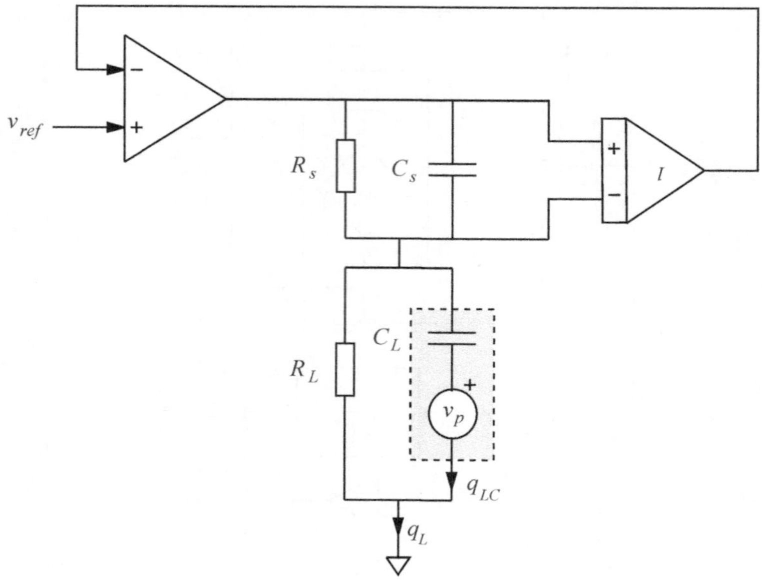

Figure 12.17. DC accurate charge source for grounded capacitive loads

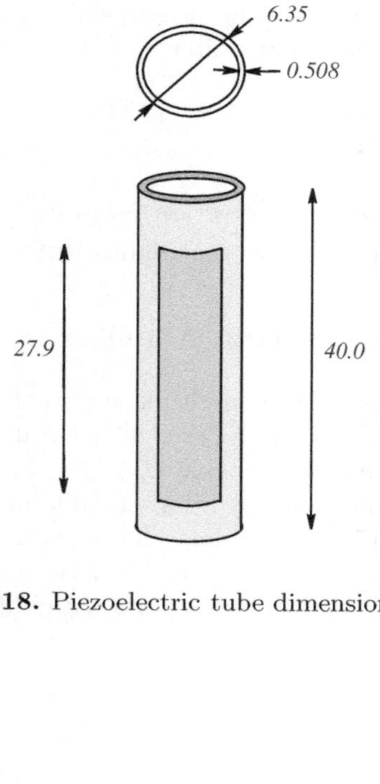

Figure 12.18. Piezoelectric tube dimensions (in mm)

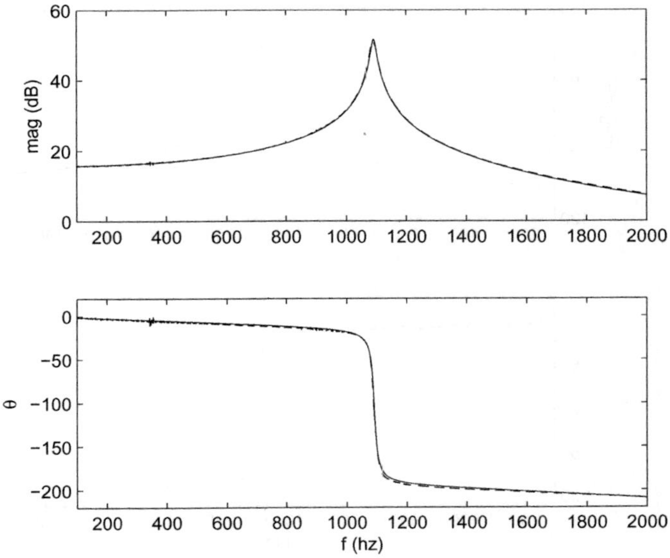

Figure 12.19. The transfer function G_{dq} measured from the charge input q (C) to the measured displacement d (m) . (—) Identified model and (- -) measured

equate the voltage measured across the sensing impedance to the reference voltage v_{ref}.

Analogous to the design presented in Section 6.5, DC accuracy is achieved by setting $C_L R_L = C_s R_s$, *i.e.*

$$\frac{R_L}{R_s} = \frac{C_s}{C_L}. \tag{12.25}$$

12.6 Experimental Results

In this section, the prototype shunt circuit and charge amplifier are employed to drive a piezoelectric tube positioner in one dimension. The tube was manufactured and patterned by Boston Piezo-Optics, physical dimensions can be found in Figure 12.18. The tube was glued (with 24-hour cure epoxy) vertically into a recessed (by 1.5 mm) aluminum block. An ADE Tech capacitive sensor was used to measure the displacement with sensitivity 10 $V/\mu m$ and bandwidth 10 kHz. An aluminum cube (1 cm × 1 cm × 1 cm) is glued onto the tube tip and grounded to provide a return for the capacitive sensor. Both the tube block and sensor mount are affixed to a stabilized optical table. Parameters of the shunt impedance and amplifier are shown in Table 12.1.

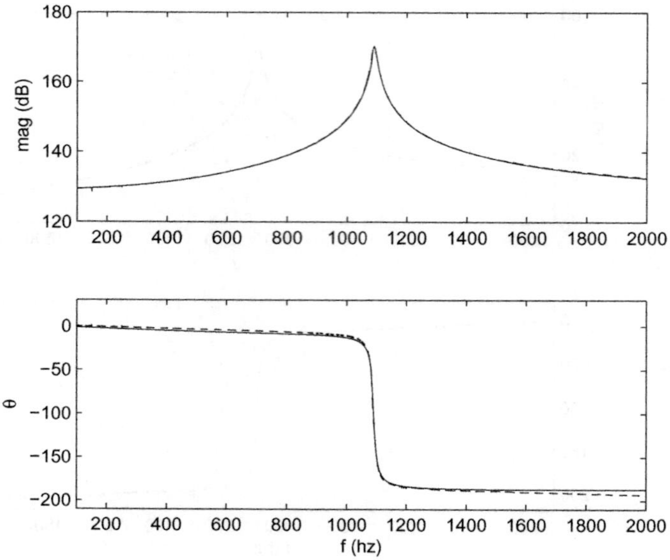

Figure 12.20. The transfer function G_{vq} measured from the charge input q (C) to the measured strain voltage v_p (V). (—) Identified model, (- -) measured.

12.6.1 Tube Dynamics

The experimental results are presented by first examining the natural response of the piezoelectric tube. The measured and identified transfer functions from charge input to strain-voltage and displacement are shown in Figures 12.19 and 12.20. The system model G, shown in Figure 12.9, was obtained through MIMO subspace frequency domain system identification [128]. The identification required 12 MIMO data points to return a single input, two output model of order 2. An excellent fit is observed in the frequency domain.

The nominal first resonance frequency and DC charge sensitivity of the tube were measured to be 1088 Hz and 5.7 m/C $(= 5.7\ \mu m/\mu C)$.

Table 12.1. Parameters of the charge amplifier and shunt impedance

Charge Gain	77.8 nC/V
Voltage Measurement Gain	0.1 V/V
L	2.9 H
C	50 nF
R	3.3 $K\Omega$

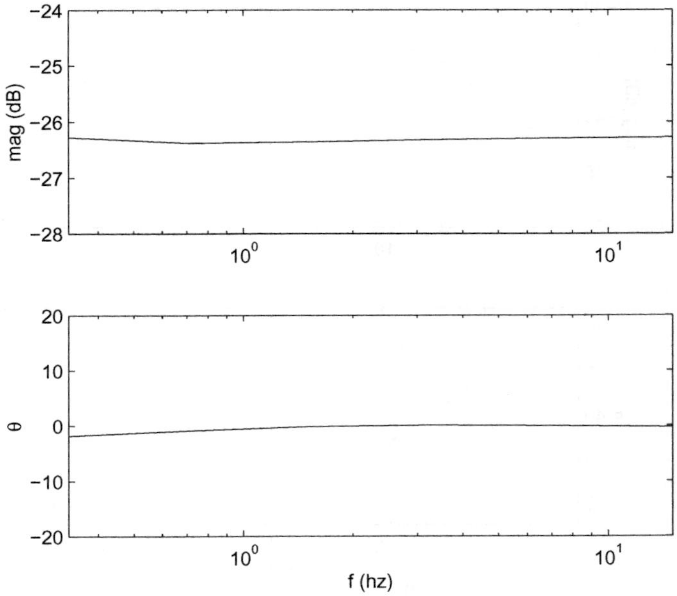

Figure 12.21. Charge amplifier low-frequency tracking performance. Measured from the charge reference signal (V) to the instrumented load voltage across a 5 nF dummy load.

12.6.2 Amplifier Performance

Both the low-frequency scanning and high-frequency vibration damping depend on the performance of the charge amplifier and related instrumentation. In the following we examine the two characteristics of foremost importance: low-frequency charge regulation - the ability of the amplifier to reproduce low-frequency inputs without drift, and the bandwidth of charge dominance - the frequency range where hysteresis will be reduced due to dominant charge feedback.

The (low-frequency) transfer function measured from an applied reference signal to the actual charge deposited on a 5 nF dummy load is shown in Figure 12.21. Excellent low-frequency tracking from 15 mHz to 15 Hz is exhibited by the amplifier and instrumentation. As discussed in Section 6.5, the bandwidth of charge dominance was ascertained by zeroing the charge reference and introducing an internal load voltage. The transfer function measured from the internal voltage to the voltage measured across the load is shown in Figure 12.22. We observe a charge dominance bandwidth of 0.8 Hz. Frequencies above this bandwidth will experience the full linearity benefit of charge actuation.

To justify the use of charge actuation we demonstrate the benefit in Figures 12.23 and 12.24. Hysteresis is reduced by approximately 89% simply through

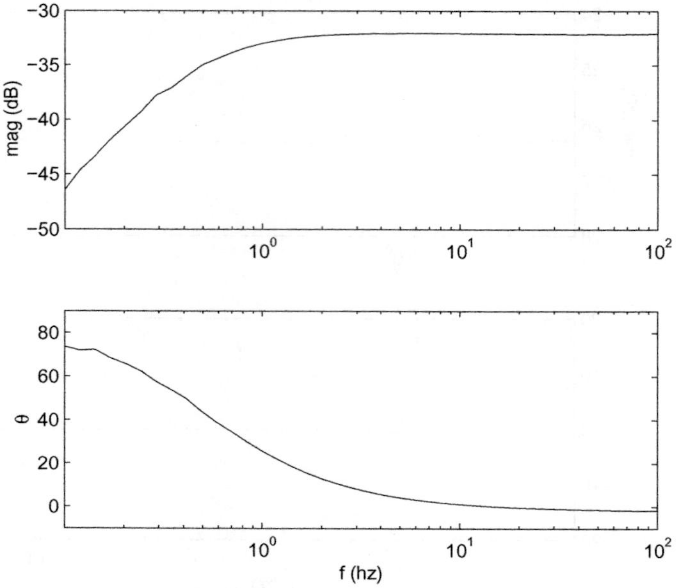

Figure 12.22. Charge dominance bandwidth. Measured from the internal tube strain voltage v_p to the load voltage.

the use of a charge amplifier. Percentage reduction is calculated by measuring the maximum excursion in the minor axis of each plot, then taking the ratio $100 \times \frac{\text{voltage}}{\text{charge}}$. It should be noted that a scan range of $\pm 3\ \mu m$ is around 20% of the full scale deflection and it is often assumed that hysteresis is negligible at such low drives. Similar plots for the same apparatus with a $\pm 8\ \mu m$ drive can be found in [63], a greater hysteresis is exhibited, and heavily reduced by through the use of a similar charge drive.

Similar to a typical voltage amplifier, the hybrid amplifier offers little or no hysteresis reduction over the frequency range of voltage dominance. For the same reason, no improvement in creep can be expected. Creep time-constants are usually greater than 10 minutes, which in this discussion, is effectively DC. At these frequencies the amplifier behaves analogously to a standard voltage amplifier.

12.6.3 Shunt Damping Performance

Scan Induced Vibration Suppression

When scanning at high frequencies, the greatest cause of tracking error is due to high frequency harmonics exciting the mechanical resonance. The influence of the shunt impedance can be observed to significantly increase the effective

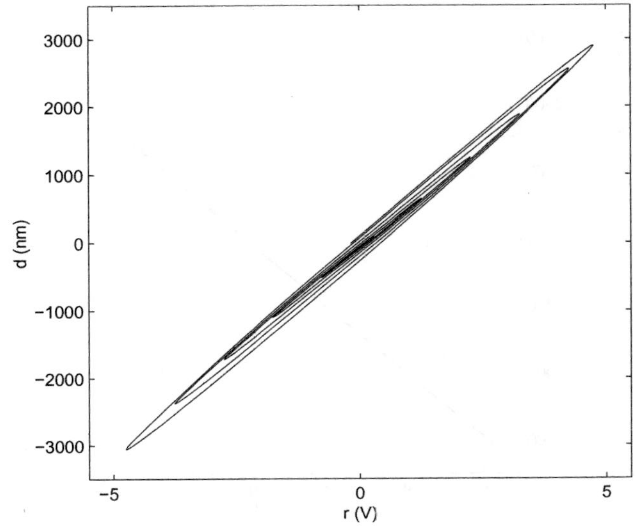

Figure 12.23. Relationship between an applied voltage and the resulting tube displacement (10 Hz ramped sinusoidal input)

damping in Figure 12.25. The simulated response shown in Figure 12.26 shows a good correlation with experimental results. The equivalent decrease in settling time can be observed in Figure 12.27.

To illustrate the improvement in triangular scanning fidelity, an unfiltered 46 Hz triangular waveform was applied to the system. The frequency and lack of filtering was chosen to illustrate the worst-case induced ripple. In practice, the triangle would be filtered or passed through a feedforward controller to reduce vibration. Regardless of the ripple magnitude, the presence of a shunt circuit provides the same decrease in settling time. At high speeds, significant increases in fast-axis resolution can be expected. In the case where feedforward vibration control [45] is applied, the damped mechanical system would allow a less severe pre-filter and provide greater immunity to modeling error.

Externally Induced Vibration

Another significant source of tracking error is external mechanical noise. Due to the highly resonant nature of the tube, high frequency noise components can excite the mechanical resonance and lead to large erroneous excursions. By applying a voltage to an opposite electrode, we can simulate the effect of a strain disturbance. A significant damping of greater than 20 dB can be observed in Figure 12.29. The effect of such damping can be observed in the time domain by applying a low frequency scanning signal. With no scan-

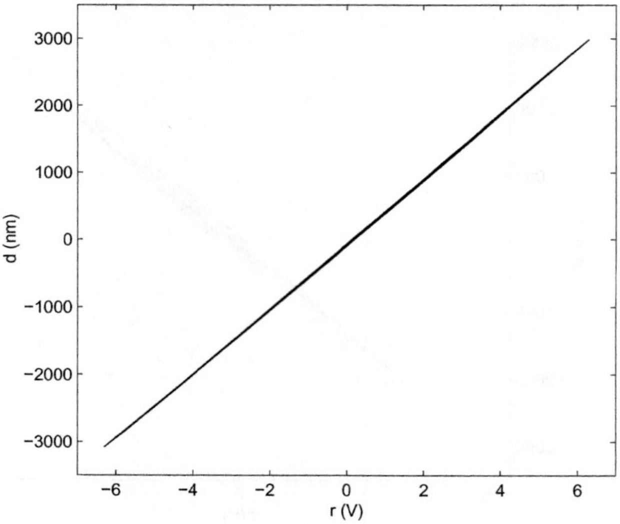

Figure 12.24. Relationship between an applied charge reference and the resulting tube displacement (10 Hz ramped sinusoidal input)

induced vibration, the external noise is dominant. The reduction of resonant vibration can be seen in Figure 12.30.

Low Frequency Scanning

The final test of such an apparatus is the ability to track DC charge offsets. In Figure 12.31 a low frequency triangle signal was applied to the charge amplifier, at time 130 sec a DC offset equivalent to around 1 μm was applied. Aside from the faithful reproduction of a 0.1 Hz triangle wave, the charge amplifier reproduces the offset without drift.

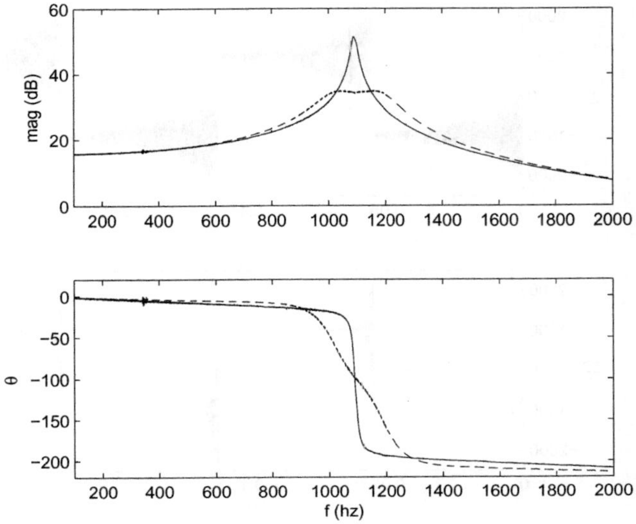

Figure 12.25. Experimental Response. The natural (–) and shunt-damped (– –) tube dynamics measured from the additive charge input q_2 (C) to the tip displacement d (m).

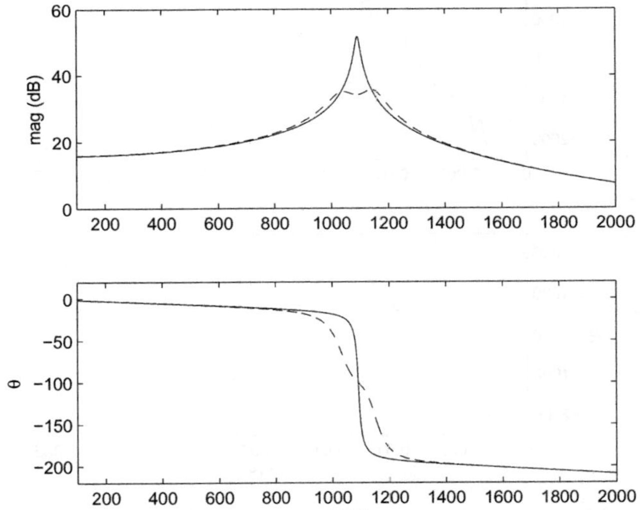

Figure 12.26. Simulated response. The natural (–) and shunt-damped (– –) tube dynamics measured from the additive charge input q_2 (C) to the tip displacement d (m).

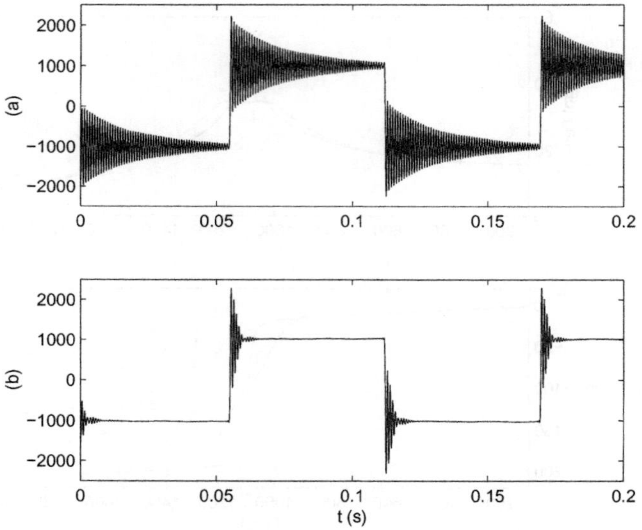

Figure 12.27. Tube deflection (in nm) resulting from square wave excitation (a) uncontrolled, and (b) with LCR shunt impedance

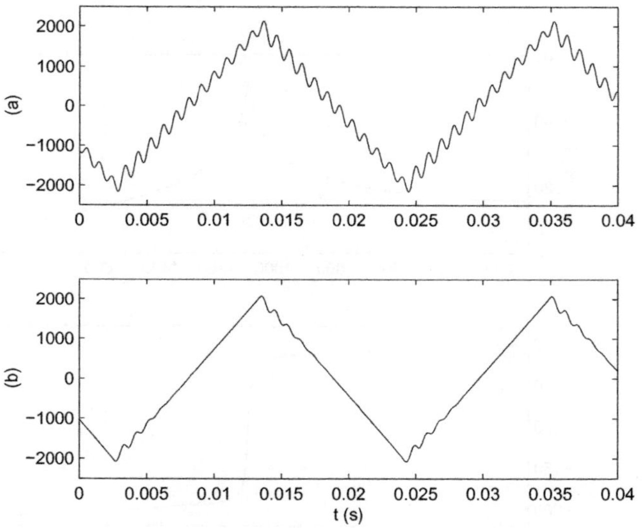

Figure 12.28. Tube deflection (in nm) resulting from a 46 Hz triangle wave excitation (a) uncontrolled, and (b) with LCR shunt impedance

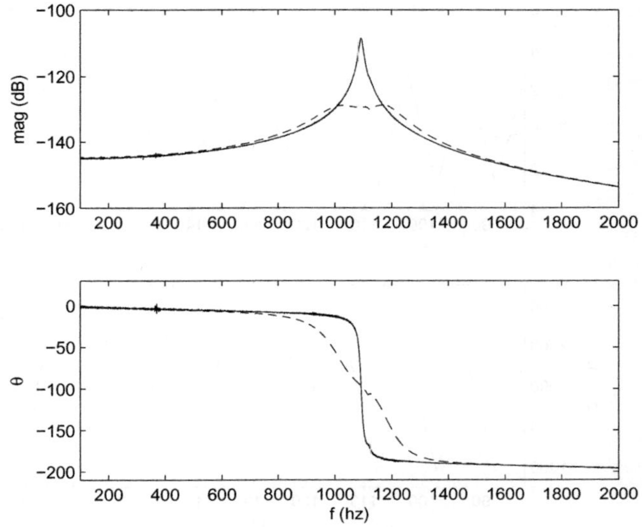

Figure 12.29. Experimental response. The natural (–) and shunt-damped (– –) tube transfer function from the applied strain disturbance (in V) to the tip displacement d (m).

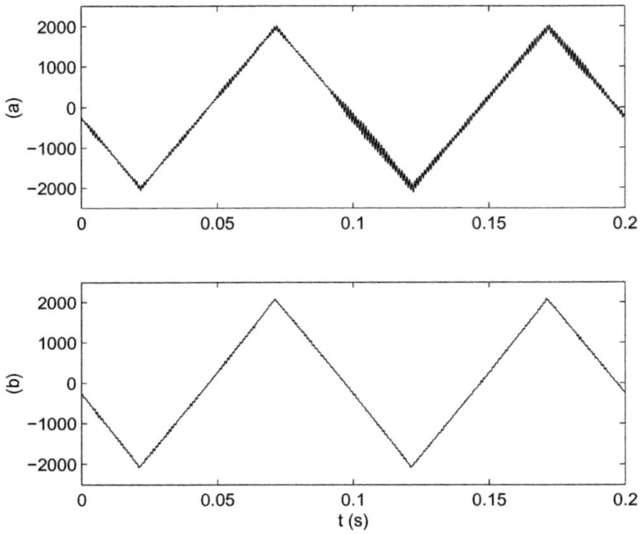

Figure 12.30. Tube deflection (in nm) resulting from a 1.6 kHz band-limited uniformly distributed random strain disturbance (a) Uncontrolled and (b) with LCR shunt impedance

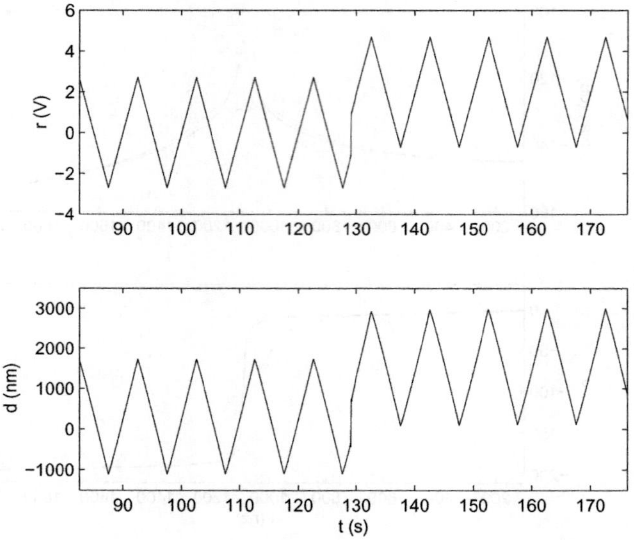

Figure 12.31. Low frequency scanning reference and resultant tube displacement with additive DC offset

References

1. S. Acrabelli and A. Tonoli. System properties of flexible structures with self-sensing piezoelectric transducers. *Journal of Sound and Vibration*, 235(1):1–23, 2000.
2. H. J. M. T. A. Adriaens, W. L. de Koning, and R. Banning. Modeling piezoelectric actuators. *IEEE/ASME transactions on mechatronics*, 5(4):331–341, December 2000.
3. B. N. Agrawal and H. Bang. Adaptive structures for large precision antennas. *ACTA Astronautica*, 38(3):175–183, 1996.
4. J. B. Aldrich, N. W. Hagood, A. von Flotow, and D. W. Vos. Design of passive piezoelectric damping for space structures. In *Proc. SPIE Conference on Smart Structures and Intelligent Systems, SPIE Vol. 1917*, pages 629–705, 1993.
5. B. D. O. Anderson and J. B. Moore. *Optimal Control: Linear Quadratic Methods.* Prentice-Hall, Englewood Cliffs, NJ, 1990.
6. B. D. O. Anderson and S. Vongpanitlerd. *Network Analysis and Synthesis.* Prentice-Hall, Englewood Cliffs, NJ, 1973.
7. B.D.O Anderson and J.B. Moore. *Optimal Filtering.* Prentice-Hall, Englewood Cliffs, N.J., 1979.
8. E. H. Anderson and N. W. Hagood. Simultaneous piezoelectric sensing actuation - analysis and application to controlled structures. *Journal of Sound and Vibration*, 174(5):617–639, 1994.
9. E. H. Anderson, N. W. Hagood, and J. M. Goodliffe. Self-sensing piezoelectric actuation: analysis and application to controlled structures. In *Proc. AIAA/ASME/ASCE/AHS/ASC Structures, structural dynamics and materials conference*, pages 2141–2155, Dallas, TX, 1992.
10. J. N. Aubrun. Theory of control of structures by low-authority controllers. *International Journal of Control*, 3(5):444–451, 1980.
11. T. Bailey and J. E. Hubbard. Distributed piezoelectric-polymer active vibration control of a cantilever beam. *Journal of Guidance, Control and Dynamics*, 8(5):605–611, 1985.
12. M. J. Balas. Active control of flexible systems. *Journal of Optimization Theory ans Applications*, 25(3):415–436, 1978.
13. M. J. Balas. Feedback control of flexible systems. *IEEE Transactions on Automatic Control*, AC-23(4), 1978.

14. A. Baz and S. Poh. Performance of an active control system with piezoelectric actuators. *Journal of Sound and Vibrations*, 126(2):327–343, 1988.
15. S. Behrens, A. J. Fleming, and S. O. R. Moheimani. A broadband controller for shunt piezoelectric damping of structural vibration. *Smart Materials and Structures*, 12(1):36–48, February 2003.
16. S. Behrens and S. O. R. Moheimani. Optimal resistive elements for multiple shunt-damping of a piezoelectric laminate beam. In *Proc. IEEE Conference on Decision and Control*, Sydney, Australia, December 2000.
17. S. Behrens, S. O. R. Moheimani, and A. J. Fleming. Multiple mode current flowing passive piezoelectric shunt controller. *Journal of Sound and Vibration*, 266(5):929–942, Oct 2003.
18. A. Belloli, D. Niederberger, S. Pietrzko, M. Morari, and P. Ermanni. Structural vibration control via r-l shunted active fiber composites. In *Proc. ICAST, Internation Conference on Adaptive Structures and Technologies*, Bar Harbor, Maine, USA, October 2004.
19. A. A. Bent. Active fiber composite material systems for structural control applications. In *Proc. SPIE Smart Structures and Materials: Industrial and Commercial Applications of Smart Structures Technologies, SPIE Vol. 3674*, pages 166–177, Newport Beach, CA, March 1999.
20. A. A. Bent and N. W. Hagood. Development of piezoelectric fiber composites for structural actuation. In *Proc. AIAA/ASME/ASCE/AHS/ASC Structures, structural dynamics and materials conference, AIAA Paper no. 93-1717*, La Jolla, CA, April 1993.
21. E. Bianchini, R. Spangler, and C. Andrus. The use of piezoelectric devices to control snowboard vibrations. In *Proc. SPIE Smart Structures and Materials: Smart Structures and integrated systems, SPIE Vol. 3329*, pages 106–114, 1998.
22. G. Binnig, C. F. Quate, and C. Gerber. Atomic force microscope. *Physical Review Letters*, 56(9):930–933, March 1986.
23. G. Binnig and H. Rohrer. Scanning tunneling microscopy. *Helvetica Physica Acta*, 55:726–735, 1982.
24. G. Binnig, H. Rohrer, Ch. Gerber, and E. Weibel. 7x7 reconstruction on Si(111) resolved in real space. *Physical Review Letters*, 50(2):120–123, January 1983.
25. G. Binnig and D. P. E. Smith. Single-tube three-dimensional scanner for scanning tunneling microscopy. *Review of Scientific Instruments*, 57(8):1688–1689, August 1986.
26. S. Bonnail, D. Tonneau, G. A. Capolino, and H. Dallaporta. Dynamic and static responses of a piezoelectric actuator at nanometer scale elongation. In *Proc. IEEE Industry Applications Conference*, volume 1, pages 293–298, October 2000.
27. S. Boyd, L. El Ghaoui, E. Feron, and V. Balakrishnan. *Linear Matrix Inequalities in System and Control Theory*. SIAM, Philadelphia, PA, 1994.
28. PI Ceramics. Catalogue of piezoelectric actuators and sensors. http://www.piceramic.com/.
29. S. Chandrasekaran and D. K. Lindner. Power flow through controlled piezoelectric actuators. *Journal of Intelligent Materials Systems and Structures*, 11(6):469–481, June 2000.
30. S. Chandrasekaran, D. K. Lindner, and R. C. Smith. Optimized design of switching amplifiers for piezoelectric actuators. *Journal of Intelligent Material Systems and Structures*, 11:887–901, November 2000.

31. C. J. Chen. Electromechanical deflections of piezoelectric tubes with quartered electrodes. *Applied Physics Letters*, 60(1):132–134, January 1992.
32. C. Choi and K. Park. Self-sensing magnetic levitation using a lc resonant circuit. *Sensors and Actuators A-Physical*, 72(2):169–177, January 1999.
33. I. Chopra. Review of the state of art of smart structures and integrated systems. *AIAA Journal*, 40(11):2145–2187, November 2002.
34. R. L. Clark. Accounting for out-of-bandwidth modes in the assumed modes approach: implications on colocated output feedback control. *Transactions of the ASME, Journal of Dynamic Systems, Measurement, and Control*, 119:390–395, 1997.
35. R. L. Clark, W. R. Saunders, and G. P. Gibbs. *Adaptive Structures: Dynamics & Control.* John Wiley & Sons Inc., 1998.
36. R. L. Clark, J. S. Vipperman, and D. G. Cole. Adaptive piezoelectric sensoriactuators. U.S. Patent 5,578,761, November 1996.
37. W. W. Clark. Vibration control with state-switched piezoelectric materials. *Journal of Intelligent Material Systems and Structures.*, 11:263–271, April 2000.
38. D. G. Cole and R. L. Clark. Adaptive compensation of piezoelectric sensoriactuators. *Journal of Intelligent Material Systems and Structures*, 5:665–672, 1994.
39. R. Comstock. Charge control of piezoelectric actuators to reduce hysteresis effect. U.S. Patent 4.263,527, 1981.
40. R. D. Cook. *Finite Element Modelling for Stress Analysis.* John Wiley and Sons, 1995.
41. L. R. Corr and W. W. Clark. Comparison of low-frequency piezoelectric switching shunt techniques for structural damping. *IOP Smart Materials and Structures*, 11:370–376, 2002.
42. E. F. Crawley and J. de Luis. Use of piezoelectric actuators as elements of intelligent structures. *AIAA Journal*, pages 1373–1385, October 1987.
43. D. Croft and S. Devasia. Vibration compensation for high speed scanning tunneling microscopy. *Review of Scientific Instruments*, 70(12):4600–4605, December 1999.
44. D. Croft, D. McAllister, and S. Devasia. High-speed scanning of piezo-probes for nano-fabrication. *Transactions of the ASME, Journal of Manufacturing Science and Technology*, 120:617–622, August 1998.
45. D. Croft, G. Shed, and S. Devasia. Creep, hysteresis, and vibration compensation for piezoactuators: Atomic force microscopy application. *Transactions of the ASME, Journal of Dynamic Systems, Measurement, and Control*, 123:35–43, March 2001.
46. D. Croft, S. Stilson, and S. Devasia. Optimal tracking of piezo-based nanopositioners. *Nanotechnology*, 10:201–208, 1999.
47. J. M. Cruz-Hernandez and V. Hayward. Phase control approach to hysteresis reduction. *IEEE Transactions on Control Systems Technology*, 9(1):17–26, 2001.
48. J. Curie and P. Curie. Development, par pression, de l'electricite polarise dans les crystaux hemiednes et fares inclines. *Comp. Rend.*, 91:294–297, 1880.
49. C. L. Davis and G. A. Lesieutre. An actively tuned solid-state vibration absorber using capacitive shunting of piezoelectric stiffness. *Journal of Sound and Vibration*, 232(3):601–617, 2000.

50. C. W. de Silva. *Vibration Fundamentals and Practice.* CRC Press, 1999.
51. K. K. Denoyer, S. F. Griffin, and D. Sciulli. Hybrid structural/acoustic control of a subscale payload fairing. In *Proc. SPIE Smart Structures and Materials: Smart Structures and Integrated Systems, SPIE Vol. 3329*, pages 237–243, 1998.
52. S. Devasia, T. Meressi, B. Paden, and E. Bayo. Piezoelectric actuator design for vibration suppression - placement and sizing. *Journal of Guidance, Control and Dynamics*, 16(5):859–864, 1993.
53. E. K. Dimitriadis, C. R. Fuller, and C. A. Rogers. Piezoelectric actuators for distributed vibration excitation of thin plates. *Transactions pf the ASME - Journal of Vibration and Acoustics*, 113:100–107, January 1991.
54. J. J. Dosch, D. J. Inman, and E. Garcia. A self-sensing piezoelectric actuator for collocated control. *Journal of Intelligent Materials and Structures*, 3(1):166–185, 1992.
55. S. Elliott. *Signal Processing for Active Control.* Academic Press, 2001.
56. D. J. Ewins. Modal testing as an aid to vibration analysis. In *Proc. Conference on Mechanical Engineering*, May 1990.
57. J. L. Fanson, E. H. Anderson, and D. Rapp. Active structures for use in precision control of large optical-systems. *Optical Engineering*, 29(11), November 1990.
58. J. L. Fanson and T. K. Caughey. Positive position feedback-control for large space structures. *AIAA Journal*, 28(4):717–724, April 1990.
59. A. J. Fleming, S. Behrens, and S. O. R. Moheimani. Synthetic impedance for implementation of piezoelectric shunt damping circuits. *Electronics Letters*, 36(18):1525–1526, 2000.
60. A. J. Fleming, S. Behrens, and S. O. R. Moheimani. Optimization and implementation of multi-mode piezoelectric shunt damping systems. *IEEE/ASME Transactions on Mechatronics*, 7(1):87–94, March 2002.
61. A. J. Fleming and S. O. R. Moheimani. Adaptive piezoelectric shunt damping. *Smart Materials and Structures*, 12:36–48, January 2003.
62. A. J. Fleming and S. O. R. Moheimani. Precision current and charge amplifiers for driving highly capacitive piezoelectric loads. *Electronics Letters*, 39(3):282–284, February 2003.
63. A. J. Fleming and S. O. R. Moheimani. Hybrid DC accurate charge amplifier for linear piezoelectric positioning. In *Proc. 3rd IFAC Symposium on Mechatronic Systems*, Sydney, Australia, September 2004.
64. A. J. Fleming and S. O. R. Moheimani. Improved current and charge amplifiers for driving piezoelectric loads, and issues in signal processing design for synthesis of shunt damping circuits. *Journal of Intelligent Material Systems and Structures*, 15(2):77–92, February 2004.
65. A. J. Fleming and S. O. R. Moheimani. Synthesis of optimal piezoelectric shunt impedances for structural vibration control. In *Proc. SPIE Symposium on Smart Structures & Materials - Damping and Isolation*, San Diego, CA, USA, March 2004.
66. A. J. Fleming and S. O. R. Moheimani. Control oriented synthesis of high performance piezoelectric shunt impedances for structural vibration control. *IEEE Transactions on Control Systems Technology*, 13(1):98–112, January 2005.
67. A. J. Fleming, S. O. R. Moheimani, and S. Behrens. Reducing the inductance requirements of piezoelectric shunt damping circuits. *Smart Materials and Structures*, 12(1):57–64, February 2003.

68. R. L. Forward. Electronic damping of vibrations in optical structures. *Journal of Applied Optics*, 18(5):690–697, 1979.
69. M. I. Friswell, D. J. Inman, and R. W. Rietz. Active damping of thermally induced vibrations. *Journal of Intelligent Material Systems and Structures*, 8(8):678–685, August 1997.
70. C. R. Fuller, S. J. Elliot, and P. A. Nelson. *Active Control of Vibration*. Academic Press, 1996.
71. K. Furutani, M. Urushibata, N. Mohri, and N. May. Improvement of control method for piezoelectric actuator by combining charge feedback with inverse transfer function compensation. In *Proc. IEEE International Conference on Robotics and Automation*, pages 1504–1509, Leuven, Belgium, 1998.
72. W. Gao, R. J. Hocken, J. A. Patten, J. Lovingood, and D. A. Lucca. Construction and testing of a nanomachining instrument. *Precision Engineering*, 24(4):320–328, October 2000.
73. E. Garcia, J. D. Dosch, and D. J. Inman. Vibration attenuation in an active antenna structure. In *Proc. of the Conference on Recent Advances in Active Control of Sound and Vibration*, pages S35–S42, Virginia Polytechnique Institute and State University, Blacksburg, April 15–17 1991.
74. P. Ge and M. Jouaneh. Modeling hysteresis in piezoceramic actuators. *Precision Engineering*, 17:211–221, 1995.
75. P. Ge and M. Jouaneh. Tracking control of a piezoceramic actuator. *IEEE Transactions on Control Systems Technology*, 4(3):209–216, 1996.
76. G. P. Gibbs and C. R. Fuller. Excitation of thin beams using asymmetric piezoelectric actuators. *Journal of Acoustical Society of America*, 92(6):3221–3227, December 1992.
77. C. J. Goh and T. K. Caughey. On the stability problem caused by finite actuator dynamics in the collocated control of large space structures. *International Journal of Control*, 41(3):787–802, 1985.
78. M. Goldfarb and N. Celanovic. A lumped parameter electromechanical model for describing the nonlinear behaviour of piezoelectric actuators. *ASME Journal of Dynamic Systems, Measurement, and Control*, 119:478–485, September 1997.
79. G. C. Goodwin, S. F. Graebe, and M. E. Salgado. *Control System Design*. Prentice-Hall, 2001.
80. M. Green and D. J. N. Limebeer. *Linear Robust Control*. Prentice-Hall, 1994.
81. N. W. Hagood and E. F. Crawley. Experimental investigation of passive enhancement of damping for space structures. *AIAA Journal of Guidance, Navigation and Control*, 14(6):1100–1109, 1991.
82. N. W. Hagood and A. von Flotow. Damping of structural vibrations with piezoelectric materials and passive electrical networks. *Journal of Sound and Vibration*, 146(2):243–268, 1991.
83. D. Halim and S. O. R. Moheimani. Spatial resonant control of flexible structures - application to a piezoelectric laminate beam. *IEEE Transactions on Control Systems Technology*, 9(1):37–53, January 2001.
84. D. Halim and S. O. R. Moheimani. Reducing the effect of truncation error in spatial and pointwise models of resonant systems with damping. *Mechanical Systems and Signal Processing*, 18(2):291–315, March 2004.
85. J. H. Han, J. Tani, and I. Lee. Multi-modal vibration control of smart composite plates. *Materials Science Research International*, 5(2):122–127, June 1999.

86. H. Hanselka. Adaptronik und Fragen zur Systemzuverlässigkeit. *ATP, Automatisierungstechnische Praxis*, 2:44–49, February 2002.
87. J. Heng, J. C. Akers, R. Venugopal, M. Lee, A. G. Sparks, P. D. Washabaugh, and D. Bernstien. Modelling, identification, and feedback control of noise in an acoustic duct. *IEEE Transactions on Control Systems Technology*, 4(3):283–291, 1996.
88. K. J. G. Hinnen, R. Fraanje, and M. Verhaegen. The application of initial state correction in iterative learning control and the experimental validation on a piezoelectric tube scanner. *Journal of Systems and Control Engineering*, 218:503–511, 2004.
89. J. J. Hollkamp. Multimodal passive vibration suppression with piezoelectric materials and resonant shunts. *Journal of Intelligent Material Systems and Structures*, 5, January 1994.
90. M. A. Hopkins, D. A. Henderson, R. W. Moses, T. Ryall, D. G. Zimcik, and R. L. Spangler. Active vibration-suppression systems applied to twin-tail buffeting. In *Proc. SPIE Smart Structures and Materials: Industrial and Commercial Applications of Smart Structures Technologies, SPIE Vol. 3326*, pages 27–33, 1998.
91. P. Horowitz and W. Hill. *The Art of Electronics*. Cambridge University Press, 1980.
92. H. Hu and R. B. Mrad. On the classical Preisach model of hysteresis in piezoceramic actuators. *Mechatronics*, 13:85–94, 2003.
93. P. C. Hughes. Space structure vibration modes: how many exist? which ones are important? *IEEE Control Systems Magazine*, 8(1):22–28, February 1987.
94. T. Ikeda. *Fundamentals of Piezoelectricity*. Oxford University Press, 1996.
95. D. J. Inman. *Engineering Vibration*. Prentice-Hall, 2nd edition, 2001.
96. J. Y. Ishihara, M. H. Terra, and R. M. Sales. The full information and state feedback H_2 optimal controllers for descriptor systems. *Automatica*, 39(3):391–402, 2003.
97. R. M. Jones. *Mechanics of Composite Materials*. Hemisphere Publishing Co., New York, 1975.
98. S. M. Joshi. *Control of Large Flexible Structures*. Springer, Berlin, 1989.
99. S. M. Joshi and S. Gupta. On a class of marginally stable positive-real systems. *IEEE Transactions on Automatic Control*, 41(1):152–155, January 1996.
100. J. Juang and M. Q. Phan. *Identification and Control of Mechanical Systems*. Cambridge University Press, 2001.
101. T. Kailath. *Linear Systems*. Printice-Hall, Upper Saddle River, NJ 07458, 1980.
102. H. Kaizuka and B. Siu. Simple way to reduce hysteresis and creep when using piezoelectric actuators. *Japan Journal of Applied Physics, part 2 - Letters*, 27(5):773–776, May 1988.
103. A. Kawamoto, K. Takaba, and T. Katayama. On the generalized algebraic Riccati equation for continuous-time descriptor systems. *Linear Algebra and its Applications*, 296(1–3):1–14, 1999.
104. H. K. Khalil. *Nonlinear Systems*. Prentice-Hall, 2nd edition, 1996.
105. J. Kim, J. Choi, and R. H. Cabell. Noise reduction performance of smart panels incorporating piezoelectric shunt damping. In *Proc. SPIE Smart Structures and Materials: Industrial and Commercial Applications of Smart Structures Technologies, SPIE Vol. 4698*, pages 143–149, 2002.

106. S. J. Kim, Han C. H, and C. Y. Yun. Improvement of aeroelastic stability of hingeless helicopter rotor blade by passive piezoelectric damping. In *Proc. SPIE Conference on Passive Damping and Isolation*, pages 131–141, Newport Beach, CA, March 1999.
107. M. J. Konak, I. G. Powlesland, S. P. van der Velden, and S. P. Galea. A self-powered discrete time piezoelectric vibration damper. In *Proc. SPIE Conference on and Integrated Systems, SPIE Vol. 3241*, pages 270–279, 1997.
108. M. J. Konak, I. G. Powlesland, S. P. van der Velden, and S. P. Galea. Analysis of a self-powered piezoelectric vibration damper. In *Proc. SPIE Smart Materials, Structures, and MEMS, Smart Structures and Devices, SPIE Vol. 4235*, pages 328–339, Melbourne, Australia, Decemeber 2000.
109. S. K. Kwak, G. Washington, and R. K. Yedavalli. Acceleration feedback-based active and passive vibration control of landing gear components. *Journal of Aerospace Engineering*, 15(1):1–9, 2002.
110. H. Kwakernaak and R. Sivan. *Linear Optimal Control Systems*. Wiley, 1972.
111. H. Kwakernaak and R. Sivan. *Modern Signals and Systems*. Prentice-Hall, 1991.
112. K.K. Leang and S. Devasia. Iterative feedforward compensation of hysteresis in piezo positioners. In *IEEE Conference on Decision and Control*, Maui, HI, December 2003.
113. D. J. Leo and D. J. Inman. Pointing control and vibration suppression of a slewing flexible frame. *Journal of Gudance, Control and Dynamics*, 17(3):529–536, 1994.
114. F. L. Lewis. *Optimal Estimation*. John Wiley & Sons Inc, 1986.
115. Y. Lim, V. V. Varadan, and V. K. Varadan. Closed-loop finite element modelling of active/passive damping in structural vibration control. In *Proc. SPIE Smart Materials and Structures 1997: Mathematics and Control in Smart Structures, SPIE Vol. 3039*, San Diego, CA, March 1997.
116. D. K. Lindner, H. Zhu, C. Song, W. Huang, and D. Cheng. Low input voltage switching amplifiers for piezoelectric actuators. In *Proc. SPIE Smart Structures and Materials: Industrial and Commercial Applications of Smart Structures Technologies. SPIE Vol. 4698*, pages 282–292., San Diego, CA, March 2002.
117. D. K. Lindner, M. Zhu, N. Vujic, and D. Leo. Comparison of linear and switching drive amplifiers for piezoelectric actuators. In *Proc. AIAA/ASME/ASCE/AHS/ASC Structures, Structural Dynamics, and Materials Conference*, Denver, CO, April 2002.
118. G. J. Lippmann. Principe de la conservation de l'electricite ou second principe de la th'eorie des phenomnes electriques. *Annales de chimie et de physique*, 24:145–177, 1881.
119. L. Ljung. *System Identification Toolbox User's Guide Version 4*. The Math Works, Inc., 24 Prime Park Way, Natick, MA 01760-1500, April 1998. http://www.mathworks.com.
120. L. Ljung. *System Identification: Theory for the Users*. Prentice-Hall, Englewood Cliffs, N.J., 1999.
121. J. M. Maciejowski. *Multivariable Feedback Design*. Addison-Wesley, 1989.
122. J. A. Main and E. Garcia. Design impact of piezoelectric actuator nonlinearities. *AIAA Journal of Guidance, Control and Dynamics*, 20(2):327–332, 1997.

123. J. A. Main and E. Garcia. Piezoelectric stack actuators and control system design: Strategies and pitfalls. *AIAA Journal of Guidance, Control and Dynamics*, 20(3):479–485, 1997.
124. J. A. Main, E. Garcia, and D. V. Newton. Precision position control of piezoelectric actuators using charge feedback. *AIAA Journal of Guidance, Control and Dynamics*, 18(5):1068–1073, 1995.
125. J. A. Main, D. V. Newton, L. Massengill, and E. Garcia. Efficient power amplifiers for piezoelectric applications. *Smart Materials and Structures*, 5:766–755, December 1996.
126. G. D. Martin. *On the control of flexible mechanical systems*. PhD thesis, Stanford University, 1978.
127. I. Mayergoyz. *Mathematical Models of Hysteresis*. Springer-Verlag, 1991.
128. T. McKelvey, H. Ackay, and L. Ljung. Subspace-based identification of infinite-dimensional multi-variable systems from frequency-response data. *IEEE Transactions on Automatic Control*, AC–41:960–979, 1996.
129. T. Mckelvey, A. J. Fleming, and S. O. R. Moheimani. Subspace-based system identification for an acoustic enclosure. *Transactions of the ASME, Journal of Vibration & Acoustics*, 124(3):414–419, July 2002.
130. L. Meirovitch. *Elements of Vibration Analysis*. McGraw-Hill, Sydney, 2nd edition, 1996.
131. E. Meyer, H. J. Hug, and R. Bennewitz. *Scanning Probe Microscopy*. Springer, Heidelberg, Germany, 2004.
132. N. Mohan, T. M. Undeland, and W. P. Robbins. *Power Electronics: Converters, Applications, and Design*. Wiley, New York, 1995.
133. S. O. R. Moheimani. Experimental verification of the corrected transfer function of a piezoelectric laminate beam. *IEEE Transactions on Control Systems Technology*, 8(4):660–666, July 2000.
134. S. O. R. Moheimani. Minimizing the effect of out of bandwidth dynamics in the models of reverberant systems that arise in modal analysis: Implications on spatial $\mathcal{H}_\infty$ control. *Automatica*, 36:1023–1031, 2000.
135. S. O. R. Moheimani. Minimizing the effect of out of bandwidth modes in truncated structure models. *Transactions of the ASME - Journal of Dynamic Systems, Measurement, and Control*, 122(1):237–239, March 2000.
136. S. O. R. Moheimani. Recent developments in structural vibration control using piezoelectric transducers. In *Proc. International Congress on Mechatronics*, Prague, 2004.
137. S. O. R. Moheimani and S. Behrens. Multi-mode piezoelectric shunt damping with a highly resonant impedance. *IEEE Transactions on Control Systems Technology*, 12(3):484–491, May 2004.
138. S. O. R. Moheimani and R. L. Clark. Minimizing the truncation error in assumed modes models of structures. *Transactions of the ASME, Journal of Vibration & Acoustics*, 122(3):332–335, July 2000.
139. S. O. R. Moheimani, A. J. Fleming, and S. Behrens. A highly resonant controller for piezoelectric shunt damping. *Electronics Letters*, 37(25):1505–1506, 2001.
140. S. O. R. Moheimani, A. J. Fleming, and S. Behrens. On the feedback structure of wideband piezoelectric shunt damping systems. *Smart Materials and Structures*, 12(1):49–56, February 2003.

141. S. O. R. Moheimani, A. J. Fleming, and S. Behrens. Dynamics, stability and control of multivariable piezoelectric shunts. *IEEE/ASME Transactions on Mechatronics*, 9(1):87–99, March 2004.
142. S. O. R. Moheimani, D. Halim, and A. J. Fleming. *Spatial Control of Vibration: Theory and Experiments.* World Scientific, Singapore, 2003. ISBN: 981-238-337-9.
143. S. O. R. Moheimani and B. J. G. Vautier. Resonant control of structural vibration using charge-driven piezoelectric actuators. To appear in *IEEE Transactions on Control Systems Technology.*
144. S. O. R. Moheimani and B. J. G. Vautier. Resonant control of structural vibration using charge-driven piezoelectric actuators. In *Proc. IEEE Conference on Decision and Control*, Bahamas, 2004.
145. M. Morari and E. Zafiriou. *Robust Process Control.* Prentice-Hall, 1989.
146. R. B. Mrad and H. Hu. Dynamic modeling of hysteresis in piezoceramics. In *IEEE/ASME International conference on advanced intelligent mechatronics proceedings*, Como, Italy, July 2001.
147. C. V. Newcomb and I. Flinn. Improving the linearity of piezoelectric ceramic actuators. *IEE Electronics Letters*, 18(11):442–443, May 1982.
148. D. Niederberger. *Smart Damping Materials using Shunt Control.* PhD thesis, The Swiss Federal Institute of Technology (ETH), Zurich, 2005.
149. D. Niederberger, S. Behrens, A. J. Fleming, S. O. R Moheimani, and M. Morari. Adaptive electromagnetic system for shunt damping. To appear in *IEEE/ASME Transactions on Mechatronics.*
150. D. Niederberger, A. J. Fleming, S. O. R. Moheimani, and M. Morari. Adaptive multi-mode resonant piezoelectric shunt damping. *IOP Smart Materials and Structures*, 13(5):1025–1035, October 2005.
151. D. Niederberger, M. Morari, and S. Pietrzko. Adaptive resonant shunted piezoelectric devices for vibration suppression. In *Proc. SPIE Smart Structures and Materials - Smart Structures and Integrated Systems, Vol.5056*, pages 213–224, San Diego, CA USA, March 2003.
152. D. Niederberger, M. Morari, and S. Pietrzko. Implementation of an adaptive R-L shunt circuit in analog technology. Technical Report AUT03-16, Automatic Control Laboratory, Swiss Federal Institute of Technology, 2003.
153. C. Niezrecki and H. H. Cudney. Feasibility to control launch vehicle internal acoustics using piezoelectric actuators. *Journal of Intelligent Material Systems and Structures*, 12:647–660, September 2001.
154. Institute of Electrical and Electronics Engineers Inc. IEEE standard on piezoelectricity. ANSI/IEEE Std. 176–1987, 1988.
155. H. Perez, Q. Zou, and S. Devasia. Design and control of optimal scan trajectories: Scanning tunneling microscope example. *Journal of Dynamic Systems, Measurment, and Control*, 126:187–197, March 2004.
156. I. R. Petersen and A. V. Savkin. *Robust Kalman Filtering for Signals and Systems with Large Uncertainties.* Birkhauser, 1999.
157. L. Petit, L. E. Lefeuvre, C. Richard, and D. Guyomar. A broad-band semi passive piezoelectric technique for structural damping. In *Proc. SPIE Smart Structures and Materials*, San Diego, CA, March 2004.
158. R. Pintelon and J. Schoukens. *Frequency Domain System Identification Toolbox.* The Math Works, Inc., 24 Prime Park Way, Natick, MA 01760-1500. http://www.mathworks.com.

159. R. Pintelon and J. Schoukens. *System Identification: A Frequency Domain Approach.* IEEE Press, Piscataway, USA, 2001.
160. H. R. Pota and T. E. Alberts. Multivariable transfer functions for a slewing piezoelectric laminate beam. *ASME Journal of Dynamic Systems, Measurement, and Control*, 117:353–359, September 1995.
161. A. Preumont. *Vibration Control of Active Structures.* Kluwer, 1997.
162. K. H. Rew, J. H. Han, and I. Lee. Multi-modal vibration control using adaptive positive position feedback. *Joural of Intelligent Material Systems and Structures*, 13(1):13–22, January 2002.
163. C. Richard, D. Guyomar, D. Audigier, and H. Bassaler. Enhanced semi-passive damping using continuous switching of a piezoelectric devices on an inductor. In *Proc. SPIE Smart Structures and Materials, Damping and Isolation, SPIE Vol. 3989*, pages 288–299, Newport Beach, CA, March 2000.
164. R. H. S. Riordan. Simulated inductors using differential amplifiers. *Electronics Letters*, 3(2):50–51, 1967.
165. S. Salapaka. Control of the nanopositioning devices. In *Proc. IEEE Conference on Decision and Control*, Maui, Hawaii, December 2003.
166. S. Salapaka, A. Sebastian, J. P. Cleveland, and M. V. Salapaka. Design, identification and control of a fast nanopositioning device. In *Proc. American Control Conference*, pages 1966–1971, May 2002.
167. S. Salapaka, A. Sebastian, J. P. Cleveland, and M. V. Salapaka. High bandwidth nano-positioner: A robust control approach. *Review of Scientific Instruments*, 75(9):3232–3241, September 2002.
168. G. Schitter, R. W. Stark, and A. Stemmer. Fast contact-mode atomic force microscopy on biological specimens by model-based control. *Ultramicroscopy*, 100:253–257, 2004.
169. G. Schitter and A. Stemmer. Model-based signal conditioning for high-speed atomic force and friction force microscopy. *Microelectronic Engineering*, 67-68:938–944, 2003.
170. M. J. Schulz, P. F. Pai, and D. J. Inman. Health monitoring and active control of composite structures using piezoceramic patches. *Composites Part B-Engineering*, 30(7):713–725, 1999.
171. W. H. Shields, J. Ro, and A. M. Baz. Control of sound radiation from a plate into an acoustic cavity using active piezoelectric-damping composites. In *Proc. SPIE Smart Structures and Materials: Mathematics and Control in Smart Structures, SPIE Vol. 3039*, pages 70–90, 1997.
172. J. Simpson and J. Schweiger. Industrial approach to piezoelectric damping of large fighter aircraft components. In *Proc. SPIE Smart Structures and Materials: Industrial and Commercial Applications of Smart Structures Technologies, SPIE Vol. 3326*, pages 34–46, 1998.
173. R. C. Smith, Z. Ounaies, and R. Wieman. A model for rate-dependent hysteresis in piezoceramic materials operating at low frequencies. NASA/CR-2001-211062 ICASE report 2001-26, 2001.
174. H. A. Sodano, G. Park, and D. J. Inman. An investigation into the performance of macro-fiber composites for sensing and structural vibration applications. *Mechanical Systems and Signal Processing*, 18(3):683–697, May 2004.
175. D. Song and C. J. Li. Modeling of piezo actuator's nonlinear and frequency dependent dynamics. *Mechatronics*, 9:391–410, 1999.

176. G. Song, P. Z. Qiao, W. K. Binienda, and G. P. Zou. Active vibration damping of composite beam using smart sensors and actuators. *JOURNAL OF AEROSPACE ENGINEERING*, 15(3):97–103, July 2002.
177. G. Song, S. P. Schmidt, and B. N. Agrawal. Experimental robustness study of positive position feedback control for active vibration suppression. *Journal of Guidance, Control and Dynamics*, 25(1):179–182, 2002.
178. C. Stiebel and H. Janocha. New concept of a hybrid amplifier for driving piezoelectric actuators. In *Proc. 1st IFAC-Conference on Mechatronic Systems*, pages 379–384, Darmstadt, Germany, 2000.
179. J. E. Storer. *Passive network synthesis.* McGraw-Hill, 1957.
180. K. Takaba and T. Katayama. H_2 output feedback control for descriptor systems. *Automatica*, 34(7):841–850, 1998.
181. W. Tan, T. P. Leung, and Q. L. Tu. H_∞ control for singularly perturbed systems. *Automatica*, 34(2):255–260, 1998.
182. J. Tang and K. W. Wang. Active-passive hybrid piezoelectric networks for vibration control: comparisons and improvement. *Smart Materials and Structures*, 10(4):794–806, 2001.
183. L. Vaillon and C. Philippe. Passive and active microvibration control for very high pointing accuracy space systems. *Smart Materials and Structures*, 8(6):719–728, December 1999.
184. M. E. Van Valkenburg. *Analog Filter Design.* CBS College Publishing, New York, NY, 1982.
185. P. Vallone. High-performance piezo-based self-sensor for structural vibration control. In *Proc. SPIE Conference on Smart Structures and Integrated Systems, SPIE Vol. 2443*, pages 643–655, 1995.
186. P. Van Overschee and B. De Moor. Continuous-time frequency domain subspace system identification. *Signal Processing*, 52:179–194, 1996.
187. B. J. G. Vautier. Charge-driven piezoelectric actuators in structural vibration control applications. Master's thesis, The University of Newcastle, 2004.
188. B. J. G. Vautier and S. O. R. Moheimani. Charge-driven piezoelectric actuators for structural vibration control: Issues and implementation. *Smart Material and Structures*, 14(4):575–586, 2005.
189. M. Viberg. Subspace-based methods for the identification of linear time invariant systems. *Automatica*, 31(12):1835–1851, 1995.
190. M. Vidyasagar. *Control System Synthesis, A Factorization Approach.* MIT Press, 1987.
191. J. S. Vipperman and R. L. Clark. Hybrid analog and digital adaptive compensation of piezoelectric sensoriactuators. In *Proc. AIAA/ASME Adaptive Structures Forum*, pages 2854–2859, New Orleans, LA, 1995.
192. D. J. Warkentin and E. F. Crawley. Power flow and amplifier design for piezoelectric actuators in intelligent structures. In *Proc. SPIE Smart Structures and Materials, Smart Structures and Intelligent Systems. SPIE Vol. 2190*, pages 283–94, 1994.
193. D. J. Warkentin and N. W. Hagood. Nonlinear piezoelectric shunting for structural damping. In *Proc. SPIE Smart Structures and Materials, Smart Structures and Integrated Systems. SPIE Vol. 3041*, San Deigo, CA, 1997.
194. J. T. Wen. Time domain and frequency domain conditions for strict positive realness. *IEEE Transactions on Automatic Control*, 33(10):988–992, October 1998.

195. B. Widrow and S. D. Stearns. *Adaptive Signal Processing.* Prentice-Hall, 1985.
196. S. Wu, T. L. Turner, and S. A. Rizzi. Piezoelectric shunt vibration damping of an F-15 panel under high-acoustic excitation. In *Proc. SPIE Smart Structures and Materials: Damping and Isolation, SPIE Vol. 3989*, pages 276–287, 2000.
197. S. Y. Wu. Piezoelectric shunts with parallel R-L circuit for smart structural damping and vibration control. In *Proc. SPIE Symposium on Smart Structures and Materials – Passive Damping and Isolation*, pages 259–269, March 1996.
198. S. Y. Wu. Method for multiple mode shunt damping of structural vibration using a single PZT transducer. In *Proc. SPIE Symposium on Smart Structures and Materials – Smart Structures and Intelligent Systems*, pages 159–167, March 1998.
199. S. Y. Wu. Piezoelectric shunts for simultaneous vibration reduction and damping of multiple vibration modes. U.S. patent 5,783,898, July 1998.
200. S. Y. Wu. Multiple PZT transducers implemented with multiple-mode piezoelectric shunting for passive vibration damping. In *Proc. SPIE Smart Structures and Materials – Passive Damping and Isolation, SPIE Vol. 3672*, pages 112–122, Huntington Beach, CA, March 1999.
201. S. Y. Wu. Broadband piezoelectric shunts for structural vibration control. Patent No. 6,075,309, June 2000.
202. S. Y. Wu. Broadband piezoelectric shunts for passive structural vibration control. In *Proc. SPIE Smart Structures and Materials 2001: Damping and Isolation, SPIE Vol. 4331*, pages 251–261, Newport Beach CA, USA, March 2001.
203. S. Y. Wu and A. S. Bicos. Structural vibration damping experiments using improved piezoelectric shunts. In *Proc. SPIE Conference on Passive Damping and Isolation, SPIE Vol. 3045*, pages 40–50, 1997.
204. K. A. Yi and R. J. Veillete. A charge controller for linear operation of a piezoelectric stack actuator. *IEEE Transactions on Control Systems Technology*, 13(4):517–526, 2005.
205. J. M. Zhang, W. Chang, V. K. Varadan, and V. V. Varadan. Passive underwater acoustic damping using shunted piezoelectric coatings. *IOP Journal of Smart Materials and Structures*, 10:414–420, 2001.
206. K. Zhou, J. C. Doyle, and K. Glover. *Robust and Optimal Control.* Prentice-Hall, 1996.
207. X. Zhu and T. A. Alberts. Appending a synthetic mode to compensate for truncated modes in colocated control. In *Proc. of AIAA GNC*, Boston, 1998.

Index

9781846283314